The Illustrated Flora of Illinois

The Illustrated Flora of Illinois
Robert H. Mohlenbrock, *General Editor*

The Illustrated Flora of Illinois

Sedges: Carex

Second Edition

Robert H. Mohlenbrock

Illustrated by Paul W. Nelson

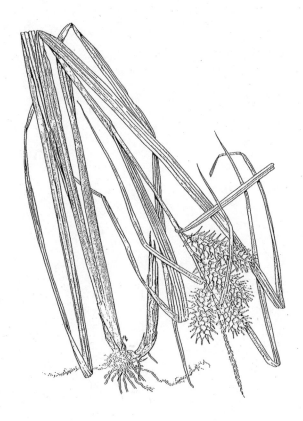

Southern
Illinois
University Press

Carbondale

To my loving wife, Beverly Ann,

for her dedication to me and my

endeavors for fifty-four years

Contents

Illustrations

[Following are additions to the list that appears in the first edition.]

Preface to the Second Edition

Since the publication of the first edition of *Sedges: Carex* in 1999, thirty-four additional species of plants covered by this book are being recognized in Illinois. Some of these are discoveries from recent field work, some are from more thorough searches of herbaria, and some are from different taxonomic philosophies. In addition, a substantial number of county records have been added for existing species.

A completely new key has been written for the species of *Carex* in Illinois.

Original illustrations for all the newly discovered species have been prepared by Paul W. Nelson.

Preface to the First Edition

This is the second volume devoted to sedges in the Illustrated Flora of Illinois series. It follows the publication of a volume on the ferns of Illinois, five volumes on the monocotyledonous plants of Illinois, six volumes on the dicotyledonous plants of Illinois, and a volume on diatoms.

The Illustrated Flora of Illinois is a multivolumed flora of the state of Illinois, designed to present every group of plants. A description will be provided for each species. Distribution maps and ecological notes will be included. Keys to aid in easy identification of the plants will be presented.

An advisory board was created to screen, criticize, and make suggestions for each volume of the series during its preparation. The board is composed of botanists eminent in their area of specialty—Floyd Swink, the Morton Arboretum (flowering plants); Robert F. Thorne, Rancho Santa Ana Botanical Garden (flowering plants); and Rolla M. Tryon Jr., University of South Florida (ferns).

The author is editor of the series and will prepare many of the volumes. Specialists in various groups will be preparing the volumes on plants of their special interest.

There is no definite sequence for publication of the Illustrated Flora of Illinois. Rather, volumes will appear as they are completed.

Acknowledgments

Specimens studied for this volume are from the Gray Herbarium of Harvard University, the Morton Arboretum, the United States National Herbarium, and also from herbaria located at Eastern Illinois University, Field Museum of Natural History, Illinois Natural History Survey, Illinois State Museum, Missouri Botanical Garden, New York Botanical Garden, Shawnee National Forest, Southern Illinois University at Carbondale, University of Illinois at Urbana, and Western Illinois University. In addition, a few private collections have been examined. To the curators and staff of the herbaria, extreme gratefulness is accorded for their courtesies.

I extend my thanks to Dan K. Evans of Marshall University, who assisted me in the early preparation of this book and who wrote the draft of species 1–15, and to Joyce Hoffman Peterson who wrote the drafts of species 149–154. I am also indebted to Dr. Gerould Wilhelm for providing most of the descriptions of the sheaths. My daughter Wendy Preece prepared the maps, while my son Trent Mohlenbrock labeled the illustrations for publication. My wife Beverly typed the several drafts of the difficult manuscript. Paul Nelson, Director of Operations of the Resource Management Program, Missouri Department of Natural Resources, meticulously prepared all of the illustrations. Special thanks must go to the herbarium and library staff of the Missouri Botanical Garden. The Gaylord and Dorothy Donnelley Foundation furnished funds for travel and for the illustrator. Without their assistance, this volume would not have been produced.

The Illustrated Flora of Illinois

County Map of Illinois

Introduction

This is the second of two volumes devoted to sedges, a group of plants belonging to the family Cyperaceae. This book deals only with the treatment of the genus *Carex* in Illinois, where 159 species are currently known. The first volume on sedges, published in 1976, contains the genera *Cyperus, Dulichium, Eleocharis, Fimbristylis, Bulbostylis, Lipocarpha, Fuirena, Scirpus, Eriophorum, Rhynchospora, Cladium,* and *Scleria.*

The plants in the family Cyperaceae produce flowers, although the flowers are greatly reduced and do not possess any colorful petallike structures. To many amateurs, the sedges often are not distinguished from the similarly appearing grasses.

Since sedges produce flowers, they are classified as Angiosperms, or flowering plants, and since only a single seed-leaf, or cotyledon, is formed upon germination, they are further distinguished as Monocotyledoneae, or monocots. Other monocots are arrowheads, lilies, irises, orchids, rushes, etc., which are treated in other volumes in the Illustrated Flora of Illinois series (Mohlenbrock 1970, 1970a). Grasses, likewise, are monocots, and have appeared in previous volumes in the series (Mohlenbrock 1972, 1973).

The nomenclature followed in this volume is based largely on the most recent monographic treatments available. Synonyms that have applied to species in Illinois are given under each species, and complete author citations are provided. A description based primarily on Illinois material is provided for each species. The description, while not necessarily intended to be complete, covers the most important features of each species.

Common names other than "sedge" are generally not used for members of the Cyperaceae. Local names when they are used, however, are indicated. The habitat designation is not always the habitat throughout the range of the species, but only for it in Illinois. The overall range for each species is given from the northeastern to the northwestern extremities, south to the southwestern limit, then eastward to the southeastern limit. The range has been compiled from various sources, including examination of herbarium material. A general statement is given concerning the range of each species in Illinois, and dot maps showing county distribution of each sedge in Illinois are provided. Each dot represents a voucher specimen deposited in some herbarium (with the exception of the record for *Carex folliculata*). There has been no attempt to locate each dot with reference to the actual locality within each county. The distribution in Illinois has been compiled from field study as well as herbarium study. Each species has been illustrated by Paul Nelson, showing the habit as well as some of the distinguishing features in detail.

Morphology of Illinois Sedges

The term sedges refers to those plants assigned to the family Cyperaceae. These are plants that have grasslike leaves and clustered flowers devoid of a perianth. Each flower is subtended, instead, by a single scale. One of the chief technical differences between sedges and grasses is in the fruit, which in the grasses has the seed adjoined to the pericarp, or ovary wall, while in the sedges, the seed is not fused with the pericarp. Other differences, primarily concerning vegetative characters, include such things as a generally hollow stem in grasses, solid in sedges; open sheaths in grasses, closed in sedges; 2-ranked leaves in grasses, 3-ranked in sedges; and anthers versatile in grasses, basifixed in sedges.

Although some sedges may be annuals, all species of *Carex* are perennials. These perennials may be tufted, but they usually have rhizomes; stolons; or short, thick, subterranean rootstocks. The stem that bears the leaves and inflorescence is called the culm. The culm is usually solid and is more or less 3-angled in all species of *Carex*. The culms are usually simple and erect, although a few species of *Carex* have reclining or even prostrate culms.

Sedge leaves are borne at the nodes in three ranks along the culm. The leaf is normally composed of a blade and a sheath, although in a few species of *Carex*, the leaf is reduced to a bladeless sheath. The sheath invariably wraps around and encloses a portion of the culm, a condition known as a closed sheath. The blade is the free part of the leaf. It is parallel-veined and generally elongated. The blades are usually flat, but they may be folded (plicate), deeply channeled (canaliculate), or inrolled into a slender tube (involute). Along the inner face of the leaf, where the blade adjoins the sheath, there is often a distinguishing structure of varying texture, shape, and size known as a ligule.

The inflorescence is composed of one to many spikes and is often subtended by a setaceous or leaflike bract. The basic unit of the *Carex* inflorescence is the spike. The spike usually contains several to numerous flowers, although in a few species, the pistillate spike is sometimes reduced to a single flower. The spike is composed of an axis along which are borne small bracts known as scales. Each scale in a spike usually subtends either a staminate flower or a pistillate flower, but never both. The scales may be keeled or rounded on the back.

The staminate flowers are always borne separate from the pistillate flowers. Each spike may be entirely staminate, entirely pistillate, or with both staminate and pistillate flowers in the same spike. In the last case, if the staminate flowers are above the pistillate flowers, the condition is known as androgynous. If the pistillate flowers are above the staminate flowers, the condition is known as gynecandrous. Although staminate and pistillate flowers usually occur on the same plant, there are rare cases where the plants may be dioecious.

The sedge flower is much reduced from the flower of the Liliaceae and other more showy flowering plants. The staminate flower consists of one to three stamens. The pistillate flower consists of a single pistil. Each stamen bears a 2-celled anther. Each pistil is composed of a 1-celled superior ovary with but one ovule. The styles are 2- or 3-cleft. The pistillate flower and the subsequent achene are nearly entirely enclosed by a saclike structure known as a perigynium. For absolute certainty in identifying a species of *Carex*, a mature perigynium must be seen. Perigynia show a wide range of variation in the genus in shape, size, texture, and vein pattern. Some species have a perigynium that tapers to a terminal beak.

The fruit of *Carex* is an achene that is dry, 1-seeded, and indehiscent. It is flat (lenticular) if the style is 2-cleft, or 3-angled (trigonous) if the style is 3-cleft. It is nearly entirely enclosed by the perigynium.

Distribution of *Carex*

The largest percentage of *Carex* species occupies habitats decidedly on the moist side, although only a few species can be considered true aquatics. On the other hand, some species of *Carex* grow in dry woodlands, on exposed bluffs and ledges, and in dry prairies. Most species of *Carex* in Illinois are native and do not regularly grow in extremely disturbed situations.

A few tamarack bogs can still be found in northeastern Illinois, although most that were in that region as little as one hundred years ago have been destroyed. In those remaining, a conspicuous sedge flora prevails. Characteristic species of *Carex* that live in these bogs are *C. brunnescens, C. canescens, C. chordorrhiza, C. disperma, C. interior, C. leptalea, C. limosa, C. oligosperma,* and *C. trisperma.*

In the sand regions along Lake Michigan, species of *Carex* that are often found include *C. aurea, C. crawei, C. garberi, C. granularis, C. tetanica,* and *C. viridula.* Also restricted to northern Illinois is a seepage habitat referred to as a calcareous fen. Species of *Carex* usually found in this habitat are *C. granularis, C. hystericina, C. leptalea,* and *C. stipata.*

Many marshes occur in the northern half of Illinois. Several species of *Carex* are found here, including *C. comosa, C. conjuncta, C. cristatella, C. lacustris, C. pellita, C. retrorsa, C. scoparia, C. aquatilis* var. *substricta, C. trichocarpa, C. vesicaria* var. *monile,* and *C. vulpinoidea.*

Although many wet meadows in Illinois have been destroyed, those that remain often support a rich sedge flora, including *Carex alopecoidea, C. aquatilis* var. *substricta, C. bebbii, C. buxbaumii, C. diandra, C. interior, C. sartwellii, C. sterilis, C. stipata,* and *C. stricta.*

Woodland regions in which the forest floor contains pockets of standing water throughout most of the year are referred to here as swampy woods. These can be found throughout the length of the state, and certain sedges can be expected to occur in many of them. In this group are *Carex crinita, C. grayi, C. lupuliformis, C. lupulina, C. lurida,* and *C. muskingumensis.* Mostly limited to swampy woods in the northern counties are *Carex lacustris, C. tuckermanii,* and *C. sprengelii,* while in the southern counties are *C. hyalinolepis* and *C. oxylepis.*

In the southern tip of Illinois, in an area long ago covered by seas, can be found scattered bald cypress swamps, remnants of an extensive region of swamps in the past. These bald cypress swamps display a strong coastal plain flora with many sedges present. Some of these are *Carex crus-corvi, C. decomposita, C. louisianica,* and *C. lupuliformis.*

A few isolated seep springs appear in the woodlands of extreme southeastern Illinois. Here may be found *Carex atlantica, C. bromoides,* and *C. crinita.*

Mesic woods are found throughout Illinois, and this habitat includes some of the most widespread species of *Carex* in the state. Among these are *Carex albursina, C. amphibola, C. blanda, C. cephaloidea, C. davisii, C. festucacea, C. hirtifolia, C. jamesii, C. radiata, C. rosea,* and *C. sparganioides.*

In moist, open ground throughout the state are usually found *Carex shortiana, C. squarrosa,* and *C. vulpinoidea.*

Prairie remnants in Illinois support some sedges. Moist prairies may include

Carex buxbaumii, C. conoidea, and *C. interior,* while drier prairies often have *Carex bicknellii, C. meadii, C. suberecta,* and *C. tetanica.*

A few sedges are characteristic of dry woodlands. Throughout the state are *Carex albicans, C. cephalophora, C. glaucodea,* and *C. hirsutella.* In the southern part of Illinois, *Carex bushii* and *C. virescens* may be found in this habitat.

Relationship of the Sedges

Sedges (Cyperaceae) and grasses (Poaceae) traditionally have been placed near each other in phylogenetic schemes. Sedges and grasses are similar in that they both bear reduced flowers subtended by an assortment of scales or bracts; they have similar leaves and are of the same general habit; and they both produce 1-seeded fruits. Hutchinson (1959) and others have placed the rushes (Juncaceae) near to these families. The rushes, although similar to Cyperaceae and Poaceae in general habit and in the presence of inconspicuous flowers, possess an actual perianth.

Recent evidence seems to point to a relationship of sedges, grasses, and rushes to several other families. These groups appear to be closely related to the Flagellariaceae (of Old World tropics and subtropics), the Restionaceae (of the Southern Hemisphere and Indochina), and the Centrolepidaceae (of the Southern Hemisphere).

Thorne (1968) proposes that the Cyperaceae is a highly specialized family of the order Commelinales, surpassed in degree of specialization only by the Poaceae. Other Illinois families placed in the Commelinales and considered somewhat less specialized than the sedges and grasses are the Xyridaceae, Commelinaceae, Pontederiaceae, and Juncaceae.

Under Thorne's classification, the sedges appear as follows:

Class Angiospermae
Subclass Monocotyledoneae
Superorder Commeliniflorae
Order Commelinales
Suborder Poineae
Family Cyperaceae

How to Identify a Species of *Carex*

A key is included for the identification of the species of *Carex* in Illinois. A botanical key is a device that, when properly employed, enables the user to identify correctly the plant in question. It is the intent to use characters that are easy to observe and to avoid the more technical characters that often show best relationships. Initially, there is a general key to groups of *Carex* that have similar characteristics. Once the group is established, turn to that group and begin the next key.

The keys in this work are dichotomous—that is, with pairs of contrasting statements. Always begin by reading both members of the first pair of statements. By choosing the statement that best fits the specimen to be identified, the reader will be guided to the next proper pair of statements. Eventually, a name will be derived.

Sometimes, while studying in the field or trying to simplify going through the key, certain characteristics that some plants may possess can quickly be determined by checking the following groupings. The species in each grouping are arranged alphabetically.

Hairs Present on Leaves, Sheaths, and/or Culms. *Carex atherodes, C. bushii, C. caroliniana, C. davisii, C. formosa, C. hirta, C. hirsutella, C. hirtifolia, C. hitchcockiana, C. oxylepis, C. pallescens, C. swanii,* and *C. virescens.*

Perigynia Pubescent. *Carex abdita, C. albicans, C. bushii, C. caroliniana, C. communis, C. emmonsii, C. grayi, C. heliophila, C. hirta, C. hirtifolia, C. lasiocarpa, C. lucorum, C. nigromarginata, C. oxylepis, C. pedunculata, C. pellita, C. pensylvanica, C. physorhyncha, C. richardsonii, C. X subimpressa, C. swanii, C. tonsa, C. trichocarpa, C. umbellata,* and *C. virescens.*

Leaves Never More than 2 mm Wide. *Carex atlantica* var. *capillaris, C. brunnescens, C. chordorrhiza, C. disperma, C. eburnea, C. emmonsii, C. formosa, C. lasiocarpa* var. *americana, C. leptalea, C. radiata, C. stenophylla* var. *enervis, C. texensis,* and *C. trisperma.*

Some or All of the Leaves More than 10 mm Wide. *Carex albursina, C. atherodes, C. blanda, C. careyana, C. comosa, C. crinita, C. crus-corvi, C. flaccosperma, C. folliculata, C. gigantea, C. granularis, C. hyalinolepis, C. lacustris, C. laxiculmis, C. laxiflora, C. lupuliformis, C. nebrascensis, C. plantaginea, C. platyphylla,* and *C. utriculata.*

Some Part of the Plant Glaucous. *Carex blanda, C. buxbaumii, C. crawei, C. deweyana, C. flaccosperma, C. glaucodea, C. granularis, C. heterostachya, C. lacustris, C. laxiculmis, C. limosa, C. pedunculata, C. platyphylla,* and *C. stricta.*

Culms Purple, Red, Red-Brown, or Purple-Brown at the Base. *Carex abdita, C. albicans, C. amphibola, C. aquatilis* var. *substricta, C. atherodes, C. baileyi, C. bushii, C. buxbaumii, C. careyana, C. caroliniana, C. communis, C. crinita, C. davisii, C. debilis* var. *rudgei, C. emmonsii, C. emoryi, C. formosa, C. frankii, C. gigantea, C. gracilescens, C. gracillima, C. grayi, C. haydenii, C. heliophila, C. hirsutella, C. hirta, C. hystericina, C. intumescens, C. jamesii, C. lacustris, C. laeviconica, C. lasiocarpa* var. *americana, C. louisianica, C. lucorum, C. lupulina, C. lupuliformis, C. lurida, C. nebrascensis, C. nigromarginata, C. oligosperma, C. oxylepis, C. pallescens, C. pedunculata, C. pellita, C. pensylvanica, C. physorhyncha, C. prasina, C. stricta, C. swanii, C. tetanica, C. tonsa, C. torta, C. trichocarpa, C. tuckermanii, C. umbellata, C. utriculata, C. vesicaria* var. *monile, C. virescens, C. willdenowii,* and *C. woodii.*

Some or All of the Pistillate Scales Longer than the Perigynia. *Carex aquatilis* var. *substricta, C. brachyglossa, C. bushii, C. buxbaumii, C. conoidea, C. crinita, C. davisii, C. emoryi, C. folliculata, C. frankii, C. gracilescens, C. gravida, C. haydenii, C. heterostachya, C. hirta, C. hitchcockiana, C. hyalinolepis, C. limosa, C. lucorum, C. nigromarginata, C. oligocarpa, C. pellita, C. pensylvanica, C. praegracilis, C. prairea, C. praticola, C. richardsonii, C. sprengelii, C. stenophylla* var. *enervis, C. tonsa, C. utriculata,* and *C. vulpinoidea.*

Key to the Genera of Cyperaceae in Illinois

Genera marked with an asterisk (*) have been treated in a separate volume of the *Illustrated Flora of Illinois*.

1. Flowers bisexual; achenes not enclosed in a saclike structure (perigynium), not white in color (except for *Cyperus erythrorhizos*) 2
1. Flowers unisexual; achenes either enclosed in a saclike structure (perigynium) or white in color 14
  2. Spikelets 3- or more-flowered (1-flowered in *Cyperus densicespitosus*, which has a sweet odor) 3
  2. Spikelets 1- to 2-flowered 13
3. Spikelets flattened; scales 2-ranked 4
3. Spikelets not flattened; scales spirally arranged 5
  4. Inflorescence terminal; bristles absent.................................... 1. *Cyperus**
  4. Inflorescence axillary; bristles present .... 2. *Dulichium**
5. Achene crowned by a tubercle.. 6
5. Achene without a tubercle 7
  6. Spikelet one; leaf blades absent .... 3. *Eleocharis**
  6. Spikelets 2–many; leaf blades present.............................. 4. *Bulbostylis**
7. Achenes not subtended by bristles or dilated scales (rudimentary scales only in *Lipocarpha*) 8
7. Achenes subtended by bristles or dilated scales although, in some species of *Scirpus*, the scales sometimes falling away early... 10
  8. Bract subtending the inflorescence one .... 8. *Scirpus**
  8. Bracts subtending the inflorescence two or more.......................... 9
9. Styles the same thickness throughout .... 8. *Scirpus**
9. Styles dilated toward base .... 5. *Fimbristylis**
  10. Achenes subtended by minute, dilated rudimentary perianth scales ..... 6. *Lipocarpha**
  10. Achenes subtended by ovate-oblong scales or by bristles............................ 11
11. Achenes subtended by ovate-oblong scales .... 7. *Fuirena**
11. Achenes subtended by bristles 12
  12. Bristles pale or dark brown, reddish or blackish at maturity .... 8. *Scirpus**
  12. Bristles white at maturity, forming "cottony" heads .... 9. *Eriophorum**
13. Achenes crowned with a tubercle; bristles subtending the flower ...... 10. *Rhynchospora**
13. Achenes without a tubercle .... 11. *Cladium**
  14. Achenes white, not enclosed in a saclike structure (perigynium) .... 12. *Scleria**
  14. Achenes not white, enclosed in a saclike structure (perigynium) .... 13. *Carex*

Key to the Species of *Carex* in Illinois

1. Hairs present on leaves, sheaths, and/or culms (this lead does not include plant parts that are merely scabrous or perigynia that may be papillose or granular) . . . . Group I

1. Leaves, sheaths, and/or culms all glabrous.

  2. Perigynia pubescent . . . . Group II

  2. Perigynia glabrous.

    3. Stigmas 3; achenes trigonous.

      4. Spike 1, androgynous (with staminate flowers at tip, pistillate flowers below) . . . . . .
      . . . . 62. *C. leptalea*

      4. Spikes more than 1 (sometimes only 1 in *C. squarrosa*), some of them entirely staminate or gynecandrous (with pistillate flowers at tip, staminate flowers below), or androgynous in rare specimens of *C. sprengelii* and *C. retrorsa*).

        5. Terminal spike entirely staminate.

          6. Perigynia 10 mm long or longer . . . . Group III

          6. Perigynia up to 10 mm long, usually much shorter.

            7. Lowest scale of pistillate flower leaflike, 15–50 mm long (do not confuse this with bract that subtends the spike); pistillate spike 1 per culm, with 2–4 perigynia.

              8. Perigynia subglobose, abruptly contracted to the beak; staminate scales truncate . . . . 63. *C. jamesii*

              8. Perigynia ellipsoid to ellipsoid-fusiform, narrowed to the beak; staminate scales acute . . . . 64. *C. willdenowii*

            7. Lowest scale of pistillate flowers less than 10 mm long, not leaflike; pistillate spikes usually more than 1 per culm, usually with more than 4 perigynia.

              9. Staminate spikes more than 1 per culm . . . . Group IV

              9. Staminate spike 1 per culm.

                10. Leaves threadlike, 0.5–1.0 mm wide; perigynia 1.5–2.0 mm long . . . .
                . . . . 78. *C. eburnea*

                10. Leaves filiform or wider, some or all of them more than 1 mm wide; perigynia usually more than 2 mm long.

                  11. Flowering culms hidden among the leaves; plants to 15 cm tall . .
                  . . . . 75. *C. tonsa*

                  11. Flowering culms not hidden among the leaves; plants usually more than 15 cm tall.

                    12. Perigynium prominently 2-toothed at the tip, or prolonged into a conspicuous bidentate beak . . . . Group V

                    12. Perigynium ending abruptly at the tip, either without teeth or with merely a small notch.

                      13. Most or all of the leaves, particularly on sterile shoots, 15 mm wide or wider . . . . Group VI

                      13. Most or all the leaves less than 15 mm wide.

                        14. At least the lowest pistillate spikes on flexuous or pendulous peduncles . . . . Group VII

                        14. All pistillate spikes ascending to erect.

                          15. Beak or tip of perigynium bent or curved to one side . . . . Group VIII

15. Beak or tip of perigynium not bent or curved to
one side .... Group IX
5. Terminal spike androgynous or gynecandrous .... Group X
3. Achenes lenticular; stigmas 2.
16. Some or all the spikes pedunculate; staminate flowers usually on separate spikes ..
.... Group XI
16. All spikes sessile, essentially alike, with staminate flowers either at the tip or at the
base of each spike (exceptional specimens of *C. sterilis, C. bromoides,* and *C. praegracilis* may have a few all staminate or all pistillate flowers).
17. Culms solitary, or forming stoloniferous or rhizomatous colonies............
.... Group XII
17. Culms cespitose, not arising singly from extensive stolons and/or rhizomes.
18. Spikes androgynous .... Group XIII
18. Spikes gynecandrous.
19. Perigynia plano-convex, with rounded margins or with narrowly
rimmed margins...................................... Group XIV
19. Perigynia more or less flat and scalelike, with winged margins ....
.... Group XV

Group I
Hairs present on leaves, sheaths, and/or culms.

1. Perigynia pubescent.
2. Terminal spike entirely staminate.
3. Leaves 5–10 mm wide; staminate spike 1, sessile or nearly so; perigynia broadly ellipsoid, 3.5–5.0 mm long, with teeth of the beak minute.............. 79. *C. hirtifolia*
3. Leaves 2–4 mm wide; staminate spikes 1–3, on long peduncles; perigynia conic-ovoid,
5–9 mm long, with teeth of the beak 1–2 mm long.................... 95. *C. hirta*
2. Terminal spike gynecandrous (pistillate flowers at top, staminate flowers below).
4. Spikes linear-cylindric; perigynia ellipsoid-ovoid to lance-ellipsoid, prominently
nerved.
5. Perigynia ellipsoid-ovoid, 2.0–2.5 mm long, more or less compressed.............
.... 100. *C. virescens*
5. Perigynia lance-ellipsoid, 3.5–5.0 mm long, angled .... 104. *C. oxylepis*
4. Spikes short-cylindric to subglobose; perigynia broadly ovoid, finely nerved ....
.... 101. *C. swanii*
1. Perigynia glabrous.
6. At least the terminal spike staminate.
7. Staminate spikes 2–5; remainder of spikes usually androgynous; pistillate scales
shorter than the perigynia; perigynia numerous per spike, lance-ovoid, 7–9 mm long,
with a beak 1.5–3.0 mm long .... 139. *C. atherodes*
7. Staminate spike 1; lateral spikes entirely pistillate; pistillate scales as long as or longer
than the perigynia; perigynia up to 40 per spike, ellipsoid-triangular to obovoid, up to
5 mm long, beakless or with a beak about 1 mm long.
8. Leaves and sheaths pubescent; leaves up to 3 mm wide; perigynia up to 40 per
spike, ellipsoid-triangular, up to 3 mm long, beakless............ 96. *C. pallescens*
8. Only the sheaths pubescent; leaves 3–7 mm wide; perigynia up to 10 per spike, obovoid, 4–5 mm long, with beak about 1 mm long............ 117. *C. hitchcockiana*
6. Terminal spike gynecandrous.
9. Spikes narrowly cylindric, not clavate at the base, the lowest on long, flexuous or pendulous peduncles; perigynia ellipsoid to narrowly ovoid to oblong-ovoid, more or less
trigonous, short-beaked.

10. Leaves up to 1.8 mm wide; terminal spike up to 2 mm thick; pistillate scales obtuse to acute; perigynia up to 20 per spike, ellipsoid . . . . 105. *C. formosa*
10. Leaves 3–8 mm wide; terminal spike more than 2 mm thick; pistillate scales awned; perigynia up to 40 per spike, narrowly ovoid to oblong-ovoid.
    11. Pistillate spikes 5–7 mm thick; pistillate scales as long as or longer than the perigynia; perigynia oblong-ovoid, loose around the achene . . . . 106. *C. davisii*
    11. Pistillate spikes 3–5 mm thick; pistillate scales shorter than the perigynia; perigynia narrowly ovoid, closely enveloping the achene . . . . . 104. *C. oxylepis*
9. Spikes broadly cylindric, clavate at the base, on short, ascending peduncles; perigynia obovoid, rounded on all faces, or with a flat, inner face, pointed at the tip.
  12. Leaves very sparsely hairy . . . . 98. *C. caroliniana*
  12. Leaves moderately to copiously hairy.
    13. Pistillate scales lanceolate, acuminate to awned, longer than the perigynia; perigynia rounded on all faces, strongly nerved . . . . 99. *C. bushii*
    13. Pistillate scales ovate, acute to cuspidate, shorter than the perigynia; perigynia with a flat, inner face, weakly nerved . . . . 97. *C. hirsutella*

Group II
Leaves, stems, and sheaths all glabrous; perigynia pubescent.

1. Terminal spike androgynous (staminate flowers at the tip, pistillate flowers below) . . . .
. . . . 76. *C. pedunculata*
1. Terminal spike gynecandrous (pistillate flowers at the tip, staminate flowers below), or entirely staminate.
2. Terminal spike gynecandrous.
  3. Pistillate scales lanceolate, acuminate to awned, longer than the perigynia . . . .
. . . . 99. *C. bushii*
  3. Pistillate scales ovate, acute to acuminate, shorter than the perigynia . . . .
. . . . 98. *C. caroliniana*
2. Terminal spike entirely staminate.
  4. Staminate spikes 2 or more per culm.
    5. Lower spikes sometimes androgynous; pistillate scales lanceolate to broadly lanceolate, slightly shorter than to longer than the perigynia; perigynia 2–5 mm long, the teeth of the beak up to 1 mm long.
      6. Culms smooth; leaves up to 2 mm wide, convolute, without a midvein; pistillate scales slightly shorter than the perigynia; perigynia obscurely nerved, 3–5 mm long, the teeth of the perigynia up to 0.6 mm long . . . . 93. *C. lasiocarpa*
      6. Culms usually somewhat scabrous; leaves 2–5 mm wide, flat, with a conspicuous midvein; pistillate scales as long as or longer than the perigynia; perigynia obviously nerved, 2.0–3.5 mm long, the teeth of the perigynia about 1 mm long
. . . . 94. *C. pellita*
    5. Lower spikes almost always pistillate; pistillate scales broadly ovate, shorter than the perigynia; perigynia 4.2–10.0 mm long, the teeth of the beak 1.2–2.0 mm long.
      7. Perigynia strongly nerved, up to 40 per pistillate spike, 5–10 mm long . . . .
. . . . 142. *C. trichocarpa*
      7. Perigynia obscurely nerved, up to 135 per pistillate spike, 4.2–6.4 mm long . . . . .
. . . . 143. *C.* X *subimpressa*
  4. Staminate spike 1 per culm.
    8. Perigynia 12 mm long or longer, crowded into a globose spike . . . . 149. *C. grayi*
    8. Perigynia up to 4.5 mm long, not forming a globose spike.
      9. Plants more than 60 cm tall; terminal staminate spike up to 4 cm long; perigynia 40–75 per spike . . . . 94. *C. pellita*

9. Plants up to 60 cm tall, usually much shorter; terminal staminate spike up to 2 cm long; perigynia up to 25 per spike.
 10. Perigynia spongy at base; culms capillary to slender and flexuous; leaves pale green or glaucous.
  11. Culms and peduncles capillary; perigynia 1–8 per spike, 3.8–4.5 mm long. . . . . 76. *C. pedunculata*
  11. Culms and peduncles slender and flexuous, but not capillary; perigynia 10–25 per spike, 2.0–3.5 mm long . . . . 77. *C. richardsonii*
 10. Perigynia firm at base; culms neither capillary or flexuous; leaves green to dark green.
  12. Some or all the fertile culms hidden among the leaf bases.
   13. Plants with stolons; perigynia ellipsoid, 1/2–2/3 as wide as long; usually only 1 peduncle arising from each culm. . . . . 72. *C. nigromarginata*
   13. Plants without stolons; perigynia suborbicular to orbicular, about as wide as long; 2–4 peduncles arising from each culm.
    14. Perigynia very sparsely pubescent; beak of perigynium 1–3 mm long, often nearly as long as the body. . . . . 75. *C. tonsa*
    14. Perigynia uniformly pubescent; beak of perigynium 0.5–1.7 mm long, up to 3/4 as long as the body.
     15. Beak of perigynium about 3/4 as long as the body; pistillate scales acuminate; perigynia 2.2–3.3 mm long . . . . 73. *C. umbellata*
     15. Beak of perigynium 1/2–2/3 as long as the body; pistillate scales acute; perigynia 3.2–4.5 mm long . . . . . . 74. *C. abdita*
  12. All fertile culms usually protruding above the leaves.
   16. Perigynia ovoid to orbicular.
    17. Beak of perigynium up to 1/3 as long as the body.
     18. Perigynia more or less trigonous; beak of perigynium not more than 1/4 as long as the body.
      19. Plants with slender stolons; leaves 1.5–3.0 mm wide; perigynia 1.5–3.0 mm long . . . . 65. *C. pensylvanica*
      19. Plants without stolons; leaves up to 5 mm wide; perigynia 2.5–4.0 mm long. . . . . 68. *C. communis*
     18. Perigynia more or less terete; beak of perigynium about 1/3 as long as the body . . . . 67. *C. heliophila*
    17. Beak of perigynium 2/3 as long as the body or longer . . . . 66. *C. lucorum*
   16. Perigynia ellipsoid.
    20. Plants with slender stolons . . . . 71. *C. physorhyncha*
    20. Plants without stolons.
     21. Culms and leaves often reclining, weak; staminate scales long-acuminate; leaves up to 1.5 mm wide . . . . 69. *C. emmonsii*
     21. Culms and leaves more or less erect or ascending, not weak; staminate scales obtuse to short-acuminate; leaves often 1.5–5.0 mm wide.
      22. Leaves up to 2.5 mm wide; staminate scales obtuse to acute . . . . 70. *C. albicans*
      22. Leaves often 3–5 mm wide; staminate scales acute to short-acuminate . . . . 68. *C. communis*

Group III

Hairs absent on leaves, sheaths, culms, and perigynia; perigynia glabrous; stigmas 3; achenes trigonous; terminal spike entirely staminate; perigynia 10 mm long or longer.

1. Perigynia not inflated, lanceoloid, up to 3.5 mm wide at its widest point . . . .
. . . . 148. *C. folliculata*
1. Perigynia strongly inflated, subuloid to lanceoloid, usually 3.5 mm wide or wider at its widest point.
    2. Pistillate spikes globose or subglobose.
        3. Perigynia radiating in all directions in a spike; perigynia not shiny, cuneate at the base
        . . . . 149. *C. grayi*
        3. Perigynia radiating in all directions except retrorse; perigynia shiny, rounded at the
        base . . . . 150. *C. intumescens*
    2. Pistillate spikes thick-cylindric, longer than broad.
        4. All perigynia horizontally spreading; beak of perigynium 2–3 times longer than the
        body; achene truncate at summit, broader than long . . . . 154. *C. gigantea*
        4. Perigynia ascending; beak of perigynium somewhat shorter than or slightly longer
        than the body; achene narrowed at summit, longer than broad or about as long as
        broad.
            5. Pistillate spikes more than twice as long as broad; achenes about as long as broad,
            with conspicuously knobby angles and concave sides . . . . 153. *C. lupuliformis*
            5. Pistillate spikes up to twice as long as broad; achenes longer than broad, without
            conspicuous knobby angles and with nearly flat sides.
                6. Perigynia crowded, 13–20 mm long; pistillate scales long-acuminate to awned;
                leaves 5 mm wide or wider; culms several in a tuft . . . . 152. *C. lupulina*
                6. Perigynia loosely arranged, 10–14 mm long; pistillate scales obtuse to acute;
                leaves up to 6 mm wide; culms few or solitary, never in a tuft . . 151. *C. louisianica*

Group IV

Leaves, sheaths, and/or culms all glabrous; perigynia glabrous; stigmas 3; achenes trigonous; terminal spike entirely staminate; perigynia up to 10 mm long, usually much shorter; staminate spikes more than 1 per culm

1. Style disarticulating from mature achene; perigynium abruptly contracted to a beak as long as the body; lowest spikes often on pendulous peduncles. . . . . 108. *C. sprengelii*
1. Style continuous and persistent on the mature achene; perigynium tapering to the beak, the beak usually somewhat shorter than the body of the perigynium; lowest spikes erect or ascending.
    2. Plants growing in large colonies from extensive rhizomes; perigynia thick and firm, not inflated.
        3. Culms red-purple at the base; leaves not glaucous or blue-green; nerves of perigynia conspicuous and slightly elevated.
            4. Teeth of the beak of the perigynium up to 1 mm long; most of the leaves 8 mm
            wide or wider . . . . 137. *C. lacustris*
            4. Teeth of the beak of the perigynium 1.0–2.2 mm long; leaves mostly 2–6 mm wide.
            . . . . 141. *C. laeviconica*
        3. Culms brown at the base; leaves glaucous or blue-green; nerves of the perigynium finely impressed, or absent.
            5. Leaves 8–15 mm wide, flat; perigynia up to 150 per spike, 6–8 mm long . . . .
            . . . . 138. *C. hyalinolepis*

  5. Leaves 1.5–3.2 mm wide, plicate; perigynia up to 30 per spike, 3.0–4.5 mm long . . .
. . . . 140. *C. heterostachya*
 2. Plants growing in dense clumps, usually without extensive rhizomes; perigynia thin and papery, inflated.
  6. Perigynia horizontally spreading or even reflexed at maturity; sheaths loose . . . .
. . . . 155. *C. retrorsa*
  6. Perigynia spreading to ascending; sheaths tight.
   7. Culms spongy at the base; some or all of the leaves 7–12 mm wide . . . .
. . . . 156. *C. utriculata*
   7. Culms firm at the base; leaves 2–7 mm wide.
    8. Achene invaginated on one side; perigynia 7–10 mm long . . . .
. . . . 157. *C. tuckermanii*
    8. Achenes not invaginated on one side; perigynia 5–8 mm long . . . .
. . . . 159. *C. vesicaria*

Group V

Leaves, sheaths, and/or culms glabrous; perigynia glabrous; stigmas 3; achenes trigonous; terminal spike entirely staminate; perigynia up to 10 mm long, usually much shorter; staminate spike 1 per culm; perigynia prominently 2-toothed at the tip, or prolonged into a conspicuous, bidentate beak.

1. Pistillate spikes 2–4 mm thick.
 2. Pistillate spikes linear-cylindric, up to 8 cm long; lowest spikes on pendulous peduncles; perigynia 4–7 mm long, all ascending . . . . 107. *C. debilis*
 2. Pistillate spikes short-cylindric, up to 1.5 cm long; lowest spikes ascending; perigynia 2–3 mm long, spreading to reflexed . . . . 133. *C. viridula*
1. Pistillate spikes usually at least 1 cm thick (sometimes slightly more slender in *C. sprengelii*).
 3. Some of the perigynia reflexed.
  4. Perigynia up to 35 per spike, 3.2–4.5 mm long, yellowish, the beak about as long as the body; leaves up to 4.5 mm wide . . . . 132. *C. cryptolepis*
  4. Perigynia usually 50–100 per spike, 5–10 mm long, yellow-green, the beak up to half as long as the body; leaves more than 4.5 mm wide.
   5. Lowest spikes on pendulous peduncles; perigynia 5–7 mm long, the beak 1.5–2.0 mm long; pistillate scales serrulate-awned . . . . 144. *C. comosa*
   5. Lowest spikes ascending; upper perigynia 7–10 mm long, the beak 2.5–3.5 mm long; pistillate scales acuminate . . . . 155. *C. retrorsa*
 3. All perigynia spreading to ascending, never reflexed.
  6. Culms filiform; leaves 1–3 mm wide; perigynia up to 15 per spike . . . .
. . . . 158. *C. oligosperma*
  6. Culms broader than filiform; leaves 2.5–12.0 mm wide (if narrower, then plicate in *C. heterostachya*); perigynia 30–100 per spike.
   7. Culms purple-red at the base; perigynia 50–100 per spike.
    8. Pistillate scales setaceous, much longer than the perigynia; spikes subtended by bracts many times their length; perigynia obconic, broadest above the middle . . .
. . . . 134. *C. frankii*
    8. Pistillate scales serrulate-awned, not longer than the perigynia; spikes not subtended by exceptionally long bracts; perigynia ovoid, broadest near the base.
     9. Lowest spikes on slender, often pendulous peduncles; perigynia 4–6 mm long, 1.5–2.0 mm broad, the beak 1.8–2.2 mm long . . . . 145. *C. hystericina*
     9. Lowest spikes ascending; perigynia 5–9 mm long, 2–4 mm broad, the beak 3–4 mm long.

10. Leaves up to 7 mm wide; perigynia 3–4 mm wide, the beak about equalling the body. . . . . 146. *C. lurida*
    10. Leaves 2–4 mm wide; perigynia 2.0–2.5 mm wide, the beak longer than the body. . . . . 147. *C. baileyi*
7. Culms brown at the base; perigynia 30–40 per spike.
    11. Perigynia 5.0–6.5 mm long, the beak equalling the body; leaves 2.5–4.0 mm wide, usually flat. . . . . 108. *C. sprengelii*
    11. Perigynia 3.0–4.5 mm long, the beak much shorter than the body; leaves 1.5–3.2 mm wide, often plicate . . . . 140. *C. heterostachya*

Group VI

Leaves, sheaths, and/or culms all glabrous; perigynia glabrous; stigmas 3; achenes trigonous; terminal spike entirely staminate; perigynia up to 10 mm long, usually much shorter; staminate spike 1 per culm; perigynia ending abruptly at the tip, either without teeth or with merely a small notch; most or all of the leaves, particularly on sterile shoots, 15 mm wide or wider.

1. Culms leafless, bearing only long-tubular purplish sheaths; staminate spike purple . . . . 121. *C. plantaginea*
1. Culms leafy; staminate spike brown to brown-purple.
    2. Blade of uppermost leaf at most only three times longer than its sheath.
        3. Basal leaves green; culms purple at base; perigynia 5.0–6.5 mm long. . . . . 122. *C. careyana*
        3. Basal leaves glaucous; culms brown at base; perigynia 2.5–4.5 mm long . . . . 123. *C. platyphylla*
    2. Blade of uppermost leaf many times longer than its sheath.
        4. Uppermost bract much longer than the inflorescence; culms somewhat winged; pistillate scales obtuse to acute . . . . 126. *C. albursina*
        4. Uppermost bract about as long as or barely longer than the inflorescence; culms wingless; pistillate scales cuspidate to awned . . . . 127. *C. laxiflora*

Group VII

Leaves, sheaths, and/or culms all glabrous; perigynia glabrous; stigmas 3; achenes trigonous; terminal spike entirely staminate; perigynia up to 10 mm long, usually much shorter; staminate spike 1 per culm; perigynia ending abruptly at the tip, either without teeth or with merely a small notch; most or all the leaves less than 15 mm wide; at least the lowest pistillate spikes on flexuous or pendulous peduncles.

1. Perigynia up to 15 per spike; culms brown or cinnamon brown at the base.
    2. Perigynia fusiform, with rounded angles; leaves up to 7 mm wide, green; lowest pistillate scale fertile . . . . 131. *C. styloflexa*
    2. Perigynia ovoid to oblongoid, with pointed angles; leaves 5–12 mm wide, pale green to glaucous; lowest pistillate scale empty . . . . 125. *C. laxiculmis*
1. Perigynia 15–50 per spike; culms usually purple to purple-red at the base.
    3. Leaves up to 2.5 mm wide; plants stoloniferous; perigynia glaucous-green. . . . . 90. *C. limosa*
    3. Leaves 3–6 mm wide; plants not stoloniferous; perigynia green . . . . 102. *C. prasina*

Group VIII

Leaves, sheaths, and/or culms all glabrous; perigynia glabrous; stigmas 3; achenes trigonous; terminal spike staminate; perigynia up to 10 mm long, usually much shorter; staminate spike 1 per culm; perigynia ending abruptly at the tip, either without teeth or with merely a small notch; most or all the leaves less than 15 mm wide; all pistillate spikes ascending to erect; beak or tip of perigynium bent or curved to one side.

1. Perigynia crowded in pistillate spikes, always overlapping.
    2. Perigynia rounded at the base, olive green to brown; leaves more or less glaucous . . . . 109. *C. granularis*
    2. Perigynia tapering to the base, yellow-green to pale green; leaves pale green, green, or gray-green, but not glaucous.
        3. Perigynia beakless, but curved at the tip; pistillate spikes 5–7 mm thick; pistillate scales with purple-brown margins.
            4. Leaves gray-green, up to 7 mm wide; pistillate spikes 5–7 mm thick; pistillate scales cuspidate to short-awned; perigynia 3–5 mm long, turgid . . . . 118. *C. meadii*
            4. Leaves green, up to 5 mm wide; pistillate spikes about 5 mm thick; pistillate scales obtuse to acute; perigynia 2.5–3.5 mm long, not turgid . . . . 119. *C. tetanica*
        3. Perigynia with a short, curved beak; pistillate spikes 3–5 mm thick; pistillate scales without purple-brown margins.
            5. Culms brown at the base; staminate spike sessile or nearly so, not overtopping the pistillate spikes; staminate scales greenish white . . . . 129. *C. blanda*
            5. Culms purplish at the base; staminate spike pedunculate, elevated above the pistillate spikes; staminate scales brown- or purple-tinged . . . . 130. *C. gracilescens*
1. Perigynia loosely or remotely arranged in pistillate spikes, scarcely overlapping.
    6. Leaves up to 4 mm wide; plants stoloniferous; pistillate scales with purple-brown margins . . . . 120. *C. woodii*
    6. Leaves 5–12 mm wide; plants tufted; pistillate scales without purple-brown margins.
        7. Pistillate spikes 4–5 mm thick; perigynia 4–5 mm long . . . . 128. *C. striatula*
        7. Pistillate spikes 3–4 mm thick; perigynia 3.0–4.5 mm long . . . . 127. *C. laxiflora*

Group IX

Leaves, sheaths, and/or culms all glabrous; perigynia glabrous; stigmas 3; achenes trigonous; terminal spike entirely staminate; perigynia up to 10 mm long, usually much shorter; staminate spike 1 per culm; perigynia ending abruptly at the tip, either without teeth or with merely a small notch; most or all the leaves less than 15 mm wide; all pistillate spikes ascending to erect; beak or tip of perigynium not bent or curved to one side.

1. Pistillate scales dark red-purple; perigynia more biconvex than trigonous . . . . 89. *C. buxbaumii*
1. Pistillate scales not dark red-purple; perigynia trigonous.
    2. Staminate spike elevated above the pistillate spikes.
        3. Pistillate spikes loosely or remotely flowered.
            4. Perigynia with rounded angles; pistillate scales cuspidate to awned; leaves 6–20 mm wide, pale green . . . . 127. *C. laxiflora*
            4. Perigynia with pointed angles; pistillate scales acute; leaves up to 5 mm wide, dark green . . . . 124. *C. digitalis*
        3. Pistillate spikes with crowded perigynia.
            5. Leaves 1.5–3.5 mm wide, glaucous; pistillate scales red-brown, acute to cuspidate; perigynia ovoid, 3.0–3.5 mm long, strongly nerved . . . . 110. *C. crawei*
            5. Leaves 3–7 mm wide, not glaucous; pistillate scales not red-brown, awned; perigynia oblongoid to obovoid, 3.0–4.5 mm long, finely impressed nerved.

  6. Culms brownish at the base; perigynia 10–25 per spike; no spikes near base of plant .... 111. *C. conoidea*

  6. Culms purplish at the base; perigynia up to 12 per spike; lowest spikes usually near base of plant .... 112. *C. amphibola*

2. Staminate spike not strongly elevated above the pistillate spikes.

  7. Perigynia up to 10 (–12) per spike; leaves green or glaucous 2–8 mm wide.

   8. Pistillate spikes loosely flowered, up to 1.5 cm long .... 116. *C. oligocarpa*

   8. Pistillate spikes with crowded perigynia, (1.0–) 1.5–2.5 cm long.

    9. Leaves more or less glaucous.

     10. Pistillate scales up to 1/2 as long as the perigynia; perigynia 2.0–2.3 mm wide; plants more or less glaucous; plants usually of rather wet habitats .... 114. *C. flaccosperma*

     10. Pistillate scales nearly as long as the perigynia; perigynia 1.5–1.8 mm wide; plants strongly glaucous; plants usually of rather dry habitats .... 115. *C. glaucodea*

    9. Leaves green.

     11. Perigynia 3.5–4.5 mm long; one or more pistillate spikes usually near base of plant .... 112. *C. amphibola*

     11. Perigynia 4.5–5.5 mm long; no pistillate spikes near base of plant .... 113. *C. grisea*

  7. Perigynia 10–60 per spike; leaves more or less glaucous (except in *C. grisea*), up to 10 mm wide.

   12. Pistillate scales 1/2 as long as the perigynia or nearly as long; leaves more or less glaucous.

    13. Perigynia 2.3–4.0 mm long, strongly nerved; pistillate spikes 3–6 mm thick; sheaths tight .... 109. *C. granularis*

    13. Perigynia 3.5–6.0 mm long, finely impressed-nerved; pistillate scales 5–8 mm thick; sheaths loose.

     14. Pistillate scales up to 1/2 as long as the perigynia; perigynia 2.0–2.3 mm wide; plants more or less glaucous; plants of rather wet habitats .... 114. *C. flaccosperma*

     14. Pistillate scales nearly as long as the perigynia; perigynia 1.5–1.8 mm wide; plants strongly glaucous; plants of usually dry habitats .... 115. *C. glaucodea*

   12. Pistillate scales about as long as the perigynia or slightly longer; leaves green .... 113. *C. grisea*

Group X

Leaves, sheaths, and/or culms all glabrous; perigynia glabrous; stigmas 3; achenes trigonous; terminal spike androgynous or gynecandrous.

1. Terminal spike androgynous.

 2. Spike one per culm; leaves usually 2 per culm, filiform, 1–3 mm wide; perigynia 1–10 per spike, flattened, 2.5–4.0 mm long, beakless .... 62. *C. leptalea*

 2. Spikes several per culm; leaves more than 2 per culm, not filiform, 2.5–10.0 mm wide; perigynia 20–many per spike, trigonous, 5–10 mm long, with a conspicuous beak.

  3. Pistillate spikes up to 3.5 cm long, up to 1 cm thick, the lowest on pendulous peduncles; perigynia loosely arranged, 5.0–6.5 mm long, 1.8–2.0 mm wide, 2-ribbed but otherwise nerveless, the beak as long as the body; leaves 2.5–4.0 mm wide, shorter than the culms .... 108. *C. sprengelii*

  3. Pistillate spikes up to 8 cm long, up to 2 cm thick, none of them on pendulous pedun-

cles; perigynia crowded, 7–10 mm long, 2.5–3.5 mm wide, strongly nerved, the beak prominent but usually not half as long as the body; leaves mostly 4–10 mm wide, longer than the culms . . . . 155. *C. retrorsa*

1. Terminal spike gynecandrous.
    4. Perigynia obconic; pistillate scales setaceous, much longer than the perigynia; spikes located about midway on plant, much surpassed by leaves and leaflike bracts . . . .
    . . . . 134. *C. frankii*
    4. Perigynia narrowly lanceoloid to narrowly ellipsoid to ovoid to obovoid (sometimes obconic in *C. typhina*); pistillate scales lanceolate to ovate, shorter than to about as long as the perigynia; most spikes in upper part of plant or even surpassing the leaves and bracts.
        5. Pistillate spikes (or terminal spike if only one spike present) 10–22 mm thick.
            6. Perigynia horizontally spreading, or the lowest ones reflexed; pistillate scales acute to acuminate to cuspidate. . . . . 135. *C. squarrosa*
            6. Perigynia ascending, none of them reflexed; pistillate scales usually obtuse, less commonly acute. . . . . 136. *C. typhina*
        5. Pistillate spike up to 10 mm thick.
            7. At least the lowest spikes pendulous or spreading on long, slender peduncles.
                8. Perigynia narrowly lanceoloid, 4–7 mm long . . . . 107. *C. debilis*
                8. Perigynia narrowly ellipsoid to ovoid, 2–4 mm long.
                    9. Perigynia narrowly ellipsoid, beakless; leaves up to 8 mm wide . . . .
                    . . . . 103. *C. gracillima*
                    9. Perigynia ovoid, with a beak up to 2 mm long; leaves 3–6 mm wide . . . .
                    . . . . 102. *C. prasina*
            7. All spikes sessile, or on short, ascending peduncles.
                10. Culms red-purple at the base; pistillate scales dark red-purple; perigynia more or less biconvex, glaucous-green; leaves glaucous below. . . . . . 89. *C. buxbaumii*
                10. Culms usually brownish at the base; pistillate scales brown, red-brown, or greenish; perigynia trigonous, yellow, olive green, or brown; leaves not glaucous.
                    11. Perigynia squarrose, none of them reflexed; leaves 4–9 mm wide; perigynia nerveless.
                        12. Pistillate spikes 4–6 mm thick; beak of perigynium absent or up to 0.5 mm long. . . . . 91. *C. shortiana*
                        12. Pistillate spikes 7–8 mm thick; beak of perigynium 1.0–1.5 mm long . .
                        . . . . 92. *C. X deamii*
                    11. Perigynia not squarrose, the lowest ones usually reflexed; leaves up to 4.5 mm wide; perigynia with a few nerves.
                        13. Pistillate spikes 6–10 mm thick; pistillate scales acute to acuminate, shorter than the perigynia; perigynia 3.2–4.5 mm long, narrowly ovoid, yellowish, the beak about as long as the body . . . .
                        . . . . 132. *C. cryptolepis*
                        13. Pistillate spikes 2–3 mm thick; pistillate scales obtuse, about as long as the perigynia; perigynia 2–3 mm long, ovoid, green or yellow-green, the beak about 1/3 as long as the body . . . . 133. *C. viridula*

Group XI

Leaves, sheaths, and/or culms all glabrous; perigynia glabrous; stigmas 2; achenes lenticular; some or all the spikes pedunculate; staminate flowers usually on separate spikes.

1. Terminal spike gynecandrous.
    2. Culms to 50 cm tall, slender; leaves 1.0–3.4 mm wide; pistillate scales obtuse to acute,

shorter than to about as long as the perigynia; perigynia up to 30 per spike, yellow to orange to white-papillate; pistillate spikes up to 2 cm long.

    3. Perigynia yellow to orange. . . . . 81. *C. aurea*

    3. Perigynia white-papillate . . . . 80. *C. garberi*

  2. Culms to 1.5 m tall, stout; leaves 4–14 mm wide; pistillate scales long-awned, much longer than the perigynia; perigynia more than 30 per spike, green to pale brown to stramineous; pistillate spikes up to 12 cm long. . . . . 82. *C. crinita*

1. Terminal spike entirely staminate.

    4. Culms to 50 cm tall, slender; leaves 1–3 mm wide; perigynia yellow to orange . . . .
    . . . . 81. *C. aurea*

    4. Culms to 1.5 m tall, usually stout; leaves 3–14 mm wide; perigynia brown to olive green to deep green.

      5. Pistillate scales long-awned, much longer than the perigynia . . . . 82. *C. crinita*

      5. Pistillate scales obtuse to acuminate, shorter than to barely longer than the perigynia.

        6. Lowest leaf sheath with well-developed blades; perigynia broadest above the middle
        . . . . 83. *C. aquatilis*

        6. Lowest leaf sheath bladeless; perigynia broadest at or below the middle.

          7. Perigynia with a distinct beak 0.3–1.0 mm long; pistillate scales purple-black to purple-brown.

            8. Beak of the perigynium twisted, 0.3–1.0 mm long; pistillate scales purple-black, shorter than the perigynia; leaves 3–5 mm wide. . . . . 88. *C. torta*

            8. Beak of perigynium straight or nearly so, 0.3–0.6 mm long; pistillate scales purple-brown, about as long as the perigynia; leaves up to 12 mm wide . . . . . .
            . . . . 85. *C. nebrascensis*

          7. Perigynia beakless, or the beak up to 0.3 mm long; pistillate scales red-brown, not purplish.

            9. Ligule longer than the width of the blade, V-shaped; perigynia nerveless or faintly nerved.

              10. Pistillate scales shorter than the perigynia; pistillate spikes to 10 cm long; perigynia more or less flat, 1.7–3.4 mm long; lowest sheath deep red or purple. . . . . 84. *C. stricta*

              10. Pistillate scales longer than the perigynia; pistillate spikes to 5 cm long; perigynia biconvex, 1.5–2.8 mm long; lowest sheath red-brown. . . . .
              . . . . 87. *C. haydenii*

            9. Ligule very short, not forming a V; perigynia distinctly nerved . . 86. *C. emoryi*

Group XII

Leaves, sheaths, and/or culms all glabrous; perigynia glabrous; stigmas 2; achenes lenticular; all spikes sessile, essentially alike, with staminate flowers either at the tip or at the base of each spike; culms solitary, or forming stoloniferous or rhizomatous colonies.

1. Culms arising from axils of last year's dead leaves along prostrate or reclining stems; culms usually smooth; tip of perigynia smooth, not serrulate . . . . 3. *C. chordorrhiza*

1. Culms arising directly from the rhizomes.

  2. Inflorescence headlike, with all spikes crowded; leaves filiform, up to 1.5 mm wide . . . . . .
  . . . . 1. *C. stenophylla*

  2. Inflorescence elongated, often the lowest spikes not contiguous with the ones above; leaves 1–3 mm wide, not filiform.

    3. Perigynia biconvex; sheaths tight; inflorescence usually 5–10 cm long; rhizomes short.

      4. Sheaths copper-colored at the summit; perigynia 2.5–4.0 mm long, 1.2–1.3 mm wide, stramineous to brown . . . . 25. *C. prairea*

4. Sheaths not copper-colored at the summit; perigynia 2–3 mm long, about 1 mm wide, dark brown to olive black . . . . 24. *C. diandra*

 3. Perigynia plano-convex; sheaths loose and open (except in *C. socialis*); inflorescence to 6 cm long, often shorter; rhizomes elongated.

  5. Inflorescence 5–12 mm wide; perigynia narrowly winged, conspicuously nerved ventrally (except in a rare variety of *C. foenea*); pistillate scales usually red-brown.

   6. Rhizomes slender, cordlike, brown; inflorescence 1–4 cm long; perigynia 4.5–6.0 mm long, the beak 2–3 mm long. . . . . 5. *C. foenea*

   6. Rhizomes thick, black; inflorescence 2.5–6.5 cm long; perigynia 2.0–4.6 mm long, the beak about 0.75 mm long. . . . . 4. *C. sartwellii*

  5. Inflorescence 3–5 mm wide; perigynia wingless or nearly so, usually nerveless or obscurely nerved ventrally; pistillate scales pale brown to green and hyaline.

   7. Rhizomes stout; pistillate scales pale brown, longer than the perigynia; perigynia 8–12 per spike, ovoid-lanceoloid, not spongy at the base . . . . 2. *C. praegracilis*

   7. Rhizomes slender; pistillate scales green and hyaline, shorter than the perigynia; perigynia 1–9 per spike, narrowly lanceoloid, spongy at the base. . . 10. *C. socialis*

Group XIII

Leaves, sheaths, and/or culms glabrous; perigynia glabrous; stigmas 2; achenes lenticular; all spikes sessile, essentially alike, with staminate flowers at the tip of each spike; culms cespitose, not arising singly from extensive stolons and/or rhizomes; spikes androgynous.

1. Perigynia 1–3 per spike; culms capillary . . . . 31. *C. disperma*

1. Perigynia 4–many per spike (sometimes 1–3 perigynia per spike in *C. socialis*); culms not capillary.

 2. Culms soft, wing-angled, easily compressed; sheaths loose and open.

  3. Beak of perigynium 1/2 as long as to nearly as long as the body; sheaths with red-brown dots; spikes 8–12 per inflorescence.

   4. Sheaths septate-nodulose; pistillate scales green and hyaline; beak of perigynium as long as the body. . . . . 27. *C. conjuncta*

   4. Sheaths usually not septate-nodulose; pistillate scales brownish; beak of perigynium about 1/2 as long as the body . . . . 26. *C. alopecoidea*

  3. Beak of perigynium 1 1/2–3 times longer than the body; sheaths without red-brown dots; spikes 10–25 per inflorescence.

   5. Pistillate scales shorter than the perigynia; perigynia yellow to brown; inflorescence up to 20 cm long, up to 6 cm thick; spikes 15–25 per inflorescence.

    6. Perigynia 3.5–6.0 mm long, strongly nerved on the inner face, the base tapering to the beak. . . . . 28. *C. stipata*

    6. Perigynia 6–8 mm long, nerveless or faintly nerved on the inner face, the base abruptly enlarged below the beak. . . . . 30. *C. crus-corvi*

   5. Pistillate scales about as long as the perigynia; perigynia greenish; inflorescence up to 6 cm long, up to 1.5 cm thick; spikes 10–15 per inflorescence . . . . 29. *C. laevivaginata*

 2. Culms firm, at least not conspicuously wing-angled and not easily compressed; sheaths loose or tight.

  7. Inflorescence consisting of 10 or more spikes, compound at the lower nodes.

   8. Sheaths not septate-nodulose; leaves 1.0–3.1 mm wide; perigynia biconvex, 1.0–1.3 mm wide; pistillate scales acute to cuspidate.

    9. Sheaths copper-colored at the summit; perigynia 2.5–4.0 mm long, 1.2–1.3 mm wide, stramineous to brown . . . . 25. *C. prairea*

9. Sheaths not copper-colored at the summit; perigynia 2–3 mm long, about 1 mm wide, dark brown to olive black . . . . 24. *C. diandra*

8. Sheaths septate-nodulose; leaves 2–8 mm wide; perigynia plano-convex (biconvex in *C. decomposita*), 1.2–2.0 mm wide; pistillate scales awned or at least mucronate.

   10. Perigynia plano-convex, green to stramineous to yellow- or golden brown, not spongy at the base; inflorescence up to 10 cm long, up to 1.5 cm thick.

      11. Leaves longer than the culms; inflorescence up to 10 cm long, usually with 15 or more spikes; perigynia stramineous to green, the beak 0.8–1.2 mm long. . . . . 21. *C. vulpinoidea*

      11. Leaves shorter than the culms; inflorescence up to 7 cm long, comprised of 10–15 spikes; perigynia yellow- to golden brown, the beak up to 0.7 mm long. . . . . 22. *C. brachyglossa*

   10. Perigynia biconvex, olive black, spongy at the base; inflorescence up to 18 cm long, up to 4 cm thick . . . . 23. *C. decomposita*

7. Inflorescence consisting of up to 10 spikes (occasionally up to 15 in *C. gravida* and *C. sparganioides*), not compound at the lower nodes.

   12. Sheaths loose, with green mottling or green septations; leaves 4–10 mm wide (sometimes only 3 mm wide in *C. aggregata*).

      13. Pistillate scales as long as or longer than the perigynia, awn-tipped . . . . 17. *C. gravida*

      13. Pistillate scales 1/2 as long as the perigynia to nearly as long, acute (rarely awned in *C. aggregata*).

         14. Inflorescence 5–15 cm long, the lowest spikes remote from the others; leaves up to 10 mm wide; perigynia 15–50 per spike . . . 20. *C. sparganioides*

         14. Inflorescence up to 5 cm long, the spikes all contiguous; leaves 3–8 mm wide; perigynia 5–20 per spike.

            15. Bracts absent, not subtending the spikes; leaves 3–5 mm wide; beak of the perigynium 1.0–1.5 mm long; summit of the sheaths concave . . . . 18. *C. aggregata*

            15. Bracts subtending spikes setaceous; leaves 4–8 mm wide; beak of the perigynium 0.5–1.3 mm long; summit of the sheaths truncate . . . . 19. *C. cephaloidea*

   12. Sheaths tight, without green mottling or green septations; leaves up to 4 mm wide (sometimes up to 5 mm wide in *C. cephalophora*).

      16. All spikes crowded into a head; perigynia usually not spongy-thickened at the base.

         17. Leaves 5–10 per culm; perigynia 2.7–4.0 mm long, 2.0–2.5 mm wide; ligule as wide as or wider than long; plants to 1 m tall.

            18. Pistillate scales about equalling the perigynia; inflorescence up to twice as long as thick; ligule wider than long . . . . 12. *C. mesochorea*

            18. Pistillate scales shorter than the perigynia; inflorescence 2–4 times longer than thick; ligule as wide as long. . . . . 14. *C. muhlenbergii*

         17. Leaves 3–5 per culm; perigynia 2.5–3.2 mm long, 1.2–2.0 mm wide; ligule longer than wide; plants up to 75 cm tall.

            19. Leaves 2–5 mm wide; perigynia broadest above the base, 1.5–2.0 mm wide . . . . 11. *C. cephalophora*

            19. Leaves up to 3 mm wide; perigynia broadest at the base, 1.2–2.0 mm wide . . . . 13. *C. leavenworthii*

      16. Spikes in an elongated, often interrupted, inflorescence; perigynia spongy-thickened at the base (except in *C. muhlenbergii*).

         20. Perigynia 2–3 mm wide; inflorescence up to 1 cm thick, often capitate above.

21. Perigynia not spongy-thickened at the base; leaves 5–10 per culm; beak of the perigynium up to 1 mm long . . . . 14. *C. muhlenbergii*

21. Perigynia spongy-thickened at the base; leaves 3–6 per culm; beak of the perigynium usually 1.0–1.8 mm long.

    22. Bracts leaflike, up to 25 cm long; leaves 1.0–2.5 mm wide; perigynia 3.5–4.0 mm long, pale green; beak of the perigynium 1.0–1.2 mm long. . . . . 16. *C. arkansana*

    22. Bracts setaceous, up to 2 cm long; leaves 2–4 mm wide; perigynia 4.0–5.5 mm long, green-brown to black; beak of the perigynium 1.5–1.8 mm long. . . . . 15. *C. spicata*

20. Perigynia 0.7–2.0 mm wide; inflorescence up to 5 mm thick, never capitate at the tip.

  23. Beak of the perigynium smooth; bracts absent or only poorly developed; some perigynia reflexed.

    24. Perigynia broadly ovoid, 1.3–2.0 mm wide, greenish brown at the base; leaves up to 2.5 mm wide. . . . . 6. *C. retroflexa*

    24. Perigynia narrowly lanceoloid, 0.7–1.2 mm wide, pale green throughout; leaves up to 1.5 mm wide . . . . 7. *C. texensis*

  23. Beak of the perigynium serrulate; bracts usually present, setaceous, up to 15 cm long; perigynia spreading to ascending.

    25. Stolons present, the plants colonial; perigynia narrowly lanceoloid . . . . 10. *C. socialis*

    25. Stolons absent, the plants not colonial; perigynia ovoid-lanceoloid.

      26. Stigmas twisted; leaves up to 3 mm wide; beak of the perigynium 0.8–1.5 mm long . . . . 8. *C. rosea*

      26. Stigmas more or less straight; leaves up to 1.9 mm wide; beak of the perigynium 0.5–1.0 mm long. . . . . 9. *C. radiata*

Group XIV

Leaves, sheaths, and/or culms all glabrous; perigynia glabrous; stigmas 2; achenes lenticular; all spikes sessile, essentially alike, with staminate flowers at the base of each spike; culms cespitose, not arising singly from extensive stolons and/or rhizomes; spikes gynecandrous; perigynia plano-convex, with rounded margins or with narrowly rimmed margins.

1. Perigynia with rounded margins, ascending.

  2. Plants weak and often reclining; spikes 1–3 (–4) per culm; perigynia 1–5 per spike, oblongoid. . . . . 32. *C. trisperma*

  2. Plants firm or, if weak, scarcely reclining; spikes usually 4 or more per culm (rarely fewer in *C. bromoides* and *C. brunnescens*); perigynia 5 or more per spike, lanceoloid to ellipsoid to narrowly ovoid (rarely only 4 in *C. bromoides*).

    3. All spikes overlapping; perigynia 3.5–4.5 mm long; beak of the perigynium 1.25–1.50 mm long.

      4. Perigynia lanceoloid, nerved on both faces; most leaves up up to 2 mm wide . . . . 35. *C. bromoides*

      4. Perigynia ovoid-lanceoloid, nerveless or faintly nerved only on the convex face; most leaves more than 2 mm wide . . . . 36. *C. deweyana*

    3. Some or all of the spikes separated from each other; perigynia 1.8–3.0 mm long; beak of the perigynium up to 1.2 mm long.

      5. Leaves up to 2 mm wide, dark green; perigynia 5–10 per spike; ventral band of sheath red-brown dotted. . . . . 34. *C. brunnescens*

5. Leaves 2–4 mm wide, glaucous to pale green; perigynia 10–30 per spike; ventral band of sheath not red-brown dotted .... 33. *C. canescens*
1. Perigynia with narrowly rimmed margins, spreading to reflexed.
    6. Terminal spike usually unisexual, either entirely staminate or entirely pistillate; perigynia castaneous to nearly black .... 38. *C. sterilis*
    6. Terminal spike gynecandrous; perigynia green to dark brown.
        7. Ventral face of the perigynium nerveless or nearly so; staminate part of the terminal spike proportionally longer than in the next two species .... 37. *C. interior*
        7. Ventral face of the perigynium usually with up to 10–12 nerves; staminate part of the terminal spike proportionally shorter than in *C. interior.*
            8. Perigynia broadly ovoid to suborbicular, 2.0–3.5 mm long, 1.3–2.8 mm wide, the beak 0.5–1.2 mm long .... 40. *C. atlantica*
            8. Perigynia lanceoloid to ovoid, 2.5–4.5 mm long, 1–2 mm wide, the beak 1–2 mm long .... 39. *C. echinata*

Group XV
Leaves, sheaths, and/or culms all glabrous; perigynia glabrous; stigmas 2; achenes lenticular; all spikes sessile, essentially alike, with staminate flowers at the base of each spike; culms cespitose, not arising singly from extensive stolons and/or rhizomes; spikes gynecandrous; perigynia more or less flat and scalelike, with winged margins.

1. Perigynia 7–10 mm long, the beak 4.0–4.8 mm long; spikes 12–27 mm long .... 41. *C. muskingumensis*
1. Perigynia up to 7 mm long (–7.5 mm long in *C. bicknellii*), usually much shorter, the beak rarely as much as 4 mm long; spikes up to 12 (rarely –15) mm long.
    2. Tip of pistillate scales exceeding, or at least equalling, the tip of the perigynium.
        3. Tip of pistillate scales exceeding the tip of the perigynium; perigynia 4.5–6.5 mm long, 1.5–2.0 mm wide, not broadly winged .... 61. *C. praticola*
        3. Tip of pistillate scales as long as the tip of the perigynium but not exceeding it; perigynia 4–5 mm long, 2.5–4.0 mm wide, broadly winged.............. 60. *C. alata*
    2. Tip of pistillate scales not reaching the tip of the perigynium.
        4. Perigynia up to 2 mm wide.
            5. Perigynia 0.8–1.1 mm wide, the wing very narrow to almost absent .... 44. *C. crawfordii*
            5. Perigynia 1.2 mm wide or wider, obviously winged.
                6. Perigynia widest at the middle.
                    7. Perigynia evenly winged all the way to the base.
                        8. Perigynia appressed or ascending.
                            9. Perigynia lanceolate.
                              10. Perigynia usually nerved only on the dorsal face; sterile tufts of leaves usually not present; perigynia 3.8–5.5 mm long............ .... 42. *C. scoparia*
                              10. Perigynia nerved on both faces; sterile tufts of leaves present; perigynia 3.0–3.6 mm long .... 48. *C. bebbii*
                        9. Perigynia ovate to obovate.
                            11. Perigynia ovate; pistillate scales usually awn-tipped.. 46. *C. straminea*
                          11. Perigynia obovate; pistillate scales acute to acuminate, not awn-tipped... 53. *C. longii*
                      8. Perigynia spreading.
                        12. Wing of perigynium not reaching tip of perigynium; perigynia obovate; spikes usually tapering to tip .... 52. *C. albolutescens*

12. Wing of perigynium extending to tip of perigynium; perigynia ovate; spikes usually subglobose, rounded at tip . . . . 49. *C. normalis*

7. Wing of perigynium narrowed above the base.
  13. Perigynia lance-ovate to ovate, usually crimped on the "shoulder", all widely spreading; spikes globose . . . . 47. *C. cristatella*
  13. Perigynia lanceolate, not crimped on the "shoulder", at least some of them usually ascending; spikes usually not globose.
    14. All but the uppermost spikes remote from each other . . . . 45. *C. projecta*
    14. Spikes crowded, with only 1 or 2 remote spikes, if any . . . . 43. *C. tribuloides*

6. Perigynia widest above or below the middle, but not at the middle.
  15. Perigynia widest a little above the middle; wing of the perigynium not reaching the tip of the perigynium . . . . 52. *C. albolutescens*
  15. Perigynia widest a little below the middle or nearly at the base; wing of the perigynium reaching the tip of the perigynium.
    16. Perigynia lanceolate, the wing diminishing above the base of the perigynium . . . . 45. *C. projecta*
    16. Perigynia ovate to orbicular, the wing extending all the way to the base of the perigynium.
      17. All spikes except sometimes the lowest 1–2 crowded . . . . 49. *C. normalis*
      17. Several of the lower spikes usually remote.
        18. Pistillate scales ovate; perigynia nerved on both faces . . . . 50. *C. tenera*
        18. Pistillate scales lanceolate; perigynia nerved on the dorsal face but only finely nerved on the ventral face . . . . 51. *C. festucacea*

4. Some or all of the perigynia more than 2 mm wide.
  19. Pistillate scales aristate to short-awned.
    20. Perigynia ovate, 2.5–3.5 mm long, widest at the middle, evenly winged all the way to the base, nerved . . . . 46. *C. straminea*
    20. Perigynia rhombic to suborbicular, 4–5 mm long, widest above the middle, the wing diminishing before the base or, if extending to the base, then very broad, nerveless or faintly nerved only on the ventral face.
      21. Wing of perigynium very broad, usually extending to the base of the perigynium; perigynia finely nerved on the ventral face, rounded at the base . . . . 60. *C. alata*
      21. Wing of perigynium not particularly broad, diminishing before reaching the base of the perigynium; perigynia more or less nerveless, cuneate at the base . . . . 55. *C. suberecta*
  19. Pistillate scales obtuse to acute to acuminate, never aristate or short-awned.
    22. Perigynia broadest above the middle.
      23. Wing of perigynium diminishing before reaching the base of the perigynium; perigynia more or less nerveless on both faces . . . . 55. *C. suberecta*
      23. Wing of perigynium extending to the base of the perigynium; perigynia at least finely nerved on both faces.
        24. Perigynia spreading, the wing not reaching the tip of the perigynium . . . . 52. *C. albolutescens*
        24. Perigynia appressed to ascending, the wing reaching the tip of the perigynium.
          25. Perigynia finely nerved, rhombic-orbicular . . . . 54. *C. cumulata*

25. Perigynia strongly nerved, obovate . . . . 53. *C. longii*
22. Perigynia broadest at or below the middle.
 26. Perigynia less than 4 mm long.
  27. Perigynia broadest below the middle . . . . 51. *C. festucacea*
  27. Perigynia broadest at the middle.
   28. Perigynia spreading.
    29. Wing of the perigynium not reaching the tip of the perigynium
      . . . . 52. *C. albolutescens*
    29. Wing of the perigynium reaching the tip of the perigynium . . .
      . . . . 49. *C. normalis*
   28. Perigynia appressed to ascending . . . . 53. *C. longii*
 26. Some or all of the perigynia 4 mm long or longer.
  30. Perigynia nerveless on the ventral face, orbicular.
   31. Perigynia 2.5–3.5 mm wide . . . . 56. *C. brevior*
   31. Perigynia 3.5–5.0 mm wide . . . . 58. *C. reniformis*
  30. Perigynia nerved, at least finely so, on the ventral face, ovate to
   obovate to suborbicular.
   32. Spikes globose to subglobose.
    33. Perigynia strongly nerved, 3.0–4.5 mm long, 1.5–2.1 mm
     wide, ovate. . . . . 49. *C. normalis*
    33. Perigynia faintly nerved, 4.0–5.5 mm long, 2–3 mm wide,
     broadly ovate to suborbicular . . . . 57. *C. molesta*
   32. Spikes longer than broad.
    34. Perigynia 4.5–7.5 mm long, 2.8–5.0 mm wide, broadly ovate,
     the wing becoming golden brown at maturity . . . 59. *C. bicknellii*
    34. Perigynia 2.6–4.5 mm long, 1.5–2.7 mm wide, obovate, the
     wing not becoming golden brown at maturity.
     35. Perigynia spreading, the wing not reaching the tip of the
      perigynium . . . . 52. *C. albolutescens*
     35. Perigynia appressed to ascending, the wing reaching the
      tip of the perigynium . . . . 53. *C. longii*

Descriptions and Illustrations

Carex L.—Sedge

Perennial herbs, sometimes with rhizomes and/or stolons; leaves flat, less commonly involute or plicate, glabrous or pubescent, often scabrous along the margins, surpassing or shorter than the culms; inflorescence composed of 1–several spikes in an elongated or capitate cluster; spikes either entirely staminate or entirely pistillate or with staminate flowers above the pistillate flowers (androgynous) or with pistillate flowers above the staminate flowers (gynecandrous), sessile or pedunculate, ascending or spreading or pendulous; each staminate flower consists of 1–3 stamens subtended by a scale; each pistillate flower consists of one pistil subtended by a scale; each pistil enclosed in a saclike structure known as a perigynium; perigynia often beaked, sometimes spongy at the base, plano-convex or biconvex or trigonous, nerved or nerveless; ovary superior, with one ovule; styles 2- or 3-cleft; achenes lenticular or trigonous.

1. **Carex stenophylla** Wahlenb. var. **enervis** (C. A. Mey.) Kukenth. in Engl. Das Pflanzenreich 4 (20):122. 1909. Fig. 1.
Carex enervis C. A. Mey. in Ledeb. Fl. Altsica 4:29. 1833.
Carex eleocharis Bailey, Mem. Torrey Club 1:6. 1889.

Plants perennial, cespitose, from fibrous roots, with scaly, wiry, filiform rhizomes; culms slender, smooth, bluntly angled, up to 20 cm tall, slightly exceeding the leaves, with the base fibrillose; leaves 2–4, up to 12 cm long, up to 1.5 mm wide, filiform, arising near the base of the culm, canaliculate, ascending, becoming triangular at the tip, scabrous on the margins; sheath open, green-nerved, with a broad hyaline ventral band, the lowermost brown, fibrous, and bladeless, the old sheaths persistent; inflorescence in an ovoid head 8–13 mm long, 3–5 mm broad, the staminate flowers above the pistillate; spikes 4–7, scarcely separated; pistillate scales ovate, 2.5–3.5 mm long, 1.5–2.0 mm wide, acuminate, reddish brown, concealing the perigynia; staminate scales narrowly lanceolate, 3–4 mm long, 1.0–1.5 mm broad, acuminate, reddish brown; bracts scalelike, encircling the culm, the lower ones awn-tipped; perigynia 3–7 per spike, 2.5–3.0 mm long, 1.25–1.50 mm broad, plano-convex, ovoid, acuminate, serrate in the upper half, obscurely striate ventrally and dorsally, nerveless, substipitate, with the bidentate beak 1.0 mm long, serrulate, hyaline-tipped and obliquely cleft dorsally, nearly hidden by the scales; achenes 1.50–1.75 mm long, 1.25–1.50 mm broad, lenticular, ovoid-orbicular, yellow-green, finely puncticulate, jointed to a short style; stigmas 2, slender, reddish brown.

Common Name: Needleleaf Sedge.
Habitat: Gravel bluff prairie.
Range: Manitoba northwest to Yukon, south to eastern Oregon, Utah,

New Mexico, Iowa, and northern Illinois.
Illinois Distribution: Known only from Kane and Winnebago counties.

Both Eurasian and American material was originally called *C. steno-phylla* Wahl. However, Eurasian plants tend to have perigynia that are 3.0–3.5 mm long and nerved, while American plants have perigynia that are 2.5–3.0 mm long and essentially nerveless.

  Carex stenophylla var. *enervis* is primarily a western and northern sedge with Illinois at the easternmost edge of its range. This taxon is considered to be a dominant in the blue grama-needlegrass-sedge com-

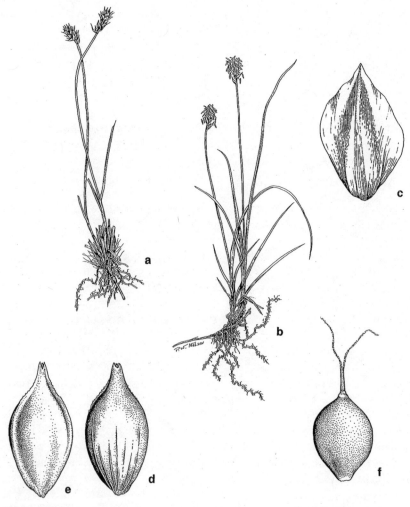

1. **Carex stenophylla var. enervis.**
a. Pistillate plant.

b. Staminate plant.
c. Pistillate scale.

d. Perigynium, dorsal view.
e. Perigynium, ventral view.
f. Achene.

munity in western North Dakota. This sedge ranks high among the species of the mixed prairie where it occurs in dense patches and provides early grazing. This variety flowers from May to July.

2. **Carex praegracilis** W. Boott in Coult. Bot. Gaz. 9:87. 1864. Fig. 2.

Plants perennial, cespitose, from fibrillose roots and stout, horizontal, brown to black, fibrous rhizomes; culms rising singly along the rhizome, slender, sharply angled and scabrous above, less angled below, 15–70 cm tall, usually equalling or slightly exceeding the upper leaves, the base scaly; leaves 2–4, 5–35 cm long, 1.5–3.0 mm wide, flattened, canaliculate, often keeled, ascending to spreading, the margins scabrous; upper sheaths pale, with a broad milky white ventral band, the lower sheaths dark, resembling those of the rhizome, the old sheaths persistent; inflorescence narrowly cylindric to slightly spreading, 2–5 cm long, 3–5 mm broad, the staminate flowers above pistillate or the staminate and pistillate flowers borne on separate plants; bracts lanceolate, the lowest enlarged and scabrous-awned, becoming smaller and less awned toward the apex, often encircling the culm with a dark ring at the base; spikes 5–12, the upper ones crowded, the lower ones interrupted; pistillate scales ovate, 3.5–4.0 mm long, 1.5–2.0 mm broad, concealing the perigynia, acuminate, pale to brown with a darker keel, the margins hyaline; staminate scales narrowly lanceolate, inconspicuous, 3–4 mm long, 1–2 mm broad, acuminate, pale, the margins hyaline; perigynia 8–12 per spike, 2.80–3.75 mm long, 1.25–1.50 mm broad, plano-convex, ovoid-lanceoloid, stipitate, sparsely nerved dorsally, nerveless ventrally, thin-winged or wingless and serrulate above, brownish black with the 1 mm long serrulate beak bidentate; achenes 1.5–2.0 mm long, 0.75–1.25 mm broad, lenticular, dark brown, polished, finely puncticulate, jointed to a short style; stigmas 2.

Common Name: Sedge.
Habitat: Low prairies; roadsides, particularly where salt has been applied during the winter months; dry sterile soil; low areas of medians and drainage swales.
Range: Manitoba northwest to Yukon, south to Mexico, Missouri, northern Illinois, and northern Michigan.
Illinois Distribution: This northern species is restricted to a few northern Illinois counties, as well as Christian County. It was first found in Illinois in 1897, but increased dramatically along roadsides during the 1970s because of the high salinity.

Carex praegracilis and *C. stenophylla* var. *enervis* are the only representatives of section Divisae in Illinois. They are both rhizomatous with usually solitary culms, androgynous spikes, two styles, a short-beaked perigynium, and lenticular achenes. They differ from each other in that *C. praegracilis* has stouter rootstocks, taller culms, longer inflorescences, and more perigynia per spike.

Immature specimens of *C. praegracilis* may be confused with *C. sartwellii* or *C. foenea*, but the brownish black perigynia without ventral nerves and the often scabrous culms usually distinguish *C. praegracilis*.

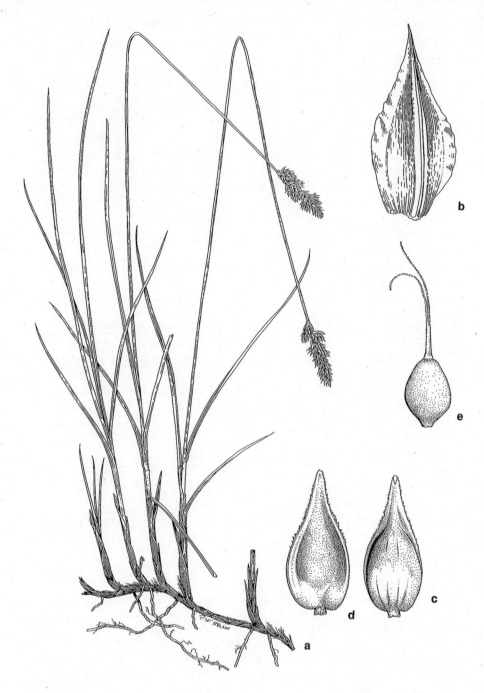

2. Carex praegracilis.
a. Habit.

b. Pistillate scale.
c. Perigynium, dorsal view.

d. Perigynium, ventral view.
e. Achene.

The spikes in *C. praegracilis* are variable in shape and color. This species flowers in Illinois from late April to early August.

3. Carex chordorrhiza Ehrh. in L. f. Suppl. Pl. Syst. Veg. 414. 1781. Fig. 3.

Plants perennial, cespitose, from cordlike roots and slender rhizomes; culms usually solitary, arising from the axils of leaves on prostrate or reclining stems, slender, usually smooth beneath the inflorescence, canaliculate, up to 35 cm tall, exceeding the leaves, reclining at the base, with the nodes bearing both fertile and sterile culms; leaves 1–4, up to 20 cm long, 1–2 mm wide, canaliculate, ascending to appressed with scabrous margins, becoming smooth near the base; sheaths open, concave at the mouth, the old sheaths persistent; inflorescence nearly head-like, 1.00–1.75 cm long, 0.5–1.0 cm broad, the staminate flowers above the pistillate; bracts (at least the lowest) awn-tipped; spikes 3–5, crowded; pistillate scales ovate, 3–4 mm long, 1.5–2.5 mm broad, acute to acuminate, reddish brown with hyaline margins, about as long as the perigynia; staminate scales conspicuous, narrowly lanceolate, 3–4 mm long, 1.0–1.5 mm broad, reddish brown to hyaline; perigynia 2–6 per spike, 2.5–3.5 mm long, 1.5–2.0 mm broad, compressed-ovoid to broadly ellipsoid, glossy, yellow-brown, stipitate, the margins smooth and thickened, strongly nerved on both sides, coriaceous and spongy throughout, quickly tapering to the slightly emarginate beak, the beak 0.5 mm long, hyaline-tipped; achenes 1.75–2.00 mm long, 1.25–1.75 mm broad, plano-convex, yellow-brown, puncticulate except along the margins, continuous with the short style; stigmas 2.

Common Name: Cordroot Sedge.
Habitat: Sphagnum swamps.
Range: Newfoundland and eastern Quebec to Alaska, south to Saskatchewan, northern Iowa, northern Illinois, northern Indiana, central New York, southwestern Vermont, and central Maine.
Illinois Distribution: Known only from Lake and McHenry counties.

Carex chordorrhiza is the only Illinois representative of section Chordorrhizeae, a group of species with cordlike rhizomes; usually solitary culms; headlike inflorescences with the staminate flowers borne above the pistillate; and wingless, spongy perigynia with distinct nerves.

This species is further characterized by its reclining base that further gives rise to secondary culms in the axils of the previous year's leaves.

The beaks of the perigynia of Illinois specimens are entire, although Fernald (1950) and Mackenzie (1940) both indicate serrulations present.

George Vasey collected this species several times in Lake County, the first time apparently in 1862. The last Illinois collection was made on April 30, 1988, at Volo Bog, Lake County. This species flowers from mid-April to early June.

4. Carex sartwellii Dewey, Am. Journ. Sci. 43:90. 1842. Fig. 4.

Plants perennial, cespitose, from fibrous roots, with thick, scaly, horizontal, black rhizomes; culms sharply triangular, scabrous, subcanaliculate, up to 1.2 m

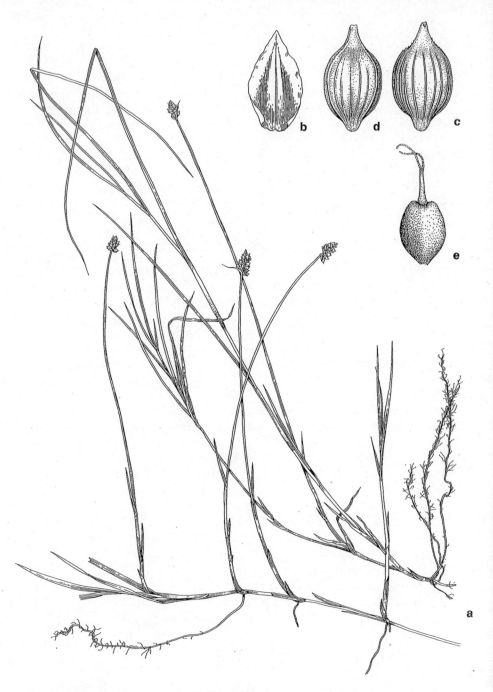

3. **Carex chordorrhiza.**
a. Habit.

b. Pistillate scale.
c. Perigynium, dorsal view.

d. Perigynium, ventral view.
e. Achene.

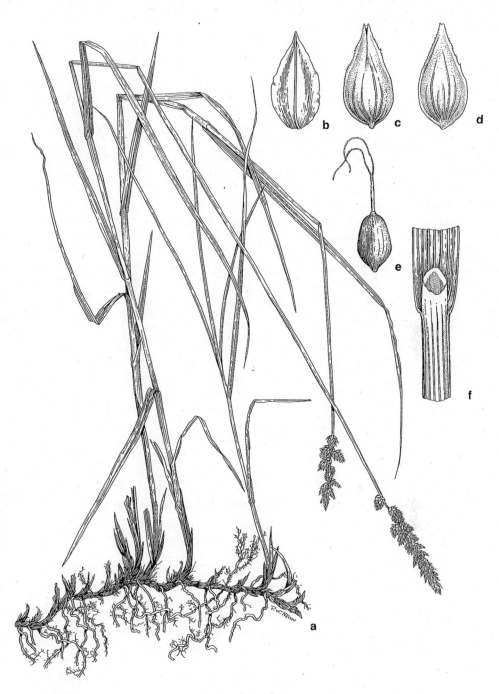

4. Carex sartwellii.
a. Habit.
b. Pistillate scale.
c. Perigynium, dorsal view.
d. Perigynium, ventral view.
e. Achene.
f. Sheath with ligule.

tall, much longer than the leaves; leaves 3–5, up to 23 cm long, 2.3–5.0 mm wide, arising from the lower half of the culm, ascending to spreading, thinly papillose above, the veins often appearing impressed on the lower surface, with the margins scabrous; sheaths open, truncate to concave at the mouth, the upper green-nerved ventrally, the lower brown to nigrescent, bladeless, with some old sheaths persisting; inflorescence narrowly cylindric, continuous or sometimes interrupted near the base, 2.5–6.5 cm long, 0.5–1.2 cm broad, stramineous, the staminate flowers above the pistillate; bracts lanceolate, the lowest scabrous-awned; spikes 10–25, the upper often entirely staminate and crowded, the lower mostly pistillate and often interrupted; pistillate scales ovate, 2.5–3.6 mm long, 1.20–1.75 mm broad, obtuse to acute to acuminate, often erose, reddish brown, the margins hyaline; staminate scales narrowly lanceolate, 2.5–4.0 mm long, 0.5–1.2 mm broad, acuminate, mostly hyaline throughout; perigynia 3–25 per spike, 2.0–4.6 mm long, 1.2–2.0 mm broad, plano-convex, ovoid-lanceoloid, substipitate, distinctly nerved dorsally and ventrally, narrowly winged to the base, serrate in the upper third, light brown, with the bidentate beak 0.75 mm long, serrulate, hyaline-tipped; achenes 1.30–1.75 mm long, 0.9–1.0 mm broad, plano-convex, reddish brown, puncticulate, the margins single-nerved, jointed to a short style; stigmas 2.

Common Name: Sartwell's Sedge.
Habitat: Low wet prairies and calcareous meadows, creek and river bottoms, marshes, dunes, peaty swamps, open cold bogs.
Range: Southwest Quebec to northern British Columbia, south to Colorado, Nebraska, Missouri, Illinois, Indiana, Ohio, and western New York.
Illinois Distribution: Mostly in the northern half of the state; also St. Clair and Washington counties.

This species and *C. foenea* are the only members of section Arenariae in Illinois. The section consists of rhizomatous perennials with mostly solitary culms; elongated spikes with staminate flowers above the pistillate; and perigynia with serrulate, bidentate beaks.

Carex sartwellii differs from *C. foenea* by its black rhizomes and smaller perigynia and achenes.

Extreme variation occurs in the inflorescence. The spikes may be very crowded or interrupted. The upper half of the inflorescence is often entirely staminate with the spikes much smaller and more sharply pointed and hyaline than the lower, subglobose, pistillate ones. In other plants, pistillate spikes occur throughout the inflorescence with a few staminate scales occurring in the tips of the spikes.

Plants with perigynia more than 4.1 mm long, known as var. *stenorrhyncha* Hermann, have not been found in Illinois.

Many of the early Illinois collections were called *C. disticha* Huds., but this binomial belongs to an apparently different species in Europe.

The first dated Illinois collection seen was made by Elihu Hall in June, 1860, although undated specimens collected from Fulton County by John Wolf and

Frederick Brendel may predate 1860. The flowers of this species bloom between mid-April and late June.

5. Carex foenea Willd. Enum. Hort. Berol. 2:957. 1809.
Carex siccata Dewey, Am. Journ. Sci. 10:278. 1826.

Plants perennial, cespitose, from fibrous roots, with scaly, slender, horizontal, cordlike rhizomes; culms slender, wiry, firm, mostly fertile, up to 80 cm tall, sharply angled and scabrous at the apex, becoming rounded, canaliculate and smooth near the base, equalling or exceeding the leaves; leaves 3–5, firm, up to 30 cm long, 2.0–3.5 mm wide, flattened, becoming triangular at the tip, ascending to spreading, with the margins scabrous; sheaths open, truncate to concave at the mouth, the lower acute, strongly nerved, brownish, bladeless, the upper green-nerved with hyaline internal tissue, the ventral band hyaline, slightly thickened at the summit, the ligule broadly V-shaped; inflorescence narrowly cylindric, pointed, 1–4 cm long, 0.5–1.0 cm broad, the staminate flowers above the pistillate or sometimes the staminate and pistillate flowers on separate plants; bracts lanceolate, the lowest scabrous-awned and encircling the culm; spikes 2–8, crowded at the apex, somewhat interrupted to widely separated at the base, the lower 1–3 spikes usually pistillate, the middle spikes staminate, and the terminal spikes larger and usually with female flowers above the male flowers; pistillate scales lanceolate, 3.50–4.75 mm long, 1.5–2.0 mm broad, acute to acuminate to aristate, light reddish brown with hyaline margins, often shorter than the perigynia; staminate scales acuminate, 3.5–4.2 mm long, 1.0–1.5 mm broad, light brown to hyaline; perigynia 3–12 per spike, 4.5–6.0 mm long, 1.5–1.8 mm broad, plano-convex, ovoid-lanceoloid, cuneate, substipitate, light reddish brown, coriaceous, distinctly nerved on both faces or rarely nerveless ventrally, the upper two-thirds narrowly winged and serrulate, with the bidentate beak 2–3 mm long, serrulate; achenes 1.90–2.25 mm long, 1.25–1.50 mm broad, lenticular, substipitate, yellow-brown, finely striate, puncticulate, jointed to the style; stigmas 2.

Two varieties of *Carex foenea* have been found in Illinois. They are distinguished by the following key.

1. Perigynia nerved on both faces, tapering gradually to the beak . . . . . . 5a. *C. foenea* var. *foenea*
1. Perigynia nerveless on the ventral face, tapering abruptly to the beak . . . .
. . . . 5b. *C. foenea* var. *enervis*

5a. Carex foenea Willd. var. **foenea** Fig. 5.

Perigynia nerved on both faces, tapering gradually to the beak.

Common Name: Sedge.
Habitat: Prairie or sandy soil, sometimes sandy woods and roadsides.
Range: Southwest Quebec to Mackenzie, south to Arizona, New Mexico, Illinois, northern Indiana, Ohio, New Jersey, and New York.

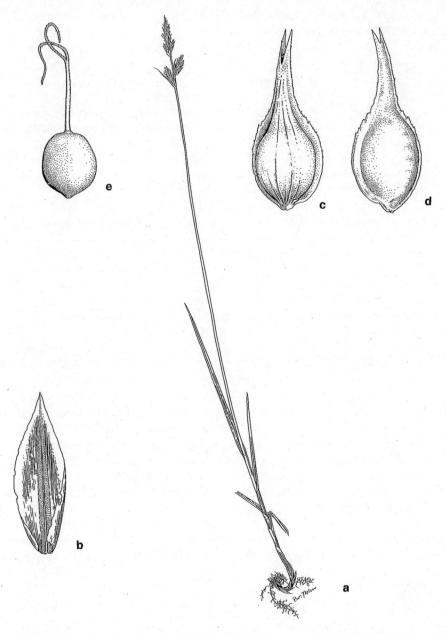

5. Carex foenea.
a. Habit.

b. Pistillate scale.
c. Perigynium, dorsal view.

d. Perigynium, ventral view.
e. Achene.

Illinois Distribution: Mostly confined to the northern sixth of the state; also Kankakee, Menard, and Peoria counties.

Carex foenea is most closely related to *C. sartwellii*, differing in its larger perigynia and achenes and in its brown rhizomes. The typical *C. foenea* strongly resembles *C. praegracilis* because of its slender inflorescence, but it is distinguished from it by its perigynium, which is many-nerved on both faces and is larger with a proportionately longer beak; by its rhizomes that are brown and scaly; and by its middle spikes that are usually wholly staminate.

Gleason and Cronquist (1991) call this species *C. siccata* Dewey, using the binomial *C. foenea* for a different species in section Ovales. This taxon flowers from mid-April through May.

5b. Carex foenea Willd. var. **enervis** Evans & Mohlenbr. Trans. Ill. Acad. Sci. 64:270. 1972. Not illustrated.

Perigynia nerveless on the ventral face, tapering abruptly to the beak.

> *Common Name:* Sedge.
> *Habitat:* Sandy plains.
> *Range:* Illinois.
> *Illinois Distribution:* Not known.

The type collection was made in 1869 by G. H. French in "sandy plains Ill." Although "Irvington, Illinois (Washington County)" is written on the label, the collection is probably not from there since Washington County is seemingly a little too far south. French, who lived for a while in Irvington, is known to have written "Irvington" on the labels of other specimens that were actually collected elsewhere in Illinois.

The absence of nerves on the ventral face of the perigynium and the more abruptly tapering beak differentiate this taxon from the typical variety. The month of the only collection is unknown.

6. Carex retroflexa Muhl. ex Willd. Sp. Pl. 4:235. 1805. Fig. 6.
Carex rosea Schkuhr var. *retroflexa* Torr. Ann. Lyc. N. Y. 3:389. 1836.
Diemisa retroflexa Raf. Good Book 27. 1840.
Carex bicostata Olney, Carices Bor.-Am. 9. 1872.

Plants densely cespitose; rhizome short, thick, dark brown, fibrous-scaly; roots wiry, abundant; culms mostly fertile, strict, becoming slightly spreading, smooth, canaliculate, with raised white striations, 1–2 mm broad just above the first developed leaf, up to 70 cm tall, mostly exceeding the leaves, becoming light to dark brown and fibrous at the base, producing leaves in the lower fourth; developed leaf blades of fertile culms 2–5, up to 40 cm long, up to 2.5 mm wide, flat, ascending to spreading, the margins scabrous except at the base, becoming triangular near the apex; sheaths tight, membranous and hyaline ventrally, strongly striate dorsally,

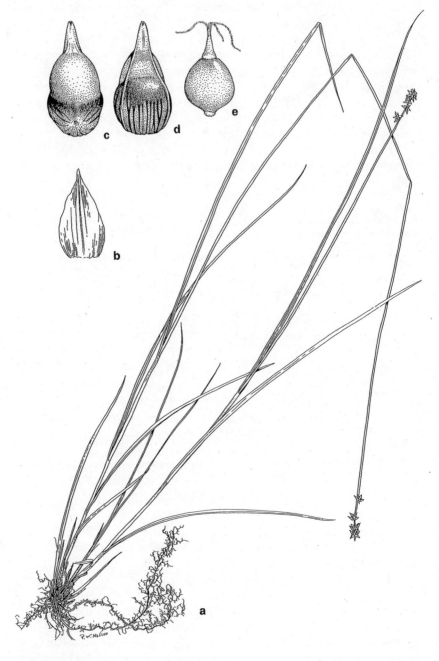

6. Carex retroflexa.
a. Habit.

b. Pistillate scale.
c. Perigynium, dorsal view.

d. Perigynium, ventral view.
e. Achene.

concave and somewhat darker and thickened at the mouth, older sheaths often persisting and becoming fibrous; inflorescence up to 4 cm long, simple; spikes 4–9, the staminate flowers above the pistillate, the upper 3–6 aggregated, the lower 2–3 interrupted; bracts poorly developed to absent; pistillate scales ovate, acute to acuminate to mucronate, 1.2–2.8 mm long, half as long to as long as the perigynia, not completely concealing the perigynia width, early deciduous; staminate scales inconspicuous except in the apical spikes, often persistent with the filaments; perigynia broadly ovoid, 2.5–3.5 mm long, 1.3–2.0 mm broad, up to 11 per spike, obscurely nerved dorsally, distinctly nerved in the gibbous-spongy, greenish brown, ventral base, reflexed, sessile, with smooth, slightly elevated, wirelike margins, the beak 0.5–1.0 mm long, hyaline dorsally, sharply bidentate, with the hyaline teeth appressed; achenes ovoid, plano-convex, contained in the upper perigynium body, 1.2–1.7 mm long, 1.2–1.5 mm broad, sharply depressed at the ventral base, light brown, finely puncticulate, polished, jointed to a short style; stigmas 2, light reddish brown, thin, flexuous.

Common Name: Sedge.
Habitat: Low wooded ridges, bluffs, dry woodlands.
Range: New Hampshire to southern Michigan and southeastern Kansas, south to Texas, Kentucky, and Virginia.
Illinois Distribution: Two primary areas of distribution, one in the southern tip of the state, the other in the east-central counties; also Henry and Woodford counties.

Carex retroflexa is readily distinguished from all species except *C. texensis* by its flattened, reflexed perigynia with smooth beaks. It is most closely related to *C. texensis*, a species that differs in its narrower perigynia, achenes, and leaves.

Evans (1976) has shown that these two species are distinct by using scanning electron microscopy to study the achenes. The flowers mature from mid-April to early June.

7. Carex texensis (G. S. Torr. ex Bailey) Bailey, Mem. Torrey Club 5:97. 1894. Fig. 7.
Carex rosea Schkuhr var. *texensis* G. S. Torr. ex Bailey, Mem. Torrey Club 1:57. 1889.
Carex retroflexa Muhl. var. *texensis* (G. S. Torr. ex Bailey) Fern. Rhodora 8:106. 1906.

Plants densely cespitose, the center of older clumps most often absent; rhizome short-continuous, wiry, knotted, brown to black, fibrous-scaly; roots wiry, abundant; culms mostly fertile, strict to spreading in taller individuals, smooth, canaliculate, with raised white striations, up to 1.5 mm broad above the first developed leaf, up to 40 cm tall, mostly far surpassing the leaves, brown and fibrous-scaly at the base, producing leaves in the lower fourth; leaf blades of fertile culms 1–5, up to 30 cm long, up to 1.5 mm wide, often crowded, flat, ascending, becoming triangu-

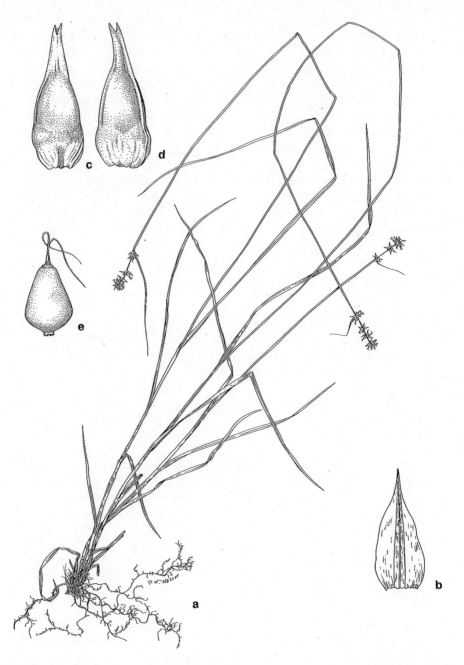

7. Carex texensis.
a. Habit.

b. Pistillate scale.
c. Perigynium, dorsal view.

d. Perigynium, ventral view.
e. Achene.

lar near the apex, the margins hyaline and scabrous, sometimes becoming smooth near the base, often inrolled; sheaths tight, membranous and hyaline ventrally, strongly striate dorsally, the mouth concave, with the older sheaths persistent and fibrous; inflorescence up to 3 cm long, simple; bracts poorly developed to absent; spikes 2–8, the staminate flowers above the pistillate, the upper 3–5 aggregated, the lowest 1–2 interrupted; pistillate scales lanceolate to ovate, acuminate to mucronate, 1.5–2.8 mm long, half to as long as the perigynia, not completely concealing the perigynium width, deciduous before the perigynia; staminate scales apical, linear-lanceolate, obscure except in the apical spikes, often persistent with the filaments; perigynia narrowly lanceoloid, up to 10 per spike, 2.5–3.2 mm long, 0.7–1.2 mm broad, obscurely wrinkled-striate dorsally, distinctly nerved on the gibbous-spongy ventral base, green to buff, reflexed, sessile, with the slightly elevated margins smooth, the beak 0.7–1.3 mm long, smooth, hyaline-striped dorsally, sharply bidentate, with the hyaline teeth appressed; achenes ovoid-lanceoloid, lenticular, sharply depressed at the ventral base, 1.2–1.6 mm long, 0.8–1.2 mm broad, contained in the upper perigynium body, light to dark brown, finely puncticulate, polished, jointed to a short, basally enlarged style; stigmas 2, light brownish red, thin, flexuous.

Common Name: Sedge.
Habitat: Disturbed soil, particularly in lawns and in cemeteries.
Range: New Jersey and Maryland to central Illinois and eastern Kansas, south to eastern Texas and South Carolina.
Illinois Distribution: Rare and scattered in Illinois.

Torrey first described this plant as a variety of *Carex rosea.* Later, Fernald treated it as a variety of *C. retroflexa.* Bailey was the first to recognize this taxon as a distinct species, and Evans (1976) concurred with this on the basis of scanning electron microscope studies. Evans has shown that this species is most closely related to *C. retroflexa. Carex texensis* is similar to *C. retroflexa* because of the smooth beak of the perigynium, but it differs in its narrower perigynia, achenes, and leaves. The flowers mature in Illinois from mid-April to mid-June.

8. **Carex rosea** Schkuhr ex Willd. Sp. Pl. 4:237. 1805. Fig. 8.
Carex rosea Schkuhr var. *pusilla* Peck in Howe, Ann. Rep. N.Y. State Mus. 48:132. 1897.
Carex convoluta Mack. Bull. Torrey Club 43:428. 1916.

Plants densely cespitose; rhizome scarcely distinguishable, discontinuous, brown to black, fibrous-scaly; roots wiry, abundant; culms mostly fertile, erect to slightly spreading, the angles antrorsely scabrous just under the inflorescence, smooth below, canaliculate, with raised white, wirelike striations, up to 80 cm tall, up to 2.2 mm broad above the first developed leaf, usually exceeding the leaves, brown-fibrous at the base, producing leaves in the lowest third; leaf blades of fertile

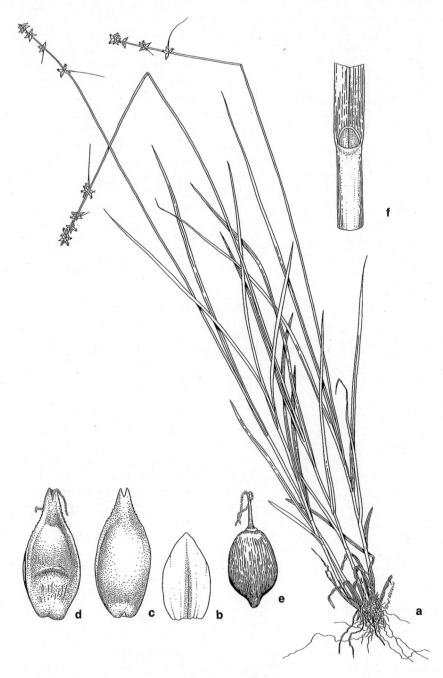

8. Carex rosea.
a. Habit.
b. Pistillate scale.

c. Perigynium, dorsal view.
d. Perigynium, ventral view.

e. Achene.
f. Sheath with ligule.

culms 3–6, up to 40 cm long, up to 3 mm wide, ascending to slightly spreading, flat, with the margins and midvein scabrous, becoming less so at the base, with few persistent old leaves; sheaths tight, ventrally membranous-hyaline, with the mouth concave to truncate; inflorescence up to 7 cm long, simple; spikes up to 8 per culm, the upper 3–5 aggregated, the lower 2–3 remote, with the staminate flowers above the pistillate; bracts bristleform, scabrous, the lowest up to 10 cm long, the others shorter or lacking; pistillate scales ovate, acute to mucronate, 1.2–2.0 mm long, about half as long as the perigynia, hyaline-buff, with the central nerve green to brown; staminate scales apical, ovate-lanceolate, inconspicuous; perigynia ovoid-lanceoloid, 1–13 per spike, averaging 7, 2.9–4.2 mm long, 1.1–1.7 mm broad, widest just below the middle, abruptly contracted into the elongated beak, green to buff, spongy at the base, becoming wrinkled with drying, widely spreading, sessile to substipitate, the margins raised and winglike, white-serrulate in the upper half, the beak 0.8–1.5 mm long, serrulate, distinctly hyaline-striped dorsally, sharply bidentate, with the white-hyaline teeth slightly divergent; achenes ovoid-orbicular, lenticular, slightly depressed at the base, 1.5–2.0 mm long, 1.1–1.5 mm broad, yellowish to dark brown, finely puncticulate, polished, jointed to the short, basally enlarged style, substipitate; stigmas 2, dark reddish brown, stout, short, flexuous to tightly twisted, deciduous with drying.

Common Name: Sedge.
Habitat: Usually mesic woodlands.
Range: Nova Scotia to Manitoba, south to Kansas, Mississippi, Georgia, and North Carolina; also in Texas.
Illinois Distribution: Occasional throughout most of the state.

This species closely resembles *C. radiata,* differing by its tightly twisted stigmas and slightly longer and broader perigynia. For a long time, this species has been known as *C. convoluta,* but Webber and Ball (1984) have indicated that *C. rosea* is the correct binomial. The plants in Illinois that are usually called *C. rosea* are now referred to as *C. radiata.* The flowering time in Illinois for this species is mid-April to mid-June.

9. **Carex radiata** (Wahlenb.) Small, Fl. S. E. U. S. 218. 1903. Fig. 9.
Carex stellulata var. *radiata* Wahlenb. Kongl. Vet. Acad. Nya Handl. 24:147. 1803.
Diemisa rosea Raf. Good Book 27. 1840.
Carex flaccidula Steud. Syn. Pl. Cyp. 99. 1855.
Carex rosea var. *minor* Boott, Ill. Carex 81, Pl. 224. 1860.
Carex rosea var. *staminata* Peck in Howe, Ann. Rep. N. Y. St. Mus. 48:132. 1897.
Carex rosea var. *minor* f. *debilis* Farwell, Papers Mich. Acad. 2:19. 1923.

Plants densely cespitose; rhizome short, stout, discontinuous, brown to black, fibrous-scaly, with 3–6 culms arising from a single node; roots abundant, wiry; culms mostly fertile, ascending to spreading, the angles antrorsely scabrous above, smooth below, canaliculate, with raised, white striations, up to 80 cm tall, up to

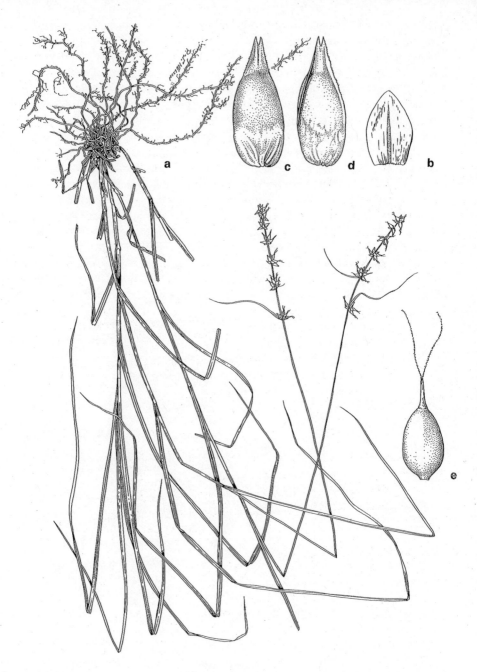

9. Carex radiata.
a. Habit.

b. Pistillate scale.
c. Perigynium, dorsal view.

d. Perigynium, ventral view.
e. Achene.

1.5 mm broad above the first developed leaf, erect to slightly spreading, equalling to exceeding the leaves, brown-fibrous at the base, producing leaves in the lowest fourth; leaf blades of fertile culms 3–6, to 40 cm long, to 1.9 mm wide, ascending to spreading, flat, with the margins and midvein scabrous, becoming smooth near the base, the old leaves often persistent; sheaths tight, membranous-hyaline ventrally, pale green to light brown, distinctly nerved dorsally, with the mouth concave and slightly thickened, old sheaths often persistent and fibrous; inflorescence up to 6.3 cm long, simple; spikes 4–9 per culm, the staminate flowers above the pistillate, the upper 3–5 aggregated, the lower 2–3 interrupted to remote; bracts elongate below, becoming rudimentary above, bristleform, the lowest up to 8 cm long, scabrous; pistillate scales ovate, obtuse to mucronate, half to two-thirds as long as the perigynia, 1.3–2.1 mm long, not concealing the perigynium width, hyaline-buff, with the central nerve darker; staminate scales apical, ovate-lanceolate, inconspicuous, persistent with the filaments; perigynia ovoid-lanceoloid, 1–9 per spike, 2.2–3.8 mm long, 1.0–1.5 mm broad, widest just below the middle, evenly tapered into the short beak, green to buff, the whitened base wrinkled-striate, spongy, withering with drying, ascending to spreading, sessile, the margins elevated, serrulate in the upper third, the beak 0.5–1.0 mm long, serrulate, obscurely hyaline-striped dorsally, bidentate, the hyaline teeth appressed; achenes ovoid, plano-convex, abruptly depressed at the ventral base, 1.4–1.9 mm long, 0.9–1.4 mm wide, yellowish to dark brown, finely puncticulate, polished, jointed to the short, basally enlarged style; stigmas 2, light to dark reddish brown, thin, straight to flexuous.

Common Name: Star Sedge.
Habitat: Woodlands, most often mesic; occasionally in disturbed areas.
Range: Nova Scotia to Alberta, south to southeastern Oklahoma, Louisiana, and South Carolina.
Illinois Distribution: Mostly throughout the state.

This species has been called *C. rosea* for many years, but that binomial should be reserved for the plant previously called *C. convoluta*, according to Webber and Ball (1984). *Carex radiata* is similar to *C. rosea* and *C. socialis* by virtue of its serrated perigynium beak. It differs from *C. socialis* in lacking extensive, creeping rhizomes and from *C. rosea* in its straight to flexuous stigmas and its slightly shorter perigynia. The mature flowers are open from late April to mid-June.

10. **Carex socialis** Mohlenbr. & Schwegm. Brittonia 21:77. 1969. Fig. 10.

Plants densely cespitose; rhizomes long-continuous, branching, wiry, brown to black, fibrous-scaly; roots wiry, abundant; culms frequently sterile, fertile culms ascending-spreading or lax, the angles antrorsely scabrous above, smooth below, with raised, white striations, up to 48 cm tall, up to 1.5 mm broad above the first developed leaf, the tallest culms equalling to exceeding the leaves, brown-fibrous at the base, producing leaves in the lower half; leaf blades of fertile culms 3–7,

10. Carex socialis.
a. Habit.

b. Pistillate scale.
c. Perigynium, dorsal view.

d. Perigynium, ventral view.
e. Achene.

1.5–2.2 mm wide, with the margins somewhat scabrous, becoming smooth in the lower fourth, the old leaves persistent; sheaths tight, delicate, membranous, hyaline, with the mouth concave, the old sheaths persistent and fibrous; inflorescence up to 4 cm long, simple; spikes 3–8, the staminate flowers above the pistillate, the upper 3–4 aggregated, the lower 2–3 interrupted to remote; bracts leaflike to bristleform, the uppermost rudimentary, the lowermost as much as 15 cm long, scabrous; pistillate scales ovate to lanceolate, acute to mucronate to subulate, 1.5–2.2 mm long, hyaline, with the central nerve green and sometimes exserted, a third to half as long as the perigynia, narrower than the perigynia; staminate scales apical, inconspicuous, often persistent with the filaments; perigynia narrowly lanceoloid, 1–9 per spike, 3.3–4.3 mm long, 0.8–1.4 mm broad, widest in the lowest fourth, evenly tapering to the elongated beak, green, with the whitish base wrinkled-striate and gibbous-spongy, becoming withered with drying, widely spreading at maturity, sessile, with the margins raised, coarsely white-serrulate in the upper half, the beak 0.8–1.3 mm long, serrulate, hyaline-striped dorsally, sharply bidentate, the hyaline teeth appressed to slightly divergent; achenes lanceoloid, plano-convex, 1.7–2.2 mm long, 0.7–1.1 mm broad, light brown, finely puncticulate, polished, jointed to the short, basally enlarged style; stigmas 2, light reddish brown, long, thin, straight to flexuous.

Common Name: Colonial Sedge.
Habitat: Wet woods, often in floodplains.
Range: Southern Indiana to southeastern Missouri, south to northeastern Texas and northwestern South Carolina.
Illinois Distribution: Confined to the southern tip of the state.

This species was first discovered in 1967 in Massac County, Illinois.
   While obviously closely related to *Carex rosea* and *C. radiata*, this species is readily distinguished by its long-continuous rootstocks resulting in the formation of colonies and by its very narrow perigynia.
   Scanning electron microscope studies of the achenes of this species by Evans (1976) indicate that this is a distinct species. The flowers bloom from mid-April to early June.

11. **Carex cephalophora** Muhl. ex Willd. Sp. Pl. 4:220. 1805. Fig. 11.

Plants perennial, cespitose, from fibrous roots and short, black rhizomes; culms slender, rough to the touch beneath the head, up to 60 cm tall, light brown at base, usually longer than the leaves; leaves 3–5, up to 40 cm long, up to 5 mm wide, confined to the lower fourth of the culm, flat, soft, pale green, rough along the margins and often along the keeled midrib beneath; sheaths open, tight, pale to greenish brown, concave and thickened at the mouth; ligule longer than wide; inflorescence capitate, subglobose to ovoid, up to 2 cm long, up to 1 cm broad;

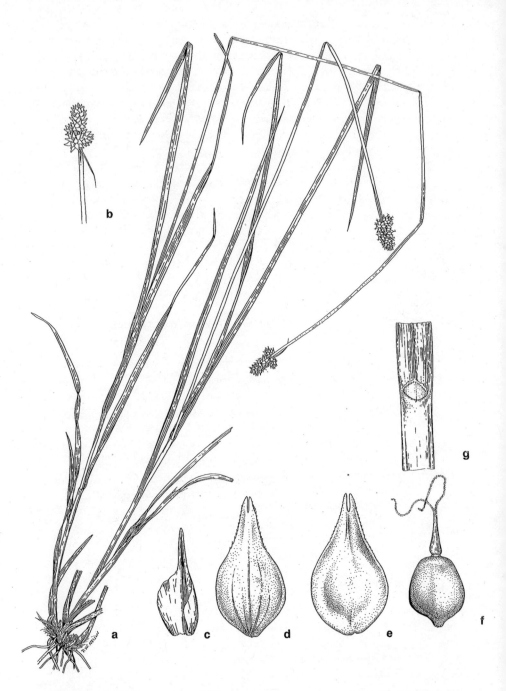

11. Carex cephalophora.
a. Habit.
b. Inflorescence.

c. Pistillate scale.
d. Perigynium, dorsal view.

e. Perigynium, ventral view.
f. Achene.
g. Sheath with ligule.

bracts setaceous, up to 5 cm long; spikes 3–8, crowded, the staminate flowers above the pistillate flowers; pistillate scales ovate, acuminate to awn-tipped, green or hyaline with a green midvein, shorter than the perigynia; staminate scales narrowly lanceolate, acuminate, pale brown, up to 1.5 mm broad; perigynia several per spike, 2.5–3.2 mm long, 1.5–2.0 mm broad, plano-convex, ovoid to ellipsoid-ovoid, pale green or light brown, nerveless or with 2–3 nerves on the dorsal face, with a 2-toothed, serrulate beak 0.5–1.1 mm long; achenes 1.0 mm long, lenticular, ovate-orbicular, apiculate, jointed to the style, the style swollen at base; stigmas 2, slender, reddish brown.

Common Name: Capitate Sedge.
Habitat: Woods, fields, lawns.
Range: Maine and Quebec to Ontario, southern Minnesota, and southeastern South Dakota, south to northeastern Texas and Florida.
Illinois Distribution: Throughout the state.

Carex cephalophora is readily distinguished by its capitate inflorescence, which consists of three to eight closely aggregated spikes. The very similar *C. mesochorea* has larger perigynia and scales that are as long as the perigynia, while *C. leavenworthii* has narrower leaves.

This is one of the more common species of *Carex* in Illinois, where it grows in a wide variety of usually dry habitats. It is often found in lawns, where it appears to survive mowing. *Carex cephalophora* flowers from early April to mid-July.

12. **Carex mesochorea** Mack. Bull. Torrey Club 37:246. 1910. Fig. 12.
Carex mediterranea Mack. Bull. Torrey Club 31:441. 1906, non C. B. Clarke (1896).

Plants perennial, cespitose, from fibrous roots and short, woody rhizomes; culms slender, rough to the touch beneath the head, up to 1 m tall, light brown at base, longer than the leaves; leaves 5–7, up to 15 cm long, up to 4.5 mm wide, confined to the lower third of the culm, flat, soft, light green, slightly rough along the margins and usually on the upper surface; sheaths open, tight, deeply concave and thickened at the mouth; ligule wider than long; inflorescence capitate, subglobose to ellipsoid, up to 2 cm long, up to 1.5 cm broad; bracts setaceous, usually not longer than 2 cm; spikes 4–8, crowded, the staminate flowers above the pistillate; pistillate scales ovate, acuminate or aristate, brownish or hyaline with a green midvein, about as long as the perigynia; staminate scales narrowly lanceolate, acuminate, pale brown, up to 1.5 mm broad; perigynia several per spike, 3–4 mm long, 2.0–2.5 mm broad, plano-convex, ellipsoid to narrowly ovoid, pale green or pale brown, weakly nerved on the dorsal face, with a 2-toothed, serrulate beak 0.8–1.0 mm long; achenes 1.5 mm long, lenticular, ellipsoid-ovoid, apiculate, jointed to the style, the style swollen at base; stigmas 2, slender, reddish brown.

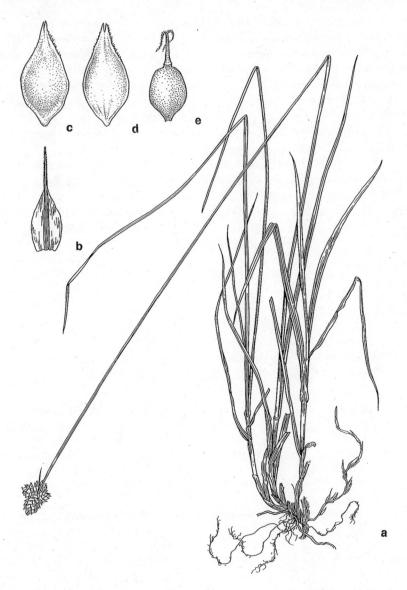

12. Carex mesochorea.
a. Habit.

b. Pistillate scale.
c. Perigynium, dorsal view.

d. Perigynium, ventral view.
e. Achene.

Common Name: Midland Sedge.
Habitat: Dry woods, fields.
Range: Massachusetts to southern Michigan, south to southern Illinois, Tennessee, and West Virginia.
Illinois Distribution: Known from Fayette, Macoupin, Shelby, and St. Clair counties.

There is some question as to the distinctness of this species from the similar *C. cephalophora*, but a number of differences occur. Because of the greater number of leaves per plant, *C. mesochorea* has a more bushy or tufted appearance than *C. cephalophora*. In all the specimens we have assigned to *C. mesochorea*, the pistillate scales reach at least to the base of the beak of the perigynium. The perigynia are slightly larger in *C. mesochorea* than in *C. cephalophora*, and the ligules are wider than they are long. The bracts that subtend the spikes are very short or sometimes lacking.

Mackenzie first named this species *C. mediterranea*, but that binomial had been used previously for a different species by C. B. Clarke. The flowers open from late May to late June.

13. Carex leavenworthii Dewey, Am. Journ. Sci. 52:246. 1846. Fig. 13.
Carex cephalophora Muhl. var. *angustifolia* Boott, Illustr. Carex 123. 1862.

Plants perennial, cespitose, from fibrous roots and short, black rhizomes; culms very slender, rough to the touch beneath the head, up to 75 cm tall, light brown at base, usually longer than the leaves; leaves 3–5, up to 25 cm long, up to 3 mm wide, confined to the lower third of the culm, flat, soft, pale green, rough along the margins; sheaths open, tight, concave and thickened at the mouth, finely russet-maculate along the margins of or throughout the hyaline ventral band; ligule longer than wide; inflorescence capitate, subglobose to oblongoid, up to 2 cm long, up to 1 cm broad; bracts setaceous, up to 2 cm long; spikes 3–8, crowded, the staminate flowers above the pistillate; pistillate scales ovate, acute to aristate, greenish or hyaline with a green midvein, shorter than the perigynia; staminate scales narrowly lanceolate, acuminate, pale brown, up to 1.5 mm broad; perigynia several per spike, 2.5–3.2 mm long, 1.2–2.0 mm broad, plano-convex, ovoid, pale green, faintly nerved on the dorsal face, with a 2-toothed, serrulate beak 0.3–0.8 mm long; achenes 1.0–1.5 mm long, lenticular, ovoid, apiculate, jointed to the style, the style swollen at base; stigmas 2, slender, reddish brown.

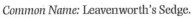

Common Name: Leavenworth's Sedge.
Habitat: Dry, open woods, either sandy or calcareous.
Range: Ontario and New Jersey to Michigan and southeastern Nebraska, south to Texas and Florida.
Illinois Distribution: Scattered in Illinois; apparently not common but probably overlooked.

13. Carex leavenworthii.
a. Habit.
b. Inflorescence.

c. Perigynium, dorsal view, with scale.

d. Perigynium, ventral view.
e. Achene.

This species is in the complex with *C. cephalophora* and *C. mesochorea*. All three species have spikes crowded into solitary heads on the culms. *Carex leavenworthii* differs from the other two in its narrower leaves and smaller stature. The flowers mature from mid-April to early June.

14. Carex muhlenbergii Schk. ex Willd. Sp. Pl. 4:231. 1805.

Plants perennial, cespitose, from fibrous roots and short, woody, black rhizomes; culms erect, rough to the touch beneath the inflorescence, up to 1 m tall, light brown at base, longer than the leaves; leaves 5–10, up to 30 cm long, 2.0–4.5 mm wide, confined to the lower fourth of the culm, flat or slightly canaliculate, pale green, rough along the margins and on the upper surface; sheaths open, tight, concave and thickened at the mouth, not as rough as the upper leaf surface, the hyaline ventral band well-developed; ligule as wide as long; inflorescence unbranched, up to 4 cm long, up to 1 cm broad, with most of the spikes crowded into a head, but with one or two of the lower spikes separated; bracts setaceous, the lowest one usually 1–2 cm long; spikes 3–10, the staminate flowers above the pistillate; pistillate scales ovate to lance-ovate, aristate, green or hyaline with a green midvein, a little shorter and narrower than the perigynia; staminate scales narrowly lanceolate, acuminate, pale brown; perigynia up to 20 per spikelet, 2.7–4.0 mm long, 2–3 mm broad, plano-convex, suborbicular to broadly ovoid, pale green, with or without nerves, with a 2-toothed, serrulate beak to 1 mm long; achenes 2.0–2.2 mm long, lenticular, ovoid-orbicular, apiculate, jointed to the style, the style swollen at base; stigmas 2, slender, reddish brown.

Three varieties occur in Illinois, distinguished by the following key:

1. Perigynia with conspicuous nerves on one or both faces, 3–4 mm long
  2. Perigynia 3.0–3.5 mm long, 2.0–2.5 mm wide; achenes 2.0–2.2 mm long . . . .
  . . . . 14a. *C. muhlenbergii* var. *muhlenbergii*
  2. Perigynia 3.5–4.0 mm long, 2.5–3.0 mm wide; achenes 2.2–2.5 mm long . . . .
  . . . . 14b. *C. muhlenbergii* var. *austrina*
1. Perigynia nerveless on both faces, 2.7–3.1 mm long . . . . 14c. *C. muhlenbergii* var. *enervis*

14a. Carex muhlenbergii Schk. var. **muhlenbergii** Fig. 14.

Perigynia with conspicuous nerves on both faces, 3.0–3.5 mm long, 2.0–2.5 mm wide; achenes 2.0–2.2 mm long.

Common Name: Muhlenberg's Sedge.
Habitat: Dry woods, scrubby black oak woods, old fields, sand prairies.
Range: Maine and Quebec to Minnesota, south to Texas and Florida.
Illinois Distribution: Throughout the state.

Most specimens of *C. muhlenbergii* in Illinois have nerves on the perigynia and are referable to the typical variety. The perigynia are 3.0–3.5 mm long and 2.0–2.5 mm wide. This is one of the more common species of *Carex* that lives in dry, often sandy habitats. It resembles *C.*

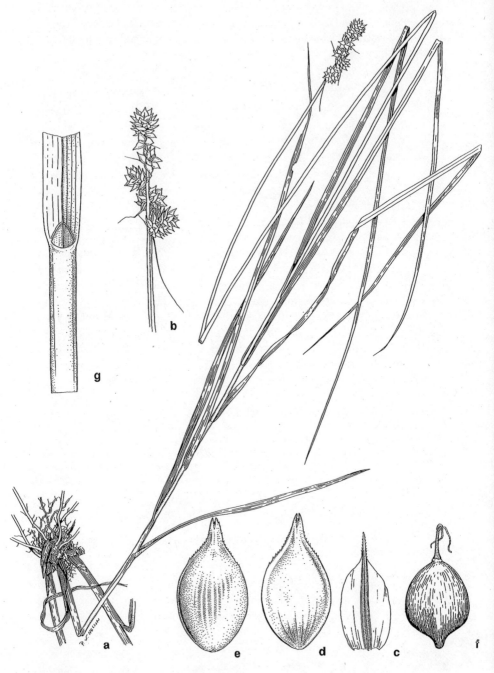

14. Carex muhlenbergii.
a. Habit.
b. Inflorescence.
c. Pistillate scale.
d. Perigynium, dorsal view.
e. Perigynium, ventral view.
f. Achene.
g. Sheath with ligule.

cephalophora except that the inflorescence is interrupted at the base. It differs from species such as *C. gravida, C. aggregata, C. cephaloidea,* and *C. sparganioides* in its tight leaf sheaths that are not septate. This taxon flowers from May through July.

14b. Carex muhlenbergii Schk. var. **austrina** Small, Fl. S. E. U. S. 218. 1903. Not illustrated.
Carex austrina (Small) Mack. Bull. Torrey Club 34:151. 1907.

Perigynia with conspicuous nerves on the dorsal face, 3.5–4.0 mm long, 2.5–3.0 mm wide; achenes 2.2–2.5 mm long.

Common Name: Southern Sedge.
Habitat: Along a railroad (in Illinois).
Range: Indiana to Kansas, south to Texas and Arkansas.
Illinois Distribution: Known only from along a railroad in Champaign County, where it is probably adventive.

This plant has larger perigynia and larger achenes than var. *muhlenbergii.* Some botanists in the past have considered it to be a distinct species.
    I reported this plant from two locations in prairies along railroads in Jackson and Perry counties. On further study, the specimens proved to be *C. gravida.*
    However, a single collection from Champaign County, originally determined as *C. spicata,* is actually *C. muhlenbergii* var. *austrina.* This variety flowers in May and June.

14c. Carex muhlenbergii Schk. var. **enervis** Boott, Illustr. Carex 3:124. 1862. Not illustrated.
Carex plana Mack. Bull. Torrey Club 50:350. 1923.

Perigynia nerveless on both faces, 2.7–3.1 mm long.

Common Name: Muhlenberg's Sedge.
Habitat: Dry woods, old fields.
Range: Vermont to Kansas, south to Texas and Mississippi.
Illinois Distribution: Scattered in Illinois.

This variety, much less common than the typical variety, often is found growing with the typical variety. Mackenzie considered this variety to be a distinct species. *Carex muhlenbergii* var. *enervis* flowers from May through July.

15. Carex spicata Huds. Fl. Angl. 349. 1762. Fig. 15.

Plants perennial, cespitose, from fibrous roots and short, woody, black rhizomes; culms erect, wiry, rough to the touch beneath the inflorescence, up to 80 cm tall,

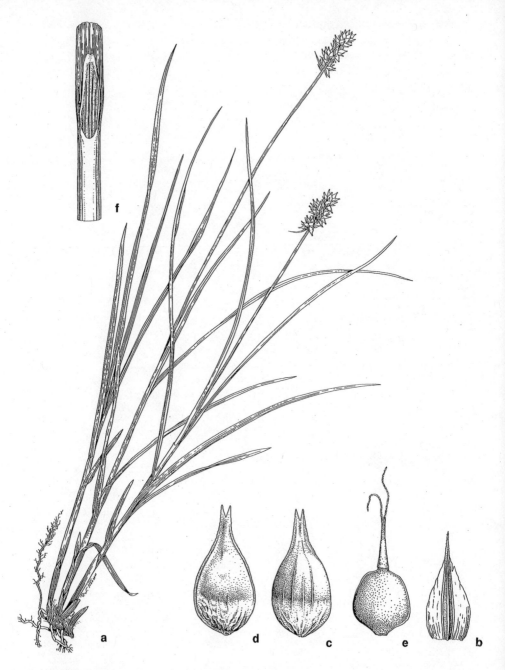

15. Carex spicata.
a. Habit.
b. Pistillate scale.

c. Perigynium, dorsal view.
d. Perigynium, ventral view.

e. Achene.
f. Sheath with ligule.

dark brown at base, longer than the leaves; leaves 3–6, up to 30 cm long, 2–4 mm wide, flat, rough along the margins and often along the midvein; sheaths open, tight, green, the lowermost red-brown; ligules much longer than broad; inflorescence unbranched, up to 4.5 cm long, up to 1 cm broad, with the spikes in a more or less interrupted head; bracts setaceous, up to 2 cm long; spikes 3–8, the staminate flowers above the pistillate; pistillate scales lanceolate to ovate, acute to acuminate to cuspidate, green or tawny with a hyaline margin, slightly shorter than the perigynia but nearly as broad; staminate scales narrower, pale brown; perigynia 5–10 in a spike, 4.0–5.5 mm long, 2.2–2.5 mm wide, spreading at maturity, lance-ovoid, plano-convex, nerveless on the ventral face, green-brown, becoming glossy black, spongy at base, with a 2-toothed, serrulate beak 1.5–1.8 mm long; achenes 2.0–2.2 mm long, lenticular, ovate-orbicular, apiculate, jointed to the style, the style swollen at base; stigmas 2, slender, red.

Common Name: Spicate Sedge.
Habitat: Disturbed, partly shaded, ground.
Range: Native of Europe and Asia; occasionally naturalized in the eastern United States.
Illinois Distribution: Collected once each in Du Page and De Kalb counties in 1982 and 1989, respectively.

This is one of the few species of *Carex* in Illinois that is not native. Lampa first found this species in Illinois at Fischer Woods Forest Preserve in Du Page County in 1982.

This species at first glance might appear to be a depauperate specimen of *C. muhlenbergii*, but the larger perigynia (4–6 mm long) are golden brown and widely spreading at maturity. The short, setaceous bracts readily distinguish *C. spicata* from *C. arkansana*, which has leafy bracts 5–25 cm long.

An earlier collection from Champaign County that had been called *C. spicata* is *C. muhlenbergii* var. *austrina*. This species flowers during May.

16. **Carex arkansana** Bailey, Bot. Gaz. 21:6. 1896. Fig. 16.

Plants perennial, cespitose, from fibrous roots and short, woody, dark rhizomes; culms erect, slender, smooth, up to 60 cm tall, brown at base, longer than the leaves; leaves 3–4, up to 20 cm long, 1.0–2.5 mm wide, confined to the lower part of the culm, flat, pale green, rough along the margins; sheaths tight, concave at the mouth; ligule shorter than broad; inflorescence unbranched, up to 4 cm long, up to 1 cm broad, with most of the spikes crowded in a head, but with the lowest spike usually separated; bracts leafy, 5–25 cm long, much exceeding the inflorescence; spikes 3–6, the staminate flowers above the pistillate; pistillate scales ovate, acute to acuminate to awned, hyaline except for the green midrib, narrower than the perigynia; staminate scales narrowly lanceolate, acuminate, pale brown; perigynia up to 20 per spike, 3.5–4.0 mm long, 2.2–3.0 mm broad, plano-convex, spongy at base, suborbicular to broadly ovoid, pale green, nerved on the back, with

16. Carex arkansana.
a. Habit.

b. Pistillate scale.
c. Perigynium, dorsal view.

d. Perigynium, ventral view.
e. Achene.

a 2-toothed, serrulate beak up to 1.2 mm long; achenes 1.5–2.0 mm long, lenticular, ovate-orbicular, apiculate, jointed to the style, the style swollen at base; stigmas 2, reddish brown.

Common Name: Arkansas Sedge.
Habitat: Moist flatwoods.
Range: Southern Illinois to southeastern Kansas, south to northeastern Texas and Arkansas.
Illinois Distribution: Known only from Saline County.

The only location for this species in Illinois is in a moist flatwoods about one mile north of Harrisburg in Saline County, where it was discovered in 1992 by John Taft. This species closely resembles *C. muhlenbergii* and *C. spicata* but differs from both in its presence of leafy bracts that range from 5–25 cm long and in its somewhat smaller achenes.

The Illinois station is the farthest east and north for this species. The flowers bloom in May and early June.

17. Carex gravida Bailey, Mem. Torrey Club 1:5. 1889.

Plants perennial, cespitose, from fibrous roots and short, dark, woody rhizomes; culms erect, up to 1 m tall, very rough to the touch beneath the inflorescence, light brown at base; leaves 4–6, up to 30 cm long, up to 8 mm wide, flat, rough on the upper surface, smooth and septate-nodulose below; sheaths open, loose, pale with green nerves, septate-nodulose, truncate or convex and slightly thickened at the mouth; inflorescence an elongated, sometimes interrupted head up to 5 cm long, up to 1.5 cm broad; bracts setaceous, the lowest as long as the inflorescence; spikes 5–15, the staminate flowers above the pistillate; pistillate scales ovate to lance-ovate, awn-tipped, green or brownish with a conspicuous midvein, about as long as or longer than the perigynia; staminate scales narrow, acuminate, pale brown; perigynia up to 15 per spike, 3.5–5.5 mm long, 2–3 mm broad, plano-convex, ovoid to suborbicular, greenish on the margins, stramineous above the spongy base, nerved on the dorsal face, with a 2-toothed, serrulate beak 0.5–1.5 mm long; achenes 2 mm long, lenticular, orbicular, apiculate, brownish yellow, jointed to the style, the style swollen at the base; stigmas 2, reddish brown to pale brown.

Two varieties of this species, which may be questionably recognized in Illinois, are separated by the following key:

1. Perigynia half as wide as long, obscurely nerved on the dorsal face ....
.... 17a. *C. gravida* var. *gravida*
1. Perigynia 2/5 to 3/4 as wide as long, conspicuously nerved on the dorsal face ....
.... 17b. *C. gravida* var. *lunelliana*

17a. Carex gravida Bailey var. gravida Fig. 17f.

Perigynia half as wide as long, obscurely nerved on the dorsal face.

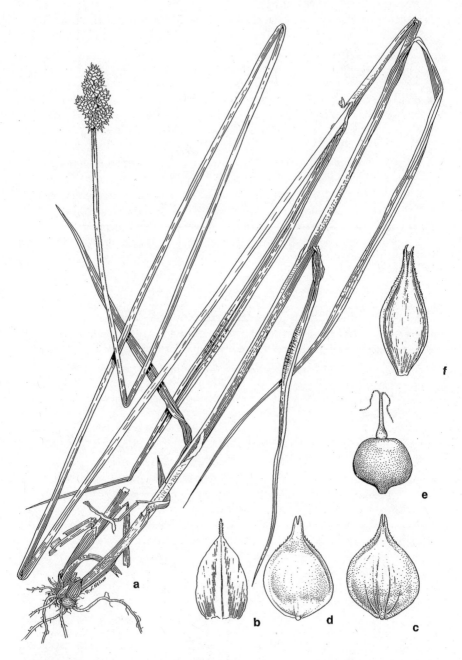

17. Carex gravida var. lunelliana.
a. Habit.

b. Pistillate scale.
c. Perigynium, dorsal view.
d. Perigynium, ventral view.

e. Achene.
var. gravida.
f. Perigynium, dorsal view.

Common Name: Sedge.
Habitat: Dry prairies, old fields, disturbed meadows.
Range: Ontario to Saskatchewan, south to Wyoming, New Mexico, southern Illinois, and western Ohio.
Illinois Distribution: Throughout most of the state

We reluctantly recognize two varieties of *C. gravida*, based on the relative width of the perigynia and the nature of the nerves on the dorsal face of the perigynium. There are some Illinois specimens in which these differences are difficult to discern.

Carex gravida is part of a complex of species in Illinois that also includes *C. aggregata*, *C. cephaloidea*, and *C. sparganioides*.

These species are similar in that they have spikes crowded into elongated inflorescences and septate-nodulose leaf sheaths. *Carex gravida* differs from all of these other species in having its perigynium with its beak less than half as long as the body and in having unwinged perigynia. Specimens from Jackson and Perry counties originally reported as *C. austrina* are actually *C. gravida*.

Carex gravida is more common in the northern half of the state, where it occurs in dry prairies, old fields, and disturbed meadows. This taxon flowers from mid-May through July.

17b. Carex gravida Bailey var. **lunelliana** (Mack.) F. J. Hermann, Am. Midl. Nat. 17:855. 1936. Fig. 17 a–e.
Carex lunelliana Mack. Bull. Torrey Club 42:615. 1915.

Perigynia two-fifths to three-fourths as wide as long, conspicuously nerved on the dorsal face.

Common Name: Sedge.
Habitat: Prairies and fields.
Range: Indiana and southern Michigan to South Dakota, south to New Mexico and Arkansas.
Illinois Distribution: Scattered in Illinois, but apparently absent in the northeastern counties.

This variety appears to occupy a more southwestern range than the typical variety of *C. gravida*.

Both varieties of *C. gravida* sometimes occur together in Illinois. The flowers bloom from mid-May through July.

18. Carex aggregata Mack. Bull. Torrey Club 37:246. 1910. Fig. 18.
Carex agglomerata Mack. Bull. Torrey Club 33:442. 1906, non C. B. Clarke (1903).
Carex sparganioides Muhl. var. *aggregata* (Mack.) Gl. Phytologia 4:22. 1952.

Plants perennial, cespitose, from fibrous roots and short, slender, woody, dark rhizomes; culms erect, up to 1 m tall, rough to the touch beneath the inflorescence,

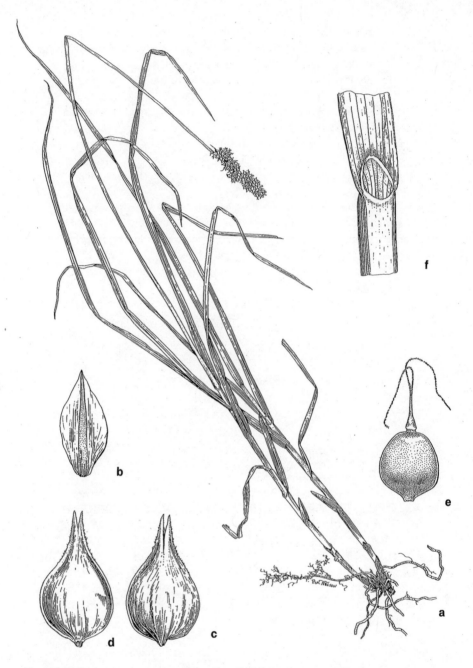

18. Carex aggregata.
a. Habit.
b. Pistillate scale.

c. Perigynium, dorsal view.
d. Perigynium, ventral view.

e. Achene.
f. Sheath with ligule.

light brown at base; leaves 3–7, up to 50 cm long, 3–5 mm wide, flat, pale green, shorter than the culms; sheaths open, loose, septate-nodulose, the summit firm, thickened and concave at the mouth, the ventral band white and nerveless; ligule about as wide as long; inflorescence an elongated, crowded or sometimes interrupted head up to 5 cm long, up to 1.2 cm broad; bracts subulate, up to 3 cm long, or lacking; spikes 5–10, the staminate flowers above the pistillate; pistillate scales lance-ovate to ovate, acute to acuminate, less commonly awned, greenish or hyaline with a green midvein, shorter and narrower than the perigynia; staminate scales narrow, acuminate, pale brown; perigynia 5–15 per spike, 3.5–4.6 mm long, 2–3 mm broad, plano-convex, ovoid, green above, fading to pale below, the base somewhat spongy thickened, obscurely nerved on the dorsal face, with a 2-toothed, serrulate beak 1.0–1.5 mm long; achenes 2 mm long, lenticular, orbicular, apiculate, jointed to the style, the style swollen at the base; stigmas 2, reddish brown.

Common Name: Clustered Sedge.
Habitat: Shaded woods, meadows.
Range: New York to southern Michigan to South Dakota, south to eastern Kansas, northern Georgia, and South Carolina.
Illinois Distribution: Scattered in the northern three-fourths of the state; also Alexander, Union, and Williamson counties.

Carex aggregata closely resembles *C. gravida* from which it differs in its longer perigynia beaks and its deeper green perigynia. It is similar to *C. cephaloidea*, a species that lacks bracts at the base of the spikes; to *C. sparganioides*, a species with a more elongated, interrupted inflorescence; and to *C. alopecoidea*, a species with slightly winged, flattened perigynia.

Gleason and Cronquist (1991) consider *C. aggregata* to be a variety of *C. sparganioides*. Mackenzie originally named this species *C. agglomerata*, apparently unaware that C. B. Clarke had used this binomial earlier for a different species. *Carex aggregata* flowers in Illinois from mid-May through July.

19. **Carex cephaloidea** (Dewey) Dewey, Rep. Pl. Mass. 262. 1840. Fig. 19.
Carex muricata var. *cephaloidea* Dewey, Am. Journ. Sci. 11:308. 1826.
Carex sparganioides Muhl. var. *cephaloidea* (Dewey) Carey in Gray, Man. Bot. 542. 1848.

Plants perennial, cespitose, from fibrous roots and short, stout, dark rhizomes; culms erect, up to 1 m tall, rough to the touch beneath the inflorescence, light brown at base; leaves 3–7, up to 35 cm long, 4–8 mm wide, flat, dark green, rough along the margins, shorter than the culm; sheaths open, loose, septate-nodulose, pale hyaline brown or with a faint flush of pink, the hyaline ventral band thin and easily torn at the summit; ligule longer than wide; inflorescence an elongated, crowded head, or with the lowest spikes not contiguous, up to 4 cm long, up to 1 cm broad; bracts subulate or lacking; spikes 5–10, the staminate flowers above the pistillate; pistillate scales ovate, acute, greenish or hyaline with a green mid-

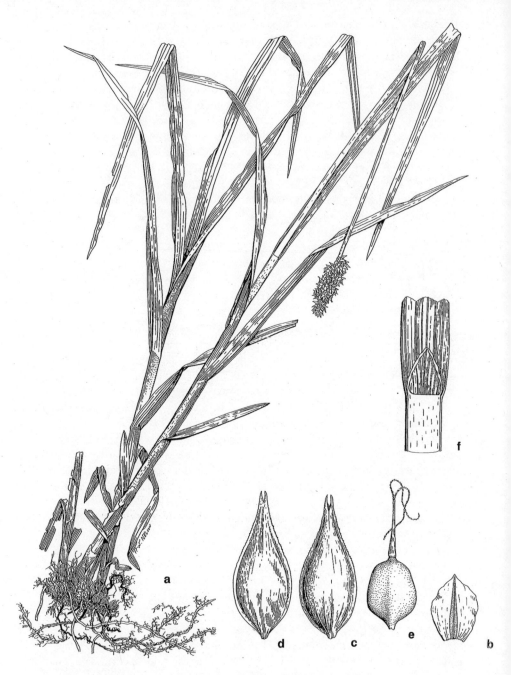

19. Carex cephaloidea.
a. Habit.
b. Pistillate scale.

c. Perigynium, dorsal view.
d. Perigynium, ventral view.

e. Achene.
f. Sheath with ligule.

vein, about half as long as the perigynia; staminate scales narrow, acuminate, pale brown; perigynia 7–20 per spike, 3.3–4.5 mm long, 1.5–2.5 mm broad, plano-convex, lance-ovoid to ovoid, pale green or becoming pale golden brown, obscurely nerved on the dorsal face, nerveless on the ventral face, with a 2-toothed, serrulate beak 0.5–1.3 mm long; achenes 1.7–2.0 mm long, lenticular, ellipsoid, apiculate, jointed to the style; stigmas 2, reddish brown.

Common Name: Sedge.
Habitat: Rich woods, sometimes disturbed; meadows.
Range: New Brunswick to northern Michigan and southern Minnesota, south to Iowa, Illinois, and New Jersey.
Illinois Distribution: Occasional in the northern three-fifths of the state, rare elsewhere.

This species is distinguished by a combination of scales that are about half as long as the perigynia, by a ligule that is much longer than it is broad, and by the summit of the leaf sheaths being truncate. The absence of bracts subtending the spikes in most specimens usually distinguishes *C. cephaloidea* from similar species, such as *C. gravida, C. aggregata,* and *C. sparganioides.*

Gleason and Cronquist (1991) recognize this plant as a variety of *C. sparganioides.* The flowers bloom from mid-May through July.

20. Carex sparganioides Muhl. ex Willd. Sp. Pl. 4:237. 1805. Fig. 20.

Plants perennial, cespitose, from fibrous roots and a short, dark, woody crown; culms erect, up to 1 m tall, wiry, rough to the touch beneath the inflorescence, light brown to yellow-brown at base; leaves 3–7, up to 40 cm long, up to 10 mm wide, flat, soft, dark green, rough along the margins, papillose above, smooth below, usually shorter than the culms; sheaths open, loose, pale with green mottling, the ventral band smooth to septate-nodulose, truncate at the mouth; ligule about as wide as long; inflorescence slender, elongated, interrupted, up to 15 cm long, up to 1.2 cm broad; bracts setaceous, nearly lacking to 4 cm long; spikes 6–15, the staminate flowers above the pistillate; pistillate scales ovate, acute to cuspidate, hyaline with a green midvein, half to nearly as long as the perigynia; perigynia up to 50 per spike, 3.0–4.3 mm long, 1.5–2.5 mm broad, plano-convex, lanceoloid to lance-ovoid, pale green or becoming golden, obscurely nerved or nerveless on the dorsal face, concave and nerveless on the inner face, narrowly winged to the base, tapering to a 2-toothed, serrulate beak 1.0–1.2 mm long; achenes 1.7–2.2 mm long, lenticular, ellipsoid, apiculate, jointed to the style, the style swollen at base; stigmas 2, reddish brown.

Common Name: Sedge.
Habitat: Dry or moist woods.
Range: Quebec and New Hampshire to South Dakota, south to Kansas, northern Georgia, and North Carolina.
Illinois Distribution: Scattered throughout the state.

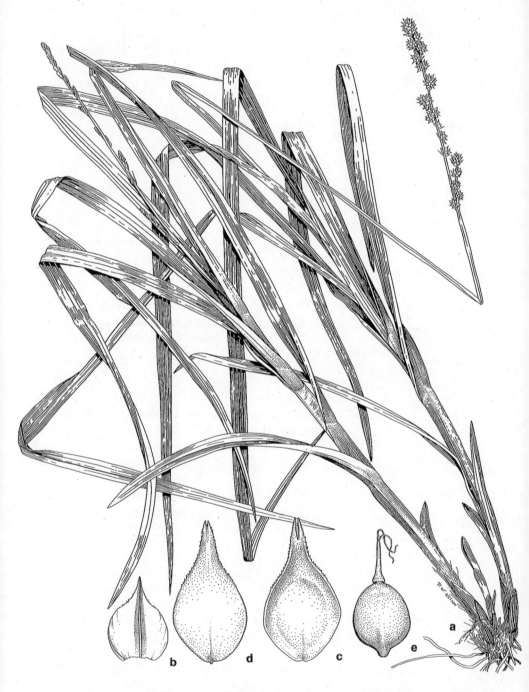

20. Carex sparganioides.
a. Habit.

b. Pistillate scale.
c. Perigynium, dorsal view.

d. Perigynium, ventral view.
e. Achene.

The abundant leaves and the slender, interrupted, graceful inflorescence make this one of the more handsome species of *Carex*. The number of spikes in an inflorescence varies considerably, but some of the lowest ones are always separated from each other. Bracts subtending the inflorescence may or may not be present. The flowers bloom from May through July.

21. **Carex vulpinoidea** Michx. Fl. Bor. Am. 2:169. 1803. Fig. 21.
Carex multiflora Muhl. ex Willd. Sp. Pl. 4:243. 1805.
Carex setacea Dewey, Am. Journ. Sci. 9:61. 1825.
Carex scabrior Sartw. in Boott, Illustr. Carex 3:125. 1862.
Carex xanthocarpa Bicknell var. *annectens* Bicknell, Bull. Torrey Club 23:22. 1896.
Carex annectens (Bicknell) Bicknell, Bull. Torrey Club 35:492. 1908.

Plants perennial, densely cespitose, from fibrous roots and short, stout, dark rhizomes; culms erect, up to 1 m tall, rough to the touch, at least beneath the inflorescence, dark brown at the base; leaves 3-several, up to 1.2 m long, 2–6 mm wide, flat or slightly canaliculate, rough along the margins, the upper ones exceeding the culms; sheaths open, tight, green-nerved, septate, with a broad, opaque, russet-maculate, often septate-nodulose ventral band, convex at the mouth; ligule broader than long; inflorescence composed of many spikes in an elongated, interrupted head up to 10 cm long and up to 1.5 cm broad; bracts setaceous, subtending most of the spikes, up to 5 cm long; spikes many, the staminate flowers above the pistillate; pistillate scales lanceolate, awned, hyaline to greenish brown and with a green midnerve, as long as or longer than the perigynia; perigynia several per spike, ovoid-lanceoloid to ovoid, 2–3 mm long, 1.2–1.8 mm broad, plano-convex, green to stramineous, nerveless or 2- to 4-nerved on the dorsal face, with a narrow, corky margin, tapering to a 2-toothed, serrulate or smooth beak about 0.8–1.2 mm long; achenes 1.2–1.4 mm long, lenticular, ovoid, red-brown, glossy, apiculate, jointed to the style, the style swollen at base; stigmas 2, reddish brown.

Common Name: Fox Sedge.
Habitat: Swamps, wet meadows, low areas, moist open ground.
Range: Newfoundland to British Columbia, south to California, Arizona, Texas, and Georgia.
Illinois Distribution: Probably in every county.

This species and *C. brachyglossa* are distinguished by their long, slender, interrupted inflorescences with conspicuous setaceous bracts subtending most of the spikes. *Carex vulpinoidea* differs from *C. brachyglossa* in its greenish or stramineous, narrower perigynia with beaks more than 0.7 mm long, and its leaves usually as long as or longer than the culm. *Carex vulpinoidea* is one of the most widespread species of *Carex* in North America. It is also probably the most common *Carex* in Illinois.

Occasional specimens with greenish or stramineous perigynia but with the leaves decidedly shorter than the culms have been referred to as *C. setacea* Dewey. There is speculation that *C. setacea* may be a hybrid between *C. vulpinoidea* and

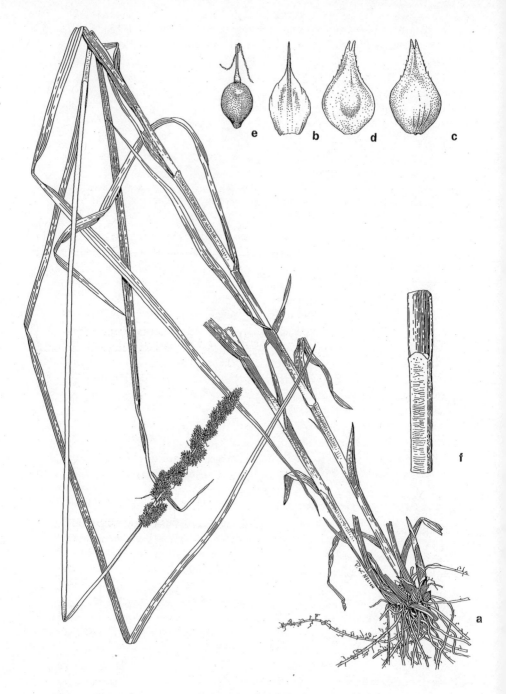

21. Carex vulpinoidea.
a. Habit.
b. Pistillate scale.

c. Perigynium, dorsal view.
d. Perigynium, ventral view.

e. Achene.
f. Sheath.

C. brachyglossa. I do not believe the differences shown by *C. setacea* are worthy of taxonomic recognition. The flowers bloom from May to mid-August.

22. Carex brachyglossa Mack. Bull. Torrey Club 50:355. 1923. Fig. 22.
Carex xanthocarpa Bicknell, Bull. Torrey Club 23:22. 1896, non Degland (1807).
Carex annectens (Bicknell) Bicknell var. *xanthocarpa* (Bicknell) Wieg. Rhodora 24:74. 1922.

Plants perennial, densely cespitose, from fibrous roots and short, stout, dark rhizomes; culms erect, up to 75 cm tall, rough to the touch beneath the inflorescence, brown at the base; leaves 3–6, up to 60 cm long, 2–5 mm wide, flat or slightly canaliculate, rough along the margins, none of them exceeding the culms; sheaths open, tight, at least the upper with the hyaline ventral band becoming septate-nodulose with age, with pale brown dots, convex at the mouth; ligule broader than long; inflorescence composed of many spikes in an elongated, interrupted head up to 7 cm long and up to 1.5 cm broad; bracts setaceous, subtending most of the spikes, up to 5 cm long; spikes 10–15, the staminate flowers above the pistillate; pistillate scales lanceolate, awn-tipped, reddish brown with hyaline margins and a green midnerve, as long as or longer than the perigynia; perigynia several per spike, ovoid to ellipsoid, 2.2–3.0 mm long, 1.5–2.0 mm broad, plano-convex, yellow to golden brown except for the margins, with 3 nerves on the dorsal face, tapering to a 2-toothed, serrulate beak up to 0.7 mm long; achenes 1.2–1.5 mm long, lenticular, broadly ellipsoid, red-brown, glossy, apiculate, jointed to the style, the style enlarged at base; stigmas 2, reddish brown.

Common Name: Yellow Fox Sedge.
Habitat: Fields, disturbed low ground.
Range: Maine to southern Michigan and Iowa, south to Kansas, northern Tennessee, and North Carolina.
Illinois Distribution: Scattered throughout the state.

Carex brachyglossa differs from *C. vulpinoidea* in its yellow or golden brown perigynia, its leaves that are shorter than the culm, and its slightly shorter beak of the perigynium. *Carex brachyglossa* grows in open areas, often with a sandy substrate.

Plants from Illinois that have been called *C. annectens* previously are now assigned to this species. The typical *C. annectens* is now considered to be synonymous with *C. vulpinoidea.* This species flowers from May through July.

23. Carex decomposita Muhl. Descr. Gram. 264. 1817. Fig. 23.

Plants cespitose; rhizomes woody, thick, blackish, fibrous; roots wiry; culms mostly fertile, spreading and arching, smooth or roughened, terete or obtusely angled, 4–8 mm wide near base, up to 1.5 m tall, slightly shorter to slightly longer than the leaves, usually with several old leaves persistent near the base, producing leaves in the lower third; leaves up to 75 cm long, up to 8 mm wide, flat or canalicu-

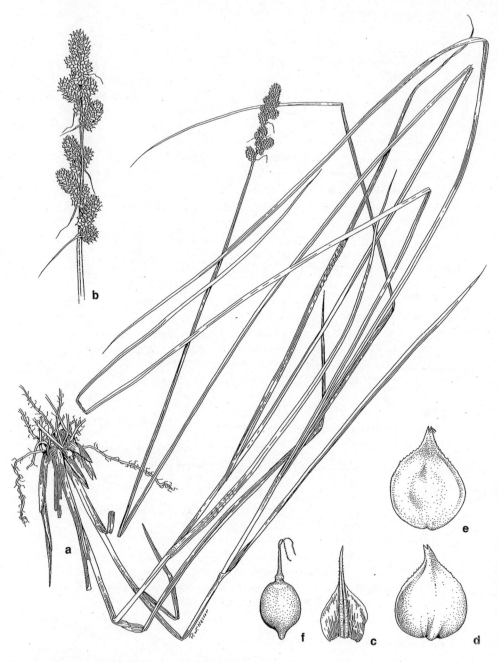

22. Carex brachyglossa.
a. Habit.
b. Inflorescence.

c. Pistillate scale.
d. Perigynium, dorsal view.

e. Perigynium, ventral view.
f. Achene.

23. Carex decomposita.
a. Habit.

b. Pistillate scale.
c. Perigynium, dorsal view.

d. Perigynium, ventral view.
e. Achene.

late, stiff, the margins and main nerves serrulate; sheaths tight, septate-nodulose, pale, sometimes red-dotted, concave at the usually reddish mouth; inflorescence up to 18 cm long, compound, the lower branches up to 4 cm long; bracts absent or setaceous, up to 5 cm long; spikes crowded, the staminate flowers above the pistillate; pistillate scales triangular-ovate, mucronate, up to 2.8 mm long, about as long as the perigynia, hyaline, with a 3-nerved green center, persistent after the perigynia have fallen; staminate scales apical, inconspicuous, crowded; perigynia obovoid, crowded, coriaceous, biconvex, shiny, spreading, 2.0–2.8 mm long, 1.50–1.75 mm broad, spongy at base, olive to blackish, obscurely few-nerved above, strongly several-nerved below, rounded at summit except for the beak, tapering to a stipitate base, the beak 0.5 mm long, flat, serrulate, bidentate; achenes about 1 mm long, closely enveloped by the perigynia, lenticular, apiculate, stipitate, jointed with the short style; stigmas 2, short, light reddish brown.

Common Name: Cypress Knee Sedge.
Habitat: Usually on fallen logs or on swollen bases of trees in cypress swamps.
Range: New York to Michigan, south to Missouri and Louisiana.
Illinois Distribution: Confined to the southern tip of the state.

The first Illinois collection was made June 27, 1951 by W. M. Bailey and J. R. Swayne from the LaRue Swamp in Union County.

The major habitat for this species in Illinois is in bald cypress swamps where it grows on the swollen bases of trees and on fallen logs.

The compound inflorescence and the black or olive perigynia are distinguishing features. The flowers mature from June to August.

24. Carex diandra Schrank, Cent. Bot. Ammerk. 57. 1781. Fig. 24.
Carex teretiuscula Gooden, Trans. Linn. Soc. 2:163. 1794.
Carex teretiuscula var. *major* Koch, Syn. Fl. Germ. 751. 1837.

Plants loosely cespitose; rhizomes short, slender, blackish, fibrous-scaly; roots wiry, abundant; culms mostly fertile, erect, becoming slightly spreading, roughened on the rounded angles, up to 1.1 m tall, exceeding the leaves, brown or dark brown at base, producing leaves in the lower fourth; blades 4–5 per culm, up to 30 cm long, 1–3 mm wide, flat, canaliculate or plicate, ascending or spreading, the margins and sometimes the tip scabrous, light green; sheaths tight, membranous and hyaline ventrally, striate dorsally, strongly red-dotted, truncate or convex at the mouth; inflorescence up to 8 cm long but usually shorter, up to 12 mm thick, subcylindric, continuous, compound and paniculate at the lowest node; bracts subulate, up to 1 mm long or absent; spikes several, crowded, dark brown, the staminate flowers above the pistillate; pistillate scales ovate-lanceolate, up to 3.5 mm long, more or less concealing the sides of the perigynia, acute to short-cuspidate, greenish but tinged with brown, the margins scarious above the middle; staminate scales apical, inconspicuous, very narrow; perigynia ovoid, biconvex, strongly rounded on the back, ascending to widely spreading, 2–3 mm long, about

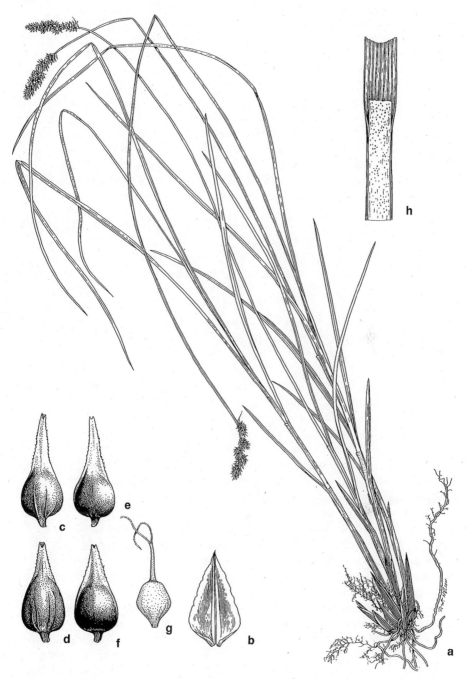

24. **Carex diandra.**
a. Habit.
b. Inflorescence.

c, d. Perigynia, dorsal views.
e,f. Perigynia, ventral views.

g. Achene.
h. Sheath.

1 mm broad, strongly nerved on the lower side, dark brown to olive black, shiny, stipitate, tapering to a green beak, the beak up to 1 mm long, bidentate, serrulate, green or whitish; achenes lenticular, up to 1 mm long, apiculate, stipitate, jointed to a short style; stigmas 2, short, reddish brown.

Common Name: Sedge.

Habitat: Calcareous wet meadows, swamps, bogs.

Range: Labrador to southern Alaska, south to California, Nebraska, northern Illinois, Ohio, Pennsylvania, and New Jersey; Europe; Asia; New Zealand.

Illinois Distribution: Confined to the northeastern corner of the state and to a few central counties near the Illinois River.

This species closely resembles *C. prairea*, differing by its more spreading, shiny, blackish perigynia and non-revolute leaves.

Until 1927, Illinois botanists called this species *C. teretiuscula*, but this binomial is a synonym for the earlier *C. diandra*. The flowers bloom from mid-May to early July.

25. Carex prairea Dewey ex Wood, Class-Book ed. 2, 414. 1847. Fig. 25.
Carex teretiuscula var. *ramosa* Boott, Illustr. Carex 43. 1867.
Carex diandra var. *ramosa* (Boott) Fern. Rhodora 10:48. 1908.

Plants loosely tufted or more or less running; rhizomes short, slender, blackish, fibrous-scaly; roots wiry, abundant; culms 1–several, mostly fertile, erect, becoming slightly spreading, roughened on the sharp angles, up to 1 m tall, up to 3 mm broad above the lowest leaf, usually slightly surpassing the leaves, brown or dark brown at base, producing leaves in the lower fourth; blades 3–4 per culm, up to 30 cm long, 2.0–3.1 mm wide, flat but with revolute margins, ascending or spreading, papillate above, the margins and sometimes the tip harshly scabrous, light green; sheaths tight, membranous and hyaline, papillate between the pale cartilaginous veins, sometimes usually minutely copper-dotted, convex and strongly copper-tinged at the mouth, prolonged 2–3 mm beyond base of blade; inflorescence up to 10 cm long, up to 2 cm thick, continuous near the tip, usually interrupted below, compound at the lower nodes; bracts setaceous or scalelike, usually never longer than the spikes; spikes several, crowded, 3–6 mm long, 2.5–4.0 mm broad, the staminate flowers above the pistillate; pistillate scales ovate-lanceolate, up to 3.5 mm long, mostly concealing the perigynia, acute to cuspidate, reddish brown except for the 1- to 3-nerved hyaline center; staminate scales apical, inconspicuous, very narrow; perigynia narrowly ovoid, biconvex, appressed to strongly ascending, 2.5–4.0 mm long, 1.2–1.3 mm broad, strongly nerved on the lower side, dark stramineous to brown, dull, stipitate, tapering to a beak, the beak up to 1 mm long, bidentate, rough along the margins, green or whitish but usually reddish tipped; achenes lenticular, up to 2 mm long, apiculate, stipitate, jointed to a short style; stigmas 2, short, reddish brown.

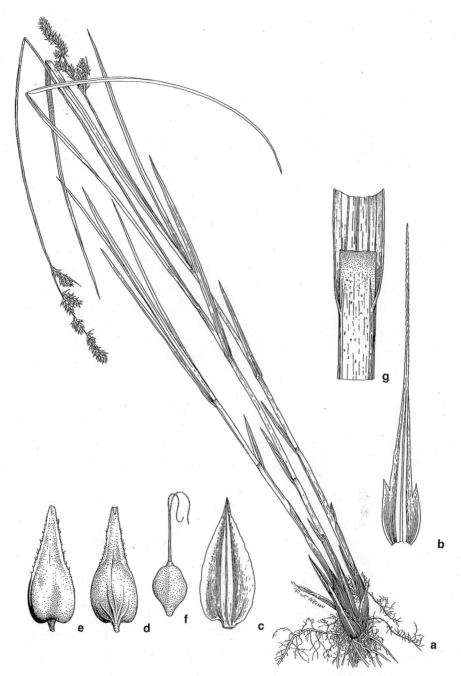

25. **Carex prairea.**
a. Habit.
b. Lowest bract.
c. Pistillate scale.
d. Perigynium, dorsal view.
e. Perigynium, ventral view.
f. Achene.
g. Sheath.

Common Name: Sedge.

Habitat: Bogs, fens, drainage swales, floating sedge mats, wet meadows, wet prairies, swamps.

Range: Quebec to Saskatchewan, south to Nebraska, Iowa, northern Illinois, Indiana, Ohio, and New Jersey.

Illinois Distribution: Occasional in northeastern Illinois and in counties along the Illinois River.

Although closely related to *C. diandra*, this species is distinguished by its dull brown achenes and the copper-tinged mouth of the sheaths.

A collection of this species was made from Illinois by George Vasey without locality during the nineteenth century. Virginius Chase then found *C. prairea* on June 4, 1900 in Stark County. This species flowers from May through July.

26. Carex alopecoidea Tuckerm. Enum. Caric. 18. 1843. Fig. 26.

Plants perennial, cespitose, from fibrous roots, with short, stout, blackish rhizomes, with scale leaves of previous year persistent; culms erect but weak, easily compressed, up to 80 cm tall, up to 4.5 mm broad at the base, sharply triangular, conspicuously winged and with concave sides, scabrous above, becoming dark brown at the base; leaves up to 60 cm long, 3.0–6.5 mm wide, usually longer than the culms, soft and flaccid, septate-nodulose; sheaths loose, hyaline, purple-dotted, not septate-nodulose, with the ligule longer than wide; inflorescence headlike or elongate-cylindrical, stramineous or light brown, 3–4 cm long, 0.8–1.5 cm thick, with the staminate flowers above the pistillate; bracts up to 1 cm long, setaceous or with a slightly dilated base; spikes 8–12 per inflorescence; staminate scales inconspicuous, brownish, very narrow; pistillate scales brownish, with a 3-nerved greenish center and hyaline margins, narrow, acuminate to cuspidate, a little shorter than to about as long as the perigynia; perigynia 8–15 per spike, ovoid, 3–4 mm long, 1.5–1.8 mm broad, plano-convex, ascending to spreading, spongy-thickened at the base, stipitate, dull brown to yellow-green, becoming tan in age, flat on the inner nerveless face, convex on the outer face, tapering into a serrulate beak, the beak 1.5–2.0 mm long, about half as long as the body of the perigynia, greenish with reddish brown teeth; achenes 1.5 mm long, lenticular, apiculate, substipitate, jointed with the short style; stigmas 2, reddish brown.

Common Name: Sedge.

Habitat: Wet meadows, moist fields, low woods; occasionally along roads.

Range: Ontario and Maine to Minnesota, south to Iowa, northern Illinois, northern Pennsylvania, and New Jersey; also Saskatchewan.

Illinois Distribution: Scattered in the northern two-thirds of the state; also Pope County.

This species belongs to a group of plants containing species with numerous spikes in crowded paniculate heads, perigynia spongy or corky at the base, lenticular achenes, and two stigmas.

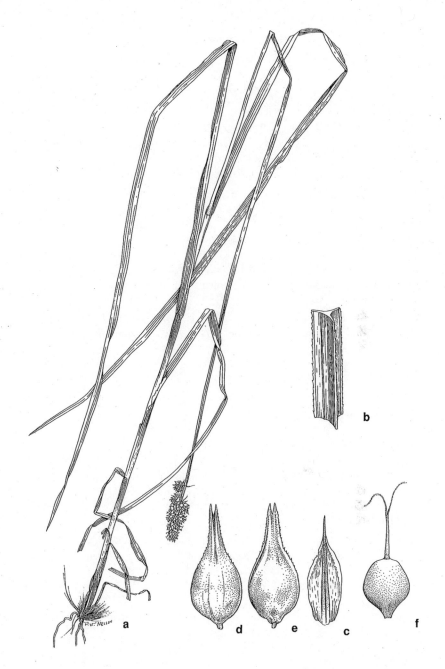

26. Carex alopecoidea.
a. Habit.
b. Cross section of fertile culm.
c. Pistillate scale.
d. Perigynium, dorsal view.
e. Perigynium, ventral view.
f. Achene.

Carex alopecoidea differs from *Carex conjuncta*, its closest relative in Illinois, in its uninterrupted inflorescence and its narrower leaves. This species flowers from May through July.

27. Carex conjuncta Boott, Illustr. Carex 122. 1862. Fig. 27.

Plants perennial, cespitose, from fibrous roots and short, dark, woody rhizomes, with scale leaves of previous year persistent; culms somewhat spongy on the sides, easily compressed, erect, up to 80 cm tall, sharply triangular, winged, harshly rough to the touch beneath the inflorescence, light brown at base, usually overtopping the leaves; leaves 3–6, up to 75 cm long, 4–8 mm wide, soft, flaccid, flat, rough along the margins and on the midvein beneath; sheaths open, loose, septate-nodulose, convex at the mouth, broadly white- to tan-hyaline, rugulose, finely purple-red dotted, easily broken; ligule longer than wide; inflorescence more or less headlike but elongated, rarely interrupted, up to 7 cm long, up to 2 cm broad; bracts setaceous, rough-awned, becoming shorter and scalelike to lacking above, up to 4 cm long; spikes 8–12, the staminate flowers above the pistillate; pistillate scales ovate to ovate-lanceolate, acuminate to cuspidate to short-awned, hyaline with three green nerves, about as long as the perigynia, but narrower than the perigynia; staminate scales narrowly lanceolate, pale brown; perigynia lance-ovoid, 3–5 mm long, 1.5–2.0 mm broad, plano-convex, round and spongy thickened at base, green except for a whitened base, nerveless or few-nerved, stipitate, tapering to a 2-toothed, serrulate beak 1.2–1.9 mm long, the teeth appressed to slightly spreading; achenes 1.50–2.25 mm long, 1.5 mm wide, lenticular, ovoid, apiculate, stipitate, yellow-brown, jointed to the style; stigmas 2, reddish brown, slender, flexuous.

Common Name: Soft Fox Sedge.
Habitat: Moist woods, swamps, wet prairies.
Range: New York to northern Indiana and southern Iowa, south to Missouri, Kentucky, and Virginia; also northern Alabama.
Illinois Distribution: Occasional throughout the state.

This species is recognized by its soft culms and flaccid leaves. It somewhat resembles *C. vulpinoidea* in appearance, except that its inflorescence is usually considerably shorter. It is closely related to *C. alopecoidea*, a species with narrower leaves, brownish pistillate scales, and sheaths lacking cross markings.

Carex stipata is similar because of its spongy culms, but the inflorescence and perigynia of *C. stipata* are larger. The flowers bloom in May and June.

28. Carex stipata Muhl. in Willd. Sp. Pl. 4:233. 1805. Fig. 28.

Plants perennial, cespitose, with fibrous roots and short, stout, dark rhizomes; culms up to 1.2 cm long, strongly 3-angled, spongy, easily compressed, rough to the touch, light brown at base, usually overtopped by the leaves; leaves 3–6, up to 1 m

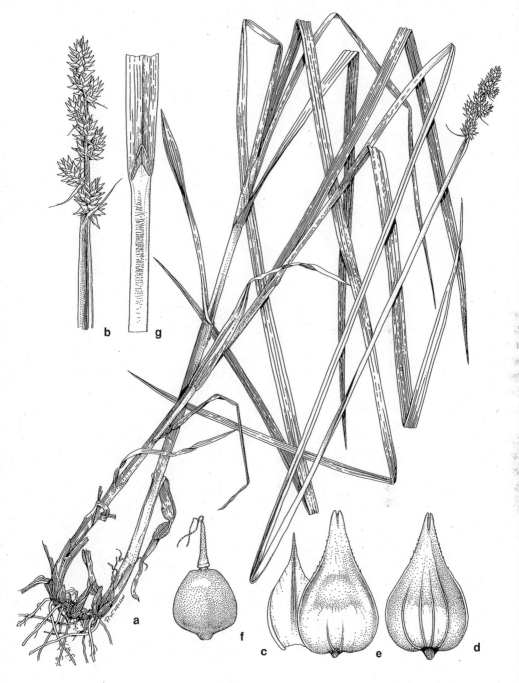

27. Carex conjuncta.
a. Habit.
b. Inflorescence.

c. Pistillate scale.
d. Perigynium, dorsal view.

e. Perigynium, ventral view.
f. Achene.
g. Sheath.

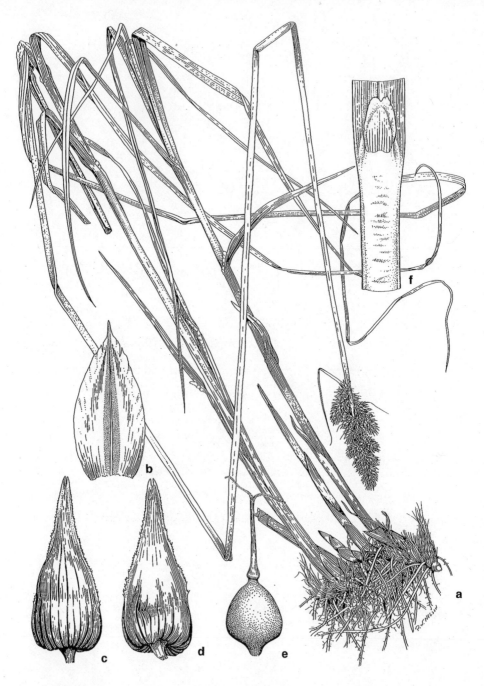

28. Carex stipata.
a. Habit.
b. Pistillate scale.

c. Perigynium, dorsal view.
d. Perigynium, ventral view.

e. Achene.
f. Sheath with ligule.

long, 4–15 mm wide, soft, flaccid, often yellowish green, rough along the margins and on the midvein beneath; sheaths open, green, prolonged at the summit, septate-nodulose; ligule longer than wide; inflorescence elongated, compound, usually continuous, up to 15 cm long, up to 4 cm broad; bracts setaceous, up to 5 cm long; spikes 15–25, the staminate flowers above the pistillate; pistillate scales lanceolate to ovate, acuminate to cuspidate to short-awned, hyaline to yellow-brown, with a green midnerve, much shorter than the perigynia; staminate scales narrowly lanceolate, pale brown; perigynia 4–10 per spike, lanceoloid, 3.5–6.0 mm long, 1.5–2.0 mm broad, plano-convex, rounded and spongy thickened at base, stipitate, yellowish or pale brown, conspicuously few-nerved, tapering gradually into a 2-toothed, serrulate green beak 2.0–3.5 mm long; achenes 1.5–2.0 mm long, lenticular, ovoid to orbicular, apiculate, substipitate, jointed to the style, the style swollen at base; stigmas 2, reddish brown.

Common Name: Sedge.
Habitat: Wet meadows, marshes, swamps, fens, wet ditches, bogs.
Range: New Brunswick to British Columbia, south to California, Utah, Colorado, Tennessee, and Virginia.
Illinois Distribution: Throughout the state.

This common species is found throughout most of the United States. It is easily identified by its long, slender, spongy-based perigynia and its spongy, strongly 3-angled culms. The larger var. *maxima*, occasionally reported from the state, does not occur in Illinois. The flowers bloom from late April through August.

29. Carex laevivaginata (Kukenth.) Mack. in Britt. & Brown. Ill. Fl., ed. 2, 1:371. 1913. Fig. 29.
Carex stipata Muhl. var. *laevivaginata* Kukenth. in Engler, Pflanzenreich 4 (20):172. 1909.

Plants perennial, densely cespitose, with fibrous roots and short, stout, black rhizomes; culms soft-spongy in fresh plants, easily compressed, stout, erect, to 80 cm tall, strongly 3-angled, with the sides concave, rough to the touch on the angles, usually overtopped by the leaves; leaves 3–6, up to 60 cm long, 3–6 mm wide, flat, flaccid, pale green, rough on the margins and usually on the midvein beneath; sheaths open, loose, pale green to yellow, strongly nerved, concave to truncate and cartilaginous at the mouth, not breaking easily, usually septate-nodulose, white-hyaline ventrally; ligule longer than wide; inflorescence a compound, elongated, usually continuous head, up to 6 cm long, up to 1.5 cm broad; bracts setaceous to scalelike, rough-awned, up to 4 cm long, seldom longer than the spike it subtends; spikes 10–15, the staminate flowers above the pistillate; pistillate scales lanceolate, acuminate to long-awned, hyaline with 3 green nerves, three-fourths as long as and narrower than the perigynia; staminate scales narrowly lanceolate, pale brown; perigynia 4–10 per spike, lanceoloid, 4.5–8.0 mm long, 1.25–2.25 mm

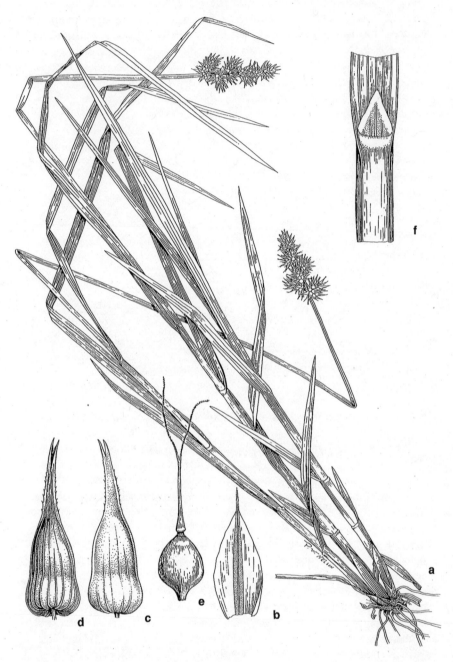

29. Carex laevivaginata.
a. Habit.

b. Pistillate scale.
c. Perigynium, dorsal view.

d. Perigynium, ventral view.
e. Achene.

broad, plano-convex, rounded and spongy thickened at base, stipitate, greenish or yellowish, conspicuously 2- to 12-nerved, tapering gradually to a 2-toothed, serrulate beak 2.0–2.5 mm long, the teeth appressed to slightly spreading; achenes 1.5–2.0 mm long, 1.25 mm wide, lenticular, ovoid to suborbicular, apiculate, substipitate, jointed to the style; stigmas 2, reddish brown, slender, flexuous.

Common Name: Sedge.
Habitat: Swamps and wet woods, wooded seeps, fens, moist limestone barrens.
Range: Massachusetts to southeastern Michigan, south to southeastern Missouri, Alabama, and Georgia.
Illinois Distribution: Scattered in Illinois, but not particularly common.

This species is very similar to the much more common *C. stipata* from which it differs in its leaf sheaths that are cartilaginous near their summit; the shorter, less-branched inflorescence; and the usually green perigynia. The flowers bloom from mid-May through August.

30. Carex crus-corvi Shuttlw. in Kunze, Riedgr. Suppl. 128. 1842. Fig. 30.
Carex siccaeformis Boott, Journ. Bost. Nat. Hist. Soc. 5:113. 1845.
Carex halei Dewey, Am. Journ. Sci. 2:248. 1846.

Plants perennial, cespitose, with fibrous roots and short, stout, blackish brown rhizomes; culms erect, stout, up to 1.2 m tall, strongly 3-angled and winged, the sides concave, not pressing completely flat, the angles rough to the touch, light brown at base, usually overtopped by the leaves; leaves 4–8 per culm, gray-green, up to 90 cm long, 5–12 mm wide, flat, pale green, septate-nodulose, rough along the margins; sheaths open, loose, concave at the summit, purple-dotted, septate-nodulose, the ventral side smooth, broadly white- to tan-hyaline, tearing easily; ligule wider than long; inflorescence compound, usually conspicuously branched, up to 20 cm long, up to 6 cm broad; bracts setaceous to scalelike to lacking, up to 8 cm long, rough-awned, rarely as long as the lowest branch of spikes; spikes 15–25, the staminate flowers above the pistillate; pistillate scales lanceolate to ovate, acuminate to aristate, green to pale brown to hyaline on the margins, the center green and 3-nerved, shorter than the perigynia; staminate scales narrowly lanceolate, pale brown; perigynia lanceoloid to subuloid, 6–8 mm long, 1.5–2.2 mm wide, plano-convex, swollen and spongy thickened at base, stipitate, yellow to brown, conspicuously nerved on the dorsal face, nearly nerveless ventrally, tapering to a long, subulate, serrulate, 2-toothed beak 4–6 mm long and about twice as long as the body of the perigynia; achenes 1.8–2.5 mm long, 1.25–1.50 mm wide, lenticular, ovoid, long-apiculate, substipitate, jointed to the style; stigmas 2, yellowish brown.

Common Name: Crowfoot Sedge.
Habitat: Swamps; wet woods, especially pin oak woods; upland swampy depressions.

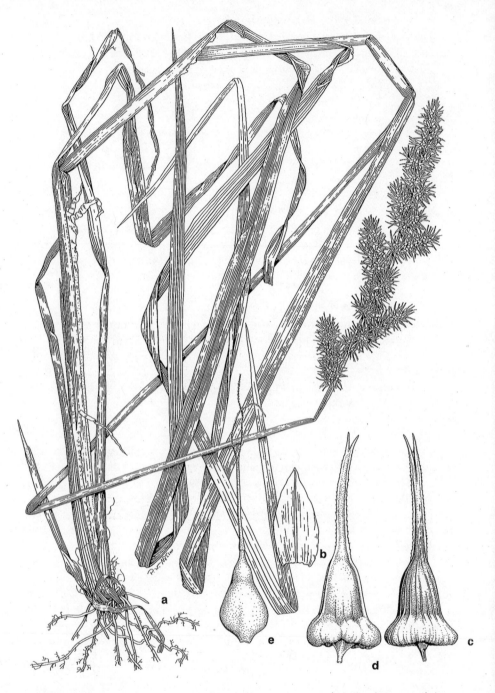

30. Carex crus-corvi.
a. Habit.

b. Pistillate scale.
c. Perigynium, dorsal view.

d. Perigynium, ventral view.
e. Achene.

Range: Southwestern Michigan to southern Minnesota, south to south-eastern Nebraska, east Texas, and Florida.
Illinois Distribution: Occasional in the southern half of the state, becoming less common in the northern half.

This is one of the most distinctive species of *Carex* in Illinois. The perigynia are unique in being greatly enlarged and spongy at the base, tapering to long-subulate beaks much longer than the body of the perigynia. The leaf sheaths are septate-nodulose and purple-spotted. The flowers bloom from mid-May through August.

31. Carex disperma Dewey, Am. Journ. Sci. 8:266. 1824. Fig. 31.
Carex tenella Schk. Riedgr. 23, pl. 104. 1801, non Thuill. (1844).

Plants perennial, loosely cespitose to mat-forming, with fibrous roots and usually slender, brown, scaly rhizomes; culms weakly erect to reclining, to 60 cm long, very slender, about 0.5 mm wide, sharply triangular, rough to the touch beneath the inflorescence, light brown at base, with several old leaves persistent at base; sterile culms common; leaves 3–6, up to 30 cm long, 1–2 mm broad, soft, dark green, rough to the touch along the margins and veins beneath, dark green, usually shorter than the culms; sheaths open, tight, concave at the summit, hyaline; ligule wider than long; inflorescence composed of a few interrupted spikes, up to 25 mm long, up to 5 mm broad; bracts setaceous to scalelike, rough-awned, enlarged at base, up to 1 cm long; spikes 2–7 per culm, aggregated above, the lower 2–4 remote, the 1–3 staminate flowers above the pistillate flowers; pistillate scales ovate, acute to acuminate to aristate, hyaline with a green midvein, shorter than to equal the length of the perigynia, narrower than the perigynia; staminate scales narrowly lanceolate; perigynia 1–6 per spike, ellipsoid-ovoid, 1.75–3.00 mm long, 1.0–1.5 mm broad, unequally biconvex, olive to yellow-green, white-puncticulate, slightly spongy at base, substipitate, the margins rounded and unwinged, finely nerved, rounded at the apex and with 2 entire teeth less than 0.3 mm long; achenes 1.5–2.0 mm long, 1 mm wide, lenticular, yellow-brown, shiny, jointed to the style; stigmas 2, reddish brown, slender.

Common Name: Soft-leaved Sedge.
Habitat: Tamarack and sphagnum swamps and bogs.
Range: Labrador and Newfoundland to Yukon and Alaska, south to California, New Mexico, northern Illinois, and Pennsylvania; Eurasia.
Illinois Distribution: Known only from Kane and Lake counties.

This northern species is distinguished by its small, weak stature; its sparsely flowered spikes; and its ellipsoid-ovoid, biconvex, nearly beakless perigynia. It is very similar to *C. trisperma*, differing primarily in its biconvex perigynia and in the fact that all its staminate flowers are terminal in the spike.

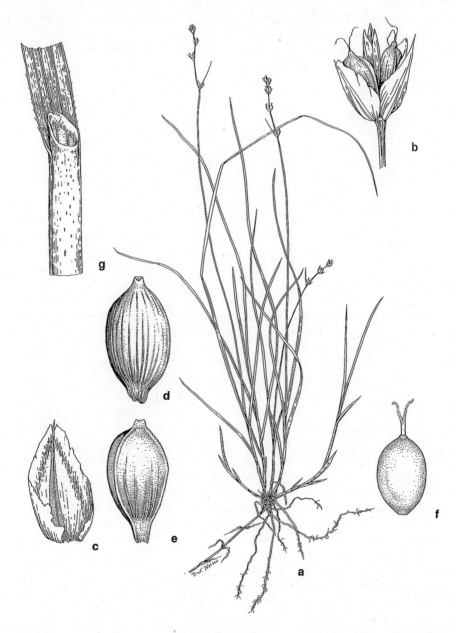

31. Carex disperma.
a. Habit.
b. Spike.
c. Pistillate scale.
d. Perigynium, dorsal view.
e. Perigynium, ventral view.
f. Achene.
g. Sheath with ligule.

Carex disperma and *C. trisperma* belong to section Heleonastes, along with *C. canescens* and *C. brunnescens*. This section is characterized by unwinged perigynia with rounded margins and with the perigynia minutely puncticulate, never reflexed, and less than 4 mm long. All but *C. disperma* have the staminate flowers scattered or at the base of the spikes. *Carex disperma* and *C. trisperma* have a perigynium beak less than 0.5 mm long and up to 6 perigynia per spike. *Carex canescens* and *C. brunnescens* have beaks at least 0.5 mm long and more than 6 perigynia per spike.

At maturity, the perigynia split open, exposing the shiny achenes. The flowers bloom from May to early August.

32. Carex trisperma Dewey, Am. Journ. Sci. 9:63. 1825. Fig. 32.

Plants perennial, loosely cespitose, with fibrous roots and slender, pale brown rhizomes; culms weakly erect to reclining, up to 70 cm tall, very slender, smooth or rough to the touch beneath the inflorescence, brown at base, with several old leaves persistent at base; leaves 3–5, up to 20 cm long, 0.75–2.00 mm wide, weak, flat or canaliculate, dark green, rough along the margins, much shorter than the culms; sheaths open, tight, concave to truncate at the summit, hyaline; ligule longer than wide; inflorescence composed of a few sparsely flowered, interrupted spikes, up to 6.5 cm long, up to 0.5 cm broad; lowest bracts filiform to setaceous, up to 10 cm long, as long as or longer than the inflorescence, the upper ones reduced to awned or aristate scales; spikes 1–4 per culm, few-flowered, the upper 2–3 often aggregated, the lowest 1–2 remote, the staminate flowers below the pistillate flowers; pistillate scales ovate to ovate-lanceolate, acute to acuminate to aristate, hyaline with 3 green nerves, shorter and barely narrower than the perigynia; staminate scales narrowly lanceolate; perigynia 1–5 per spike, oblongoid, 2.5–4.0 mm long, 1.25–2.00 mm broad, ascending, plano-convex, thick-coriaceous to spongy, substipitate, olive to light green to brown, white-puncticulate, sharp-edged along the margin, finely nerved, abruptly tapering to a very short, untoothed, green beak less than 0.50–0.75 mm long; achenes 2 mm long, lenticular, yellowish brown, shiny, apiculate, substipitate, jointed to the style; stigmas 2, reddish brown, slender.

Common Name: Three-fruited Sedge.
Habitat: Tamarack swamps and sphagnum bogs.
Range: Labrador and Newfoundland to Saskatchewan, south to Minnesota, northern Illinois, northern Indiana, Ohio, eastern Tennessee, western North Carolina, and New Jersey.
Illinois Distribution: Known only from tamarack bogs in Lake County.

Carex trisperma has a depauperate appearance because the inflorescence consists of only 1–4 spikes, each with only 1–5 perigynia. It is similar to *Carex disperma*, but the staminate flowers in *C. trisperma* are below the pistillate flowers.

32. Carex trisperma.
a. Habit.
b. Spike.

c. Pistillate scale.
d. Perigynium, dorsal view.

e. Perigynium, ventral view.
f. Achene.

The rhizomes of *C. trisperma* are usually entangled in moss and humus. The flowers bloom during June and July.

33. Carex canescens L. Sp. Pl. 974. 1753.

Plants perennial, cespitose, from short, blackish, fibrillose rootstocks; culms sharply triangular, scabrous only beneath the head, to 80 cm tall, with last year's leaves persisting at base; leaves 5–8, 2–4 mm wide, lax, flat, glaucous to pale green, pale papillate beneath, the margins scabrous toward the tip, usually shorter than the culms; sheaths tight, clear on the hyaline ventral band, the lower sheaths

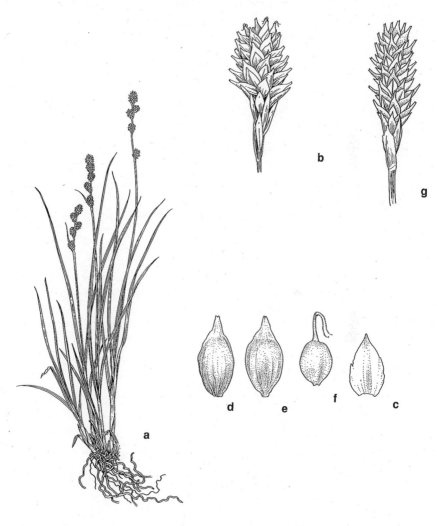

33. Carex canescens var. disjuncta.
a. Habit.

b. Spike.
c. Pistillate scale.
d. Perigynium, dorsal view.

e. Perigynium, ventral view.
f. Achene.
g. Spike.

brownish, the nerves darker, the ligule longer than broad; spikes 4–8 per culm, ovoid-cylindric, gynecandrous, overlapping or the lowest somewhat remote, 4–12 mm long, 3–5 mm wide, forming a loose head 2–10 cm long; bracts filiform to setaceous; pistillate scales broadly elliptic, obtuse to acuminate, hyaline to brownish, with a 3-nerved green center, shorter than the perigynia; perigynia 10–30 per spike, 1.8–3.0 mm long, 1.25–1.75 mm wide, spreading to ascending, narrowly ovoid to ellipsoid, plano-convex, membranous except for the spongy base, stipitate, obscurely nerved, pale green to pale brown, the beak 0.2–0.8 mm long, entire or emarginate, smooth or serrulate; achenes lenticular, about 1.5 mm long, 0.9 mm wide, ellipsoid, yellow-brown, substipitate, apiculate, jointed to the style; stigmas 2, reddish brown.

Typical var. *canescens*, with nearly all of the spikes contiguous, has not been found in Illinois.

Two varieties that have been found in Illinois may be distinguished as follows:

1. Spikes 6–12 mm long; perigynia 2.2–3.0 mm long . . . . 33a. *C. canescens* var. *disjuncta*
1. Spikes 4–7 mm long; perigynia up to 2.2 mm long . . . . 33b. *C. canescens* var. *subloliacea*

33a. Carex canescens Dewey var. **disjuncta** Fern. Proc. Am. Acad. 37:488. 1902. Fig. 33.

Spikelets 6–12 mm long; perigynia 2.2–3.0 mm long.

Common Name: Silvery Sedge.
Habitat: Sphagnum bogs.
Range: Newfoundland to Minnesota, south to northeastern Illinois, Ohio, and Virginia.
Illinois Distribution: Known only from Lake County.

Both varieties of *C. canescens* in Illinois grow together and may tend to overlap in their characteristics. This variety flowers from late April to mid-May.

33b. Carex canescens Dewey var. **subloliacea** Laestad. Nov. Act. Soc. Sci. Ups. 11:282. 1839. Not illustrated.

Spikelets 4–7 mm long; perigynia up to 2.2 mm long.

Common Name: Silvery Sedge.
Habitat: Sphagnum bogs
Range: Labrador to Alberta, south to Washington, Wyoming, Minnesota, northeastern Illinois, and New Jersey.
Illinois Distribution: Known only from Lake County.

This variety is smaller in all respects than var. *disjuncta*. Some specimens differ from any other plants of *C. canescens* in having only 7–9 perigynia per spike. This variety flowers from late April to mid-May.

34. Carex brunnescens (Pers.) Poir. in Lam. Encycl. Suppl. 3:286. 1813. Fig. 34.
Carex curta Good. var. *brunnescens* Pers. Syn. 2:539. 1807.
Carex canescens L. var. *sphaerostachya* Tuckerm. Enum. Meth. 10, 19. 1843.
Carex sphaerostachya (Tuckerm.) Dewey, Am. Journ. Sci. 49:44. 1845.
Carex canescens L. var. *vulgaris* Bailey, Bot. Gaz. 13:86. 1888.
Carex brunnescens (Pers.) Poir. var. *gracilior* Britt. in Britt. & Brown. Ill. Fl. 1:351.
1896.
Carex brunnescens (Pers.) Poir var. *sphaerostachya* (Tuckerm.) Kukenth. Pflanzenr.
38, 4, Fam. 20:220. 1909.

Plants perennial, cespitose, from short, blackish, fibrillose rootstocks; culms
triangular, scabrous, slender, weak, to 50 cm tall, longer than the leaves; leaves to
2 mm wide, lax, flat, dark green; sheaths tight, hyaline on the ventral band, with
russet spots; spikes 3–6 per culm, gynecandrous, not overlapping, 4–5 mm long,
in a more or less flexuous inflorescence up to 4 cm long; lowest bract setaceous; pistillate scales ovate, obtuse to acute, 3-nerved and green in the center, shorter than
the perigynia; perigynia 5–10 per spike, broadly ellipsoid, 2.2–2.5 mm long, 1.0–1.5
mm broad, plump, tapering to a short, serrulate beak, spongy at base, nerveless or
nearly so on the flat, ventral face, nerved on the convex dorsal face; achenes lenticular, ellipsoid, 1.3–1.5 mm long, 1.0 mm wide, yellow-brown, substipitate, apiculate, jointed to the style; stigmas 2, reddish brown.

Common Name: Brown Sedge.
Habitat: Alkaline bog.
Range: Greenland to British Columbia, south to Oregon, Colorado,
Illinois, and North Carolina.
Illinois Distribution: Known only from Lake County, where it was first
collected on June 9, 1987, by John Taft and Mary Solecki.

Fernald (1950) segregates the plants from Illinois as var. *sphaerostachya*, a variation which differs from the typical variety in its weaker
leaves and generally smaller spikes. This species flowers during May.

35. Carex bromoides Schkuhr in Willd. Sp. Pl. 4:258. 1805. Fig. 35.

Plants perennial, densely cespitose, from long, thin, blackish, fibrillous rootstocks; culms sharply triangular, the sides concave, scabrous on the angles above,
to 70 cm tall, equalling or much exceeding the leaves; leaves 3–5, 1.0–2.2 mm
wide, stiff, the margins scabrous, the old leaves persistent; sheaths tight, with the
hyaline ventral band cartilaginous and thickened at the concave summit; spikes
2–7 per culm, appressed-ascending, gynecandrous, all overlapping at the tip of the
culm, forming a loose head 2–6 cm long; bracts scalelike with rough awns, only the
lowest occasionally longer than the spike; pistillate scales lanceolate, acuminate to
short-aristate, pale or soon suffused with amber, just reaching the base of the beak

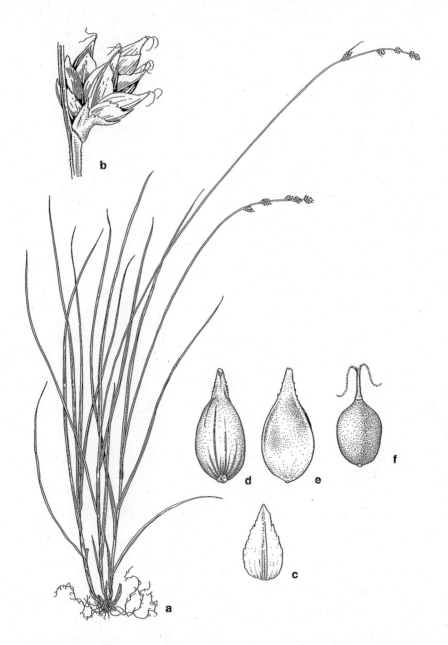

34. Carex brunnescens.
a. Habit.
b. Spike.
c. Pistillate scale.
d. Perigynium, dorsal view.
e. Perigynium, ventral view.
f. Achene.

35. Carex bromoides.
a. Habit.

b. Pistillate scale.
c. Perigynium, dorsal view.

d. Perigynium, ventral view.
e. Achene.

of the perigynium and as wide as the perigynium; perigynia 4–12 per spike, 3.5–4.5 mm long, 0.75–1.25 mm wide, appressed-ascending, lanceoloid, plano-convex, membranous except for the spongy base, substipitate, nerved dorsally and ventrally, greenish to light brown, the beak 1.25–1.50 mm long, bidentate, obliquely cut dorsally, serrulate; achenes lenticular, 1.50–1.75 mm long, 0.75 mm wide, substipitate, brown, jointed to a basally elongated style, positioned in the lengthwise center of the body of the perigynia; stigmas 2, elongated, flexuous to intertwined, reddish.

Common Name: Sedge.
Habitat: Low woods, seep springs, swamps, prairie bogs.
Range: Nova Scotia and Quebec to Wisconsin, south to Louisiana and Florida; Mexico.
Illinois Distribution: Scattered throughout the state, but less common in the southern counties.

The distinctive features of this species are the extremely narrow, lanceoloid perigynia and its stiff leaves. It differs from the similar *C. deweyana* in its narrower perigynium that is nerved on the convex face. In Illinois, *C. bromoides* flowers from early April to mid-May.

36. Carex deweyana Schwein. Ann. Lyc. N. Y. 1:65. 1824. Fig. 36.

Plants perennial, densely cespitose, from short rootstocks; culms to 1.2 m tall, slender, scabrous at least beneath the inflorescence, more or less triangular, usually longer than the leaves, brownish at the base, usually with old leaves persistent; leaves 3–6, 2–5 mm wide, lax, soft, slightly scabrous along the margins, pale green to somewhat glaucous; sheaths tight, pale, often with greenish mottling; spikes 3–7 per culm, ascending, more or less separate from each other, forming an inflorescence up to 6 cm long, the terminal spike gynecandrous, the lateral spikes entirely pistillate, 5–12 mm long, 3–5 mm thick; lowest bract setaceous, up to 4 cm long; pistillate scales ovate, as broad as the perigynia and nearly as long, obtuse to awned, hyaline with a green center; perigynia 3–15 per spike, 4.5–5.5 mm long, 1.5–2.0 mm wide, appressed-ascending, ovoid, plano-convex, membranous except at the spongy base, substipitate, obscurely nerved dorsally, nerveless ventrally, pale green, the beak about 2 mm long, bidentate, not quite half as long as the body,

serrulate; achenes lenticular, 2.0–2.5 mm long, 1.3–1.8 mm wide, substipitate, yellow-brown, jointed to the style; stigmas 2, elongated, nearly straight.

Common Name: Dewey's Sedge.
Habitat: Sandy oak woods.
Range: Labrador to British Columbia, south to Idaho, Colorado, Iowa, northern Illinois, Pennsylvania, and New England.
Illinois Distribution: Known only from a woods east of Roscoe, Winnebago County, where it was collected by E. W. Fell on May 23, 1954.

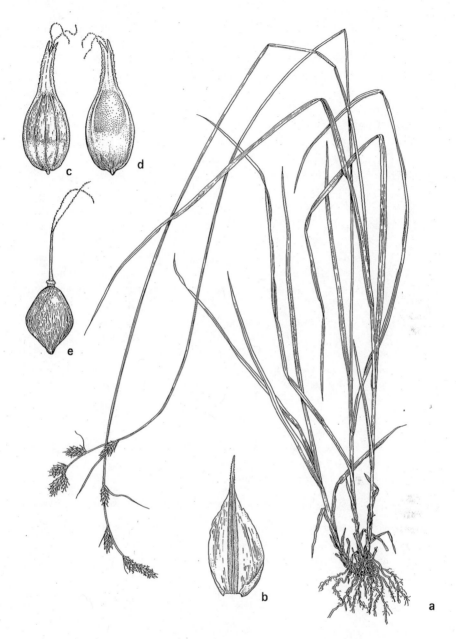

36. Carex deweyana.
a. Habit.

b. Pistillate scale.
c. Perigynium, dorsal view.

d. Perigynium, ventral view.
e. Achene.

This species, along with the somewhat similar *C. bromoides*, comprises the section Deweyanae. *Carex deweyana* differs from *C. bromoides* in its broader perigynia that are obscurely nerved, its broader spikes, and its wider leaves. This species is known in Illinois only from its original collection in Winnebago County. *Carex deweyana* flowers during late May in Illinois.

37. Carex interior Bailey, Bull. Torrey Club 20:426. 1893. Fig. 37.

Plants perennial, densely cespitose, from short rhizomes; culms triangular, smooth or scabrous only below the head, to 90 cm tall; leaves 3–5, 0.5–2.5 mm wide, flat, green, smooth or slightly scabrous along the margins, shorter than the culms; sheaths tight, smooth, the inner band hyaline and sometimes purple-dotted, concave at the apex, at least the lower ones brown to stramineous; inflorescence up to 3.5 cm long, with 2–5 sessile, separated spikes; terminal spike clavate, 5–20 mm long, gynecandrous, the staminate part 2–14 mm long, 3- to 10-flowered, the pistillate part 3–7 mm long, 4- to 16-flowered; lateral spikes pistillate or with a few staminate flowers at the base, 3–10 mm long, with scalelike bracts; pistillate scales ovate, 1.0–2.5 mm long, castaneous with a green center and hyaline margins, acute or obtuse at the tip, sometimes reaching the base of the beak of the perigynium; perigynia 2–3 mm long, 1.0–1.8 mm wide, ovoid to broadly ovoid, planoconvex, the ventral surface nerveless or with a few short nerves, spongy-thickened at the base, green to dark brown, sessile, tapering to a beak with at least the lower perigynia spreading to reflexed; beak of the perigynium 0.5–1.0 mm long, serrulate, toothed at the apex; achenes biconvex, about 1.5 mm long, up to 1.5 mm broad, substipitate, jointed to the deciduous style; stigmas 2.

Common Name: Inland Sedge.
Habitat: Bogs, wet meadows, moist prairies, wet woods, swamps, ditches, banks of streams, lakeshores.
Range: Throughout most of North America.
Illinois Distribution: Scattered throughout the state.

This small species is readily distinguished by its star-shaped spikes due to the radiating perigynia, its elongated staminate part of the terminal spike, its narrow leaves, and its often nerveless ventral face of the perigynium.

The similar *C. atlantica* and *C. echinata* differ in their nerved perigynia. This species begins to flower in late April in the southern counties and continues through May.

38. Carex sterilis Willd. Sp. Pl. 4:208. 1805. Fig. 38.
Carex muricata L. var. *sterilis* (Willd.) Gl. Phytologia 4:22. 1952.

Plants perennial, densely cespitose, from stout rhizomes; culms wiry, sharply triangular, scabrous, to 75 cm tall; leaves 3–5, 1.5–2.5 mm wide, flat, firm, green,

37. **Carex interior.**
a. Habit.
b. Inflorescence.

c. Pistillate scale.
d. Perigynium, dorsal view.

e. Perigynium, ventral view.
f. Achene.

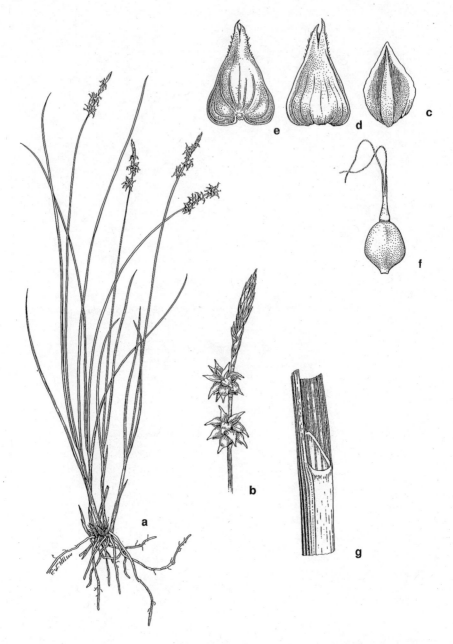

38. Carex sterilis.
a. Habit.
b. Inflorescence.

c. Pistillate scale.
d. Perigynium, dorsal view.

e. Perigynium, ventral view.
f. Achene.
g. Sheath with ligule.

scabrous along the margins, shorter than the culms; sheaths tight, smooth, green, the inner band hyaline, usually minutely papillate, concave at the apex, at least the lowermost brown; inflorescence up to 4 cm long, densely crowded above but with the lower spikes often separated, with 3–8 sessile spikes, usually dioecious, or with pistillate plants often bearing a very few staminate flowers, or staminate plants with a few scattered parigynia, or sometimes with all staminate spikes and all pistillate spikes on the same plant; terminal spike 3.5–12.0 mm long, usually unisexual; lateral spikes up to 12 mm long, subtended by scalelike bracts; pistillate scales ovate, 2–3 mm long, castaneous with a green center and hyaline margins, acute at the tip, reaching to the base of the perigynium or longer; perigynia 2.0–3.5 mm long, 1.0–2.2 mm wide, lanceoloid to deltoid, plano-convex, the ventral surface with up to 10 nerves, or nerveless, spongy-thickened at the base, castaneous to nearly black, sessile, tapering to a beak, the perigynia radiating in all directions; beak up to 1.5 mm long, serrulate, bidentate at the apex; achenes biconvex, 1.0–1.7 mm long, sessile, jointed to the deciduous style; stigmas 2.

Common Name: Sedge.
Habitat: Wet meadows, fens, marly seeps.
Range: Newfoundland to Alberta, south to northwestern Minnesota, southern Illinois, south-central Kentucky, Pennsylvania, Connecticut, and Maine.
Illinois Distribution: This species is known from the northern half of Illinois as well as Coles, St. Clair, and Washington counties.

The perigynia, which radiate in all directions, give each lateral spikelet a star-shaped appearance. This species differs from the similar *C. interior, C. echinata,* and *C. atlantica* in its unisexual terminal spikelet. *Carex sterilis* flowers from late April until the end of May.

39. Carex echinata Murray, Prod. Stirp. Gott. 76. 1770. Fig. 39.

Carex leersii Willd. Fl. Berol. Prod. 29. 1787. Nomen illeg.
Carex stellulata Goodenough, Trans. Linn. Soc. 2:144. 1794.
Carex echinata Murray var. *cephalantha* Bailey, Mem. Torrey Club 1:58. 1889.
Carex stellulata Goodenough var. *cephalantha* (Bailey) Fern. Rhodora 4:222. 1902.
Carex cephalantha (Bailey) Bicknell, Bull. Torrey Club 35:493. 1908.

Plants perennial, cespitose, from short rhizomes; culms triangular, slightly scabrous, to 90 cm tall; leaves 2–6, 1.0–3.5 mm wide, plicate, green, scabrous along the margins, a little shorter than the culms; sheaths tight, smooth, the inner band hyaline and sometimes purple-dotted, concave at the apex, the lower ones light brown; inflorescence up to 7.5 cm long, with 3–8 crowded or separate, sessile spikes; terminal spike 0.5–2.0 cm long, gynecandrous, the staminate part up to 1.5 cm long, 2- to 17-flowered, the pistillate part up to 1 cm long, 4- to 26-flowered; lateral spikes up to 1.5 cm long, gynecandrous, with scalelike bracts; pistillate scales ovate, 1.5–3.0

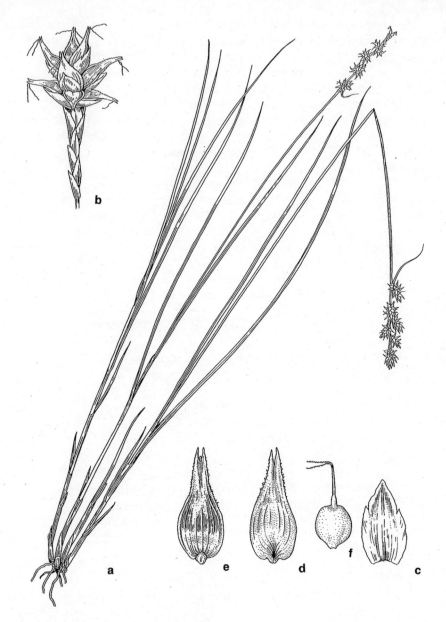

39. Carex echinata.
a. Habit.
b. Inflorescence.

c. Pistillate scale.
d. Perigynium, dorsal view.

e. Perigynium, ventral view.
f. Achene.

mm long, castaneous, with a green center and hyaline margins, acute at the tip, reaching the base of the beak of the perigynium; perigynia 2.5–4.5 mm long, 1–2 mm wide, lanceoloid to ovoid, plano-convex, the ventral surface with up to 12 veins, or nerveless, spongy-thickened at the base, green to dark brown, sessile, tapering to a beak, with the lower perigynia spreading to reflexed; beak 1–2 mm long, serrulate, with sharp, stiff teeth at the apex; achenes biconvex, 1.5–2.0 mm long, 1.0–1.5 mm broad, substipitate, jointed to the deciduous style; stigmas 2.

Common Name: Sedge.
Habitat: Wet meadows (in Illinois).
Range: Labrador to British Columbia, south to southern California, central Utah, northern Colorado, northern Iowa, eastern Indiana, southwestern North Carolina, Maryland, and Maine.
Illinois Distribution: Du Page County, collected in 1994; also Lake County.

This species is distinguished by its star-shaped spikes, gynecandrous terminal spike, narrow leaves, and perigynia 2.5–4.5 mm long. The similar *C. atlantica* and *C. interior* have shorter perigynia. This species flowers in May and June.

40. Carex atlantica Bailey, Bull. Torrey Club 20:425. 1893.

Plants perennial, cespitose, from short rhizomes; culms triangular, scabrous, to 1 m tall; leaves 3–5, 0.5–4.0 mm wide, flat, green, scabrous along the margins, shorter than or equalling the culms; sheaths tight, smooth, the inner band hyaline and sometimes purple-dotted, cartilaginous-thickened at the concave apex, light brown; inflorescence up to 5 cm long, with 2–8 sessile, crowded or separated spikes; terminal spike 0.5–2.5 cm long, gynecandrous, the staminate part to 1.5 cm long and 2- to 20-flowered, the pistillate part to 1 cm long and 4- to 35-flowered; lateral spikes gynecandrous, 0.5–1.2 cm long, with scalelike bracts; pistillate scales ovate, 1.2–2.4 mm long, green with hyaline margins, obtuse to acute at the tip, reaching or surpassing the base of the perigynium; perigynia 2.0–3.5 mm long, 1.3–2.8 mm wide, broadly ovoid to suborbicular, plano-convex, the ventral surface with up to 12 nerves, spongy-thickened at the base, green to dark brown, sessile, tapering to a beak, with the lower perigynia spreading to reflexed; beak of the perigynium 0.5–1.2 mm long, serrulate, bidentate at the apex; achenes biconvex, 1–2 mm long, substipitate, jointed to the deciduous style; stigmas 2.
Two subspecies occur in Illinois, separated by the following key:

1. Larger leaves more than 1.6 mm wide; inflorescence mostly more than 2 cm long. . . . .
. . . . 40a. *C. atlantica* ssp. *atlantica*
1. Larger leaves up to 1.6 mm wide; inflorescence up to 2 cm long. . . . .
. . . . 40b. *C. atlantica* ssp. *capillacea*

40. Carex atlantica.
a. Habit.

b. Pistillate scale.
c. Perigynium, dorsal view.

d. Perigynium, ventral view.
e. Achene.

40a. Carex atlantica Bailey ssp. **atlantica** Fig. 40.
Carex imcomperta Bicknell, Bull. Torrey Club 35:494. 1908.
Carex atlantica Bailey var. *incomperta* (Bicknell) F. J. Hermann, Rhodora 67:362.
1965.

Leaves more than 1.6 mm wide; inflorescence mostly more than 2 cm long.

Common Name: Star Sedge.
Habitat: Swampy woods.
Range: Nova Scotia to central Michigan to northwestern Indiana, south
to southeastern Missouri, southeastern Texas, and northern Florida.
Illinois Distribution: Known only from Pope County.

This subspecies is primarily a plant of the Coastal Plain. It differs from
ssp. *capillacea* in being larger in all respects, but intergradations between the two subspecies occur in the range of the two. This subspecies flowers in late April and early May.

40b. Carex atlantica Bailey ssp. **capillacea** (Bailey) Reznicek, Contr. Mich. Herb.
14:191. 1980. Not illustrated.
Carex interior Bailey var. *capillacea* Bailey, Bull. Torrey Club 20:426. 1893.
Carex howei Mack. Bull. Torrey Club 37:245. 1910.

Larger leaves up to 1.6 mm wide; inflorescence up to 2 cm long.

Common Name: Star Sedge.
Habitat: Swampy woods.
Range: Nova Scotia to southwestern Michigan and northwestern
Indiana, south to southern Illinois, southeastern Texas, and southern
Florida.
Illinois Distribution: Known only from Pulaski County.

This subspecies flowers in late April and early May.

41. Carex muskingumensis Schwein. Ann. Lyc. N. Y. 1:66. 1824. Fig. 41.

Plants perennial, densely cespitose, from short, stout, black, fibrillose rhizomes;
culms to 85 cm tall, sharply triangular, the angles very scabrous above, stiffly erect,
usually longer than the leaves, with many very leafy sterile shoots present; leaves
up to 12 per fertile culm, oriented at right angles to the culm, 2.5–4.0 mm wide
(leaves of sterile culm up to 7 mm wide), flat, the margins scabrous; sheaths tight,
striate ventrally to the V-shaped mouth, the mouth with a thickened, dark band,
the ligule short, rigid, and holding the leaf base away from the culm; spikes 6–10
per culm, 12–25 mm long, narrowly ellipsoid, light green to stramineous, ap-

41. Carex muskingumensis.
a. Habit.

b. Pistillate scale.
c. Perigynium, dorsal view.

d. Perigynium, ventral view.
e. Achene.

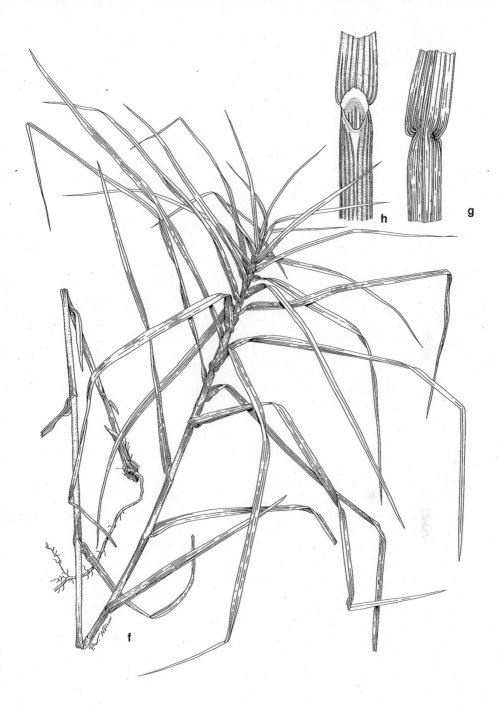

f. Sterile culm.

g. Sheath, dorsal view.

h. Sheath with ligule.

pressed-ascending, acuminate, gynecandrous, long-clavate at the base, overlapping to aggregated, forming an inflorescence 4.5–9.0 cm long, 10–20 mm wide; bracts leaflike, with prolonged, rough awns below, awnless above, rarely longer than the adjacent spike; pistillate scales lanceolate, acuminate to obtuse, two-thirds as long as and much narrower than the perigynia, tan to brown, hyaline with a darkened center; perigynia many per spike, lanceoloid to narrowly ovoid-lanceoloid, plano-convex, appressed, 7–10 mm long, 1.75–2.00 mm wide, broadest at or above the middle, substipitate, distinctly nerved dorsally and ventrally, the wing abruptly narrowed at or just below the middle, absent at the base, the beak 4.0–4.8 mm long, serrulate, sharply bidentate; achenes lenticular, 2.5–3.0 mm long, about 0.75 mm wide, apiculate, stipitate, reddish brown, jointed to and sometimes continuous with an elongated style; stigmas 2, thin, flexuous, reddish.

Name: Muskingum Sedge.
Habitat: Low, swampy woods and floodplains of major streams; wooded depressions.
Range: Manitoba, south to Kansas, Arkansas, and Kentucky.
Illinois Distribution: Occasional throughout the state, but more common in the northeastern and extreme southern counties and counties along the Wabash River.

This is one of the more easily recognized species in section Ovales because of its large, pointed spikes; leaves that stand at right angles to the culm; and extremely narrow perigynia. Many tufts of sterile culms are usually present.

This species is confined to low, swampy woods and floodplain woods. *Carex muskingumensis* flowers from mid-May to mid-June.

42. Carex scoparia Schkuhr in Willd. Sp. Pl. 4:230. 1805. Fig. 42.
Carex scoparia Schkuhr var. *moniliformis* Tuckerm. Enum. Meth. 8:17. 1843.
Carex scoparia Schkuhr var. *condensa* Fern. Proc. Am. Acad. 37:468. 1902.

Plants perennial, densely cespitose, from short, brown, fibrillose rhizomes; culms to 1 m tall, sharply triangular, the angles very scabrous above, stiffly erect, usually longer than the leaves, with sterile leafy culms uncommon; leaves up to 7 per fertile culm, ascending to spreading, 1.5–4.0 mm wide, flat, the margins and veins scabrous throughout; sheaths tight, green-nerved throughout or white-hyaline ventrally, the mouth concave, slightly thickened and sometimes darkened; spikes 3–10 per culm, 5–15 mm long, light green to stramineous, ascending, acute, gynecandrous, clavate at the base, crowded, or the lowest spike remote, or rarely all the spikes separated, forming an inflorescence 1.5–5.0 cm long, 1.0–1.5 (–2.5) cm wide; bracts setaceous; pistillate scales lanceolate, acuminate, two-thirds to three-fourths as long as and much narrower than the perigynia, tan-hyaline with the center brown to green and 3-nerved, with a central lighter nerve;

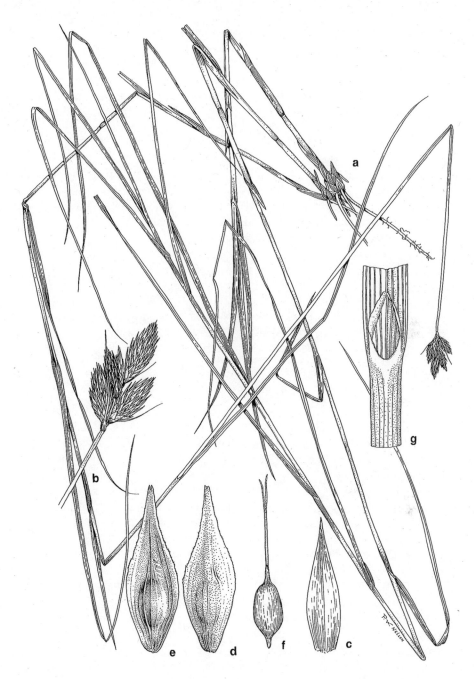

42. Carex scoparia.
a. Habit.
b. Inflorescence.

c. Pistillate scale.
d. Perigynium, dorsal view.

e. Perigynium, ventral view.
f. Achene.
g. Sheath.

perigynia many per spike, lanceolate, acuminate, plano-convex, appressed, 3.8–5.5 mm long, 1.2–2.0 mm wide, broadest at the middle, membranous, substipitate, distinctly nerved dorsally, obscurely nerved to nerveless ventrally, the wing not narrowed below the middle, stramineous, the beak about 3 mm long, serrulate, bidentate, the teeth appressed; achenes lenticular, 1.25–1.50 mm long, 0.75–1.00 mm wide, apiculate, stipitate, brown, weakly continuous with the deciduous, jointed style; stigmas 2, short, tan to reddish, flexuous.

Common Name: Sedge.
Habitat: Wet open woods, wet prairies, wet meadows, seeps, calcareous fens.
Range: Newfoundland to British Columbia, south to Oregon, New Mexico, Arkansas, and South Carolina.
Illinois Distribution: Throughout Illinois.

This species is distinguished by its very slender perigynia, narrow leaves, and pointed spikes. There is variation in the arrangement of spikes in the inflorescence.

A few specimens have densely crowded spikes forming an inflorescence 1.5–2.5 cm broad. These have been segregated as var. *condensa.* A few specimens have all spikes remote. These may be segregated as var. *moniliformis. Carex scoparia* flowers from early May to late June.

43. Carex tribuloides Wahl. Sv. Vet. Akad. Handl. 24:145. 1803. Fig. 43.

Plants perennial, densely cespitose, from short, brown to black, fibrillose rhizomes; culms to 1.2 m tall, sharply triangular, the angles scabrous, stiffly erect or somewhat lax, usually about as long as the leaves, with sterile leafy culms common; leaves 5–9 per fertile culm, ascending to spreading, 3.5–7.0 mm wide, flat, the margins scabrous in the upper half; sheaths tight, the upper greenest near the summit, the lower pale or becoming stramineous or brown, the ventral portion veiny nearly throughout; spikes 4–12 per culm, 6–12 mm long, pale green to stramineous, ascending, acute to obtuse, gynecandrous, clavate at the base, mostly aggregated, or with the lowest 1–3 spikes slightly separated, forming an inflorescence 1.5–6.0 cm long; bracts setaceous; pistillate scales lanceolate, acute to acuminate, two-thirds to three-fourths as long as and narrower than the perigynia, tan-hyaline with the center greenish to darker brown and 3-nerved; perigynia many per spike, lanceolate to narrowly lanceolate-ovate, plano-convex, appressed to slightly spreading, 3.25–5.00 mm long, 1.25–1.50 mm wide, widest at the middle, membranous, substipitate, discretely nerved dorsally and ventrally, the wing abruptly narrowed to absent below the middle, pale green to stramineous, the beak 1.0–1.5 mm long, serrulate, bidentate, the teeth appressed; achenes lenticular, 1.5–1.7 mm long, 0.75–1.00 mm wide, apiculate, stipitate, light to dark brown, weakly continuous with the deciduous style; stigmas 2, slender, elongate, reddish.

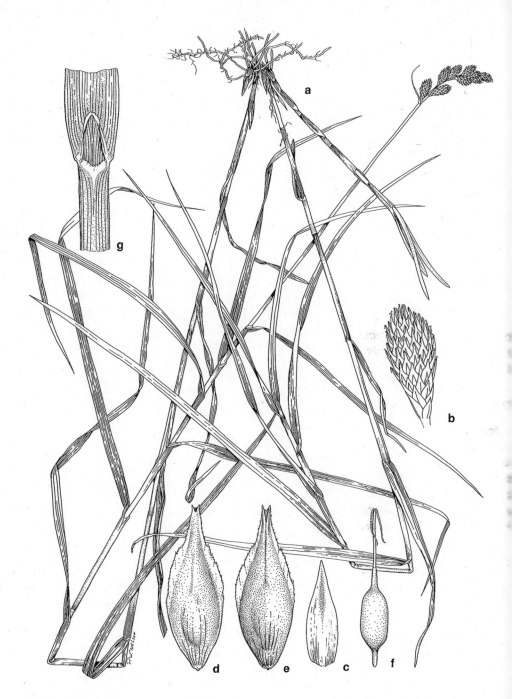

43. Carex tribuloides.
a. Habit.
b. Spike.

c. Pistillate scale.
d. Perigynium, dorsal view.

e. Perigynium, ventral view.
f. Achene.
g. Sheath with ligule.

Common Name: Sedge.

Habitat: Wet woods, swamps, wet ditches, peaty marshes, swales, wet prairies, wet meadows, peaty fens, oxbows, shores of lakes and ponds.

Range: Quebec to Minnesota, south to Louisiana and Florida.

Illinois Distribution: Throughout the state, but more frequent in the southern half.

This species is found in nearly every type of wetland habitat in the state. It is distinguished by its crowded spikelets, very narrow perigynia, and leaves that are at least 3 mm broad. The perigynia are always less than 2 mm wide, and cuneate rather than rounded at the base. The spikes are longer than broad. *Carex tribuloides* flowers from mid-May until the end of June.

44. Carex crawfordii Fern. Proc. Am. Acad. 37:469. 1902. Fig. 44.

Plants perennial, densely cespitose, from short, dark, fibrillose rhizomes; culms to 60 cm tall, sharply triangular, the angles scabrous beneath the inflorescence, stiffly erect, wiry, usually a little longer than the leaves, with sterile leafy culms uncommon; leaves 3–4 per fertile culm, ascending, 1–3 mm wide, flat or canaliculate, yellow-green, the margins scabrous in the upper half; sheaths tight, white-hyaline ventrally; spikes 7–12 per culm, 3–9 mm long, stramineous, ascending, usually acute, gynecandrous, barely clavate at the base, crowded into a densely ovoid head up to 2.5 cm long and up to 8 mm wide; bracts setaceous; pistillate scales lanceolate, acute to acuminate, shorter than the perigynia, pale brown, with a green, 3-nerved center; perigynia many per spike, lanceolate, acuminate, plano-convex, ascending, 3.0–4.5 mm long, 0.8–1.1 mm wide, membranous, stipitate, obscurely nerved or nerveless on both faces, the wing very narrow to nearly absent, brownish, the beak about 1 mm long, serrulate, bidentate, the teeth usually reddish brown; achenes lenticular, about 1 mm long, about 0.5 mm wide, apiculate, stipitate, brown, weakly continuous with the deciduous style; stigmas 2, short, red-brown.

Common Name: Crawford's Sedge.

Habitat: Degraded marsh (in Illinois).

Range: Newfoundland to British Columbia, south to Washington, Minnesota, northern Illinois, and New Jersey; also in the mountains of Tennessee.

Illinois Distribution: Known from a single location in Lake County where it was first collected in 1991.

This species, common much farther north of Illinois, was growing in a degraded marsh in northeastern Illinois. There is some speculation that it may not be native there, but since most of the associated species are native, it too is likely to be native.

44. Carex crawfordii.
a. Habit.

b. Pistillate scale.
c. Perigynium, dorsal view.

d. Perigynium, ventral view.
e. Achene.

Carex crawfordii is readily distinguished by its very slender perigynia measuring no more than 1.1 mm wide. There seems to be little evidence that *Carex crawfordii* is a hybrid between *C. bebbii* and *C. scoparia*, as suggested by Gleason and Cronquist (1963). This species flowers in late May and early June.

45. Carex projecta Mack. Bull. Torrey Club 35:264. 1908. Fig. 45.
Carex tribuloides Wahl. var. *moniliformis* Britt. in Britt & Brown. Ill. Fl. 1:336. 1896.

Plants perennial, cespitose, from short, black, fibrillose rhizomes; culms to 1 m tall, sharply triangular, the angles scabrous beneath the inflorescence, rather stiff, usually shorter than the leaves, usually with several sterile culms; leaves 4–6 per fertile culm, ascending, 3–7 mm wide, flat, lax, light green, the margins and midvein scabrous in the upper half; sheaths green, the lower ones becoming stramineous or brown, the ventral band typically hyaline only near the summit; spikes 5–15 per culm, 4–8 mm long, green or stramineous, obtuse, gynecandrous, clavate at the base, all except the very uppermost separated from each other in a moniliform inflorescence up to 7 cm long; lowest bract setaceous; pistillate scales lanceolate, obtuse to subacute, reaching the base of the beak of the perigynium, stramineous with hyaline margins and a green, 3-nerved center; perigynia up to 30 in a spike, lanceolate to linear-lanceolate, plano-convex, 2.5–5.0 mm long, 1.3–1.7 mm wide, widest at or a little below the middle, membranous, stipitate, nerved on both faces, the wing diminishing abruptly below the middle, stramineous to greenish, the beak 1–2 mm long, serrulate, bidentate; achenes lenticular, 1.5–1.7 mm long, 0.5–0.7 mm wide, apiculate, stipitate, weakly continuous with the deciduous style; stigmas 2, short, red-brown.

Common Name: Sedge.
Habitat: Riparian terraces, moist woods, swampy woods.
Range: Newfoundland to Manitoba, south to Missouri, West Virginia, and New Jersey; also British Columbia.
Illinois Distribution: Scattered throughout the state.

This species shows a strong similarity to *C. tribuloides* in characteristics of the perigynia and achenes but differs markedly in its separated spikes.

Carex projecta is more of a woodland species, whereas *C. tribuloides* occurs mostly in open areas. This species flowers from mid-May through most of June.

46. Carex straminea Willd. in Schk. Riedgr. 49. 1801. Fig. 46.
Carex tenera Dewey var. *richii* Fern. Proc. Am. Acad. 37:475. 1902.
Carex richii (Fern.) Mack. Bull. Torrey Club 49:362. 1923.

Plants perennial, cespitose, from short, black, fibrillose rootstocks; culms to 1 m tall, sharply triangular, the angles scabrous beneath the inflorescence, rather stiff,

45. Carex projecta.
a. Habit.

b. Pistillate scale.
c. Perigynium, dorsal view.

d. Perigynium, ventral view.
e. Achene.

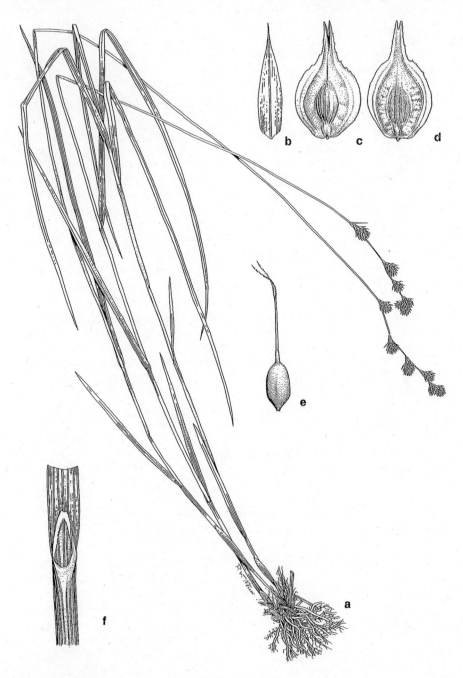

46. Carex straminea.
a. Habit.
b. Pistillate scale.
c. Perigynium, dorsal view.
d. Perigynium, ventral view.
e. Achene.
f. Sheath with ligule.

usually about as long as the leaves, usually with few or no sterile culms; leaves 3–5 per culm, ascending, 2.0–2.5 mm wide, flat, rather lax, green, the margins scabrous; sheaths tight, green-nerved throughout except for an abbreviated ventral band near the summit, the lowermost stramineous; spikes 3–8 per culm, 6–10 mm long, green or stramineous, obtuse, gynecandrous, clavate at base, the uppermost sometimes crowded, otherwise the spikes separated in an elongated inflorescence up to 8 cm long; lowest bract setaceous or scalelike; pistillate scales lanceolate, acuminate to aristate, the tip of the uppermost surpassing the base of the beak of the perigynium, tan with hyaline margins and a green center; perigynia up to 30 in a spike, 2.5–3.5 mm long, 1.5–2.7 mm wide, ovate, widest at the middle, appressed-ascending, membranous, plano-convex, strongly nerved on both faces, winged to the base, stramineous to greenish, the beak 1.5–2.0 mm long, serrulate, bidentate; achenes lenticular, 1.5–1.7 mm long, 0.75–1.00 mm wide, apiculate, stipitate, weakly jointed with the deciduous style; stigmas 2, short, reddish.

Common Name: Sedge.
Habitat: Wet savannas; along railroads.
Range: Massachusetts to Michigan, south to Illinois, Indiana, Maryland, and Delaware.
Illinois Distribution: Known only from Menard, Ogle, and Winnebago counties.

This species is distinguished by its awn-tipped pistillate scales, its ovate perigynia, and its interrupted inflorescence.

The application of the binomial *Carex straminea* has been a source of confusion for a long time. The name actually applies to a plant known in the past as *C. richii*, although it was also used by earlier botanists for the species now known as *C. tenera*. *Carex straminea* flowers in June and July in Illinois.

47. Carex cristatella Britt. in Britt. & Brown. Ill. Fl. 1:357. 1896. Fig. 47.
Carex cristata Schw. Ann. Lyc. N. Y. 1:66. 1824
Carex cristatella Britt. var. *catelliformis* Farw. Papers Mich. Acad. 2:17. 1923.
Carex cristatella Britt. f. *catelliformis* (Farw.) Fern. Rhodora 44:284. 1942.

Plants perennial, densely cespitose, from short, brownish black, fibrillose rootstocks; culms to 90 cm tall, sharply triangular with concave sides (pressing nearly flat), the angles scabrous beneath the inflorescence, stiff, usually shorter than the culms, with sterile culms common; leaves 3–6 per fertile culm, ascending to spreading, 3.0–7.5 mm wide, often revolute, green, the margins scabrous; sheaths relatively loose, wing-margined or inflated near summit, smooth, green-nerved throughout, sometimes hyaline near summit; spikes 4–12 per culm, less than 1 cm in diameter, green, globose, gynecandrous, abruptly tapering to the base, usually crowded in a headlike inflorescence up to 4 cm long, rarely separated into a moniliform inflorescence; lowest bracts setaceous, scabrous, upper bracts scalelike; pistil-

47. Carex cristatella.
a. Habit.

b. Pistillate scale.
c. Perigynium, dorsal view.

d. Perigynium, ventral view.
e. Achene.

late scales lanceolate, acute to acuminate, three-fourths as long as and narrower than the perigynia, often hidden by the recurving perigynia, the margins tan- to brown-hyaline, the center darker and 3-nerved; perigynia many per spike, 3.5–4.0 mm long, 1.4–1.8 mm wide, lance-ovate, widest at the middle, but with the body often suborbicular, the tips ascending to widely spreading, plano-convex, sometimes distended over the achenes dorsally and ventrally, membranous, distinctly nerved dorsally and ventrally, the wing abruptly narrowed just above the base, the beak 1.0–1.5 mm long, often twisted and constricted at the base, serrulate, bidentate, the teeth appressed; achenes lenticular, 1.3–1.6 mm long, 0.5–0.8 mm wide, apiculate, stipitate, weakly jointed to the deciduous style; stigmas 2, slender, elongate, reddish brown.

Common Name: Round-spikelet Sedge.
Habitat: Wet woods, marshes, swales, streambanks, ditches, meadows, bogs.
Range: Massachusetts to North Dakota, south to Nebraska, Missouri, and Virginia.
Illinois Distribution: Throughout the state, but more frequent in northern and west-central Illinois.

This species is distinguished by its globose spikes, outward curving perigynia that obscure the pistillate scales, and perigynia less than 2 mm wide that taper to the base. While this species usually has a crowded inflorescence, rare plants are found in which the spikes are separated into a moniliform inflorescence. These moniliform plants resemble *C. projecta*, but the lowest spikes are not clearly separated as they are in *C. projecta*.

This species is found in almost any type of wetland, including disturbed sites. The flowers mature from mid-May to the first of July.

48. Carex bebbii Olney, Caric. Bor. Am. 3. 1871. Fig. 48.
Carex tribuloides Wahl. var. *bebbii* Bailey, Mem. Torrey Club 1:55. 1889.

Plants perennial, densely cespitose, from short, compact, black to brown, fibrillose rootstocks; culms to 85 cm tall, sharply triangular, the angles scabrous beneath the inflorescence, stiff, longer than the leaves, often with sterile culms present; leaves 3–5 per culm, ascending, 2–4 mm wide, the apex becoming triangular, the margins and veins scabrous, at least on the upper surface; sheaths tight, with a narrow hyaline band toward the concave summit, easily broken; spikes 3–9 per culm, ascending, lance-ovate to ovate, rounded or somewhat pointed at the tip, brownish, 5–9 mm long, gynecandrous, abruptly contracted below making the staminate flowers inconspicuous, crowded (except sometimes the lowermost) into an ovoid inflorescence up to 2 cm long; lowest bracts setaceous, scabrous, upper bracts scalelike; pistillate scales lanceolate, acute to acuminate, three-fourths to

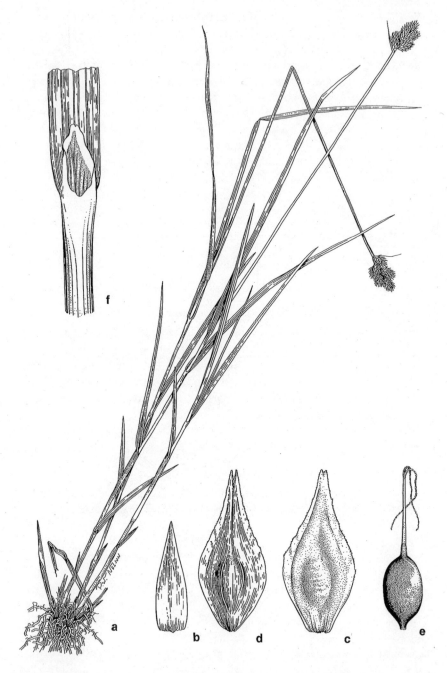

48. Carex bebbii.
a. Habit.
b. Pistillate scale.

c. Perigynium, dorsal view.
d. Perigynium, ventral view.

e. Achene.
f. Sheath with ligule.

seven-eighths as long as the perigynia and narrowed, the center green and 3-nerved and tan- to brown-hyaline margins; perigynia many per spike, 3.0–3.6 mm long, 1.2–1.5 mm wide, lanceolate to lance-ovate, widest at the middle, ascending to spreading, plano-convex, finely nerved only on the outer face, winged to the base, the beak 0.75–1.00 mm long, serrulate, bidentate; achenes lenticular, 1.2–1.4 mm long, 0.5–0.7 mm wide, apiculate, stipitate, light to dark brown, weakly jointed with the deciduous style; stigmas 2, short, slender, reddish brown.

Common Name: Bebb's Sedge.
Habitat: Wet prairies, bogs, calcareous fens, marshes.
Range: Newfoundland to British Columbia, south to Washington, Colorado, Illinois, and New Jersey.
Illinois Distribution: Occasional to frequent in the northern half of Illinois, uncommon in the southern half.

Carex bebbii is distinguished by its perigynia that are less than 2 mm wide, nerveless on the inner face, and winged all the way to the base and by its crowded spikes that are longer than they are broad. This species flowers in May and June.

49. Carex normalis Mack. Bull. Torrey Club 37:244. 1910. Fig. 49.
Carex mirabilis Dewey, Am. Journ. Sci. 30:63. 1836, non Host. (1809).
Carex mirabilis Dewey var. *perlonga* Fern. Proc. Am. Acad. 37:473. 1902.
Carex normalis Mack. var. *perlonga* Fern. Proc. Am. Acad. 37:473. 1902.
Carex normalis Mack. f. *perlonga* (Fern.) Fern. Rhodora 44:285. 1942.

Plants perennial, densely cespitose, from short, compact, brown, scaly to fibrillose rootstocks; culms to 1.5 m tall, wiry, sharply triangular, scabrous on the angles beneath the inflorescence, longer than the leaves, with several sterile culms; leaves 3–7 per culm, ascending, 2.0–6.5 mm wide, flat, lax, green, the margins scabrous toward the apex; sheaths loose, green-mottled, at least partially septate, prolonged and slightly concave at the mouth, the lowermost brownish; spikes 4–10 per culm, 6–10 mm long, green, gynecandrous, subglobose, clavate with the staminate flowers conspicuous, sometimes crowded into a head, sometimes separated in a moniliform inflorescence up to 5 cm long; lowest bracts setaceous, scabrous, upper bracts scalelike; pistillate scales ovate, obtuse to acute, barely reaching the base of the beak of the perigynium, the center green to brown and 3-nerved, the margins white- to tan-hyaline; perigynia many per spike, 3.0–4.5 mm long, 1.5–2.1 mm wide, ovate, widest at the middle, spreading, plano-convex, distinctly and finely nerved on both faces, narrowly and evenly winged to the base, substipitate, the beak 0.8–1.2 mm long, serrulate, bidentate; achenes lenticular, 1.6–1.8 mm long, 1.1–1.3 mm wide, apiculate, stipitate, weakly jointed with the deciduous style; stigmas 2, short, slender, reddish brown.

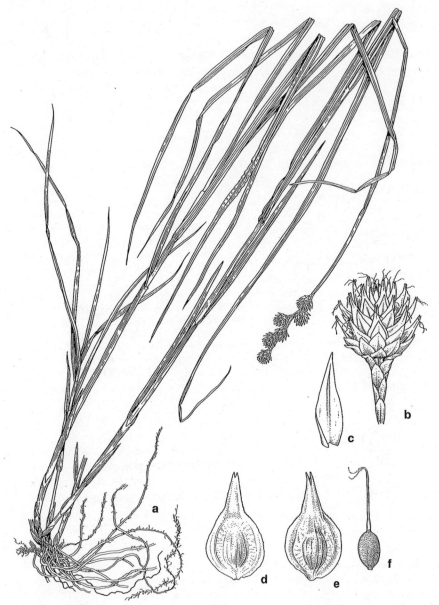

49. Carex normalis.
a. Habit.
b. Spike.

c. Pistillate scale.
d. Perigynium, dorsal view.

e. Perigynium, ventral view.
f. Achene.

Common Name: Sedge.
Habitat: Seep springs, mesic woods, flood plains, streambanks, mesic savannas, marshes, pond borders, moist fields, ditches.
Range: Maine to Manitoba, south to Oklahoma and North Carolina.
Illinois Distribution: Throughout the state, seemingly more common in the northern half of Illinois.

Carex normalis is distinguished by the numerous sterile culms, by its perigynia that are up to 2.1 mm wide and winged to the base, and by its spikes that are about as wide as they are long. This species is variable with respect to leaf width, perigynia length, and arrangement of spikes in the inflorescence. The spikes range from being crowded to somewhat separated to remote and moniliform. Specimens with moniliform spikes have been called *f. perlonga*.

The earliest collections of this species in Illinois were called *C. cristata*, but that is a different species that does not occur in Illinois. *Carex normalis* flowers during May and June.

50. Carex tenera Dewey, Am. Journ. Sci. 8:97. 1824. Fig. 50.
Carex straminea Willd. var. *echinodes* Fern. Proc. Am. Acad. 37:474. 1902. (*Carex tenera* Dewey var. *echinodes* [Fern.] Wieg. Rhodora 26:2. 1924.)

Plants perennial, densely cespitose, from short, black, scaly and fibrillose rhizomes; culms to 70 cm tall, sharply triangular, scabrous beneath the inflorescence or smooth, stiff, longer than the leaves, dark brown to black at the base, with the old bases often remaining as stubble; leaves 3–5 per culm, ascending, 1.0–2.8 mm wide, flat, green, the margins scabrous toward the apex; sheaths tight, green-nerved nearly throughout or with a narrow, hyaline, ventral band, the mouth prolonged and concave, the lowermost sheath purplish brown to blackish; spikes 3–7 per culm, 6–10 mm long, green to stramineous, obtuse, gynecandrous, somewhat clavate at base, the staminate flowers conspicuous, the spikes arranged in a moniliform inflorescence to 5 cm long, the axis of the inflorescence narrowed and flexed just above the lower spike; lowest bract usually setaceous, scabrous, the upper bracts scalelike; pistillate scales ovate to ovate-lanceolate, acute to acuminate, reaching or surpassing the base of the beak of the perigynium, the center greenish and 3-nerved, the margins hyaline; perigynia many per spike, 3.0–4.4 mm long, 1.5–2.0 mm wide, ovate, widest a little below the middle, spreading or ascending, plano-convex, strongly nerved on the outer face, more finely nerved on the inner face, narrowly and evenly winged to the base, substipitate, the beak 0.8–1.2 mm long, serrulate, bidentate; achenes lenticular, 1.5–1.8 mm long, 1.0–1.3 mm wide, apiculate, stipitate, weakly jointed with the deciduous style; stigmas 2, slender, light reddish.

Common Name: Remote Sedge.
Habitat: Floodplain woods, wet meadows, mesic prairies, swampy depressions, wet ditches.

50. Carex tenera.
a. Habit.
b. Inflorescence.

c. Pistillate scale.
d. Perigynium, dorsal view.

e. Perigynium, ventral view.
f. Achene.

Range: Quebec to Alberta, south to Missouri and North Carolina.
Illinois Distribution: Scattered throughout the state.

This *Carex* uniformly has the most remote spikes of any member of the section Ovales. Its ovate perigynia vary both in size and in having either spreading or ascending beaks.

Specimens with perigynia 4.0–4.4 mm long and with distinctly spreading beaks are sometimes known as var. *echinodes.*

Carex tenera occurs in most types of wetland habitats throughout Illinois. This species flowers from early May to late June.

51. Carex festucacea Schk. in Willd. Sp. Pl. 4:242. 1805. Fig. 51.

Plants perennial, densely cespitose, from short, black, fibrillose rhizomes; culms to 1 m tall, stout, sharply triangular, scabrous on the angles, at least beneath the inflorescence, longer than the leaves, brownish black at the base with the old leaves often remaining as stubble; leaves 3–5 per culm, ascending, 1–5 mm wide, flat, green, the margins scabrous toward the apex; sheaths tight, green-nerved nearly throughout or with a very narrow hyaline ventral band, often septate, the mouth prolonged, the lowermost sometimes becoming purplish; spikes 3–10 per culm, 6–16 mm long, gray-green, obtuse, gynecandrous, strongly clavate at the base, the staminate flowers conspicuous, the spikes sometimes rather crowded or more often separated, sometimes in a moniliform inflorescence 2.5–6.0 cm long, the axis of the inflorescence not flexed but straight; lowest bract sometimes setaceous, scalelike, the upper bracts scalelike; pistillate scales lanceolate, acute to acuminate, reaching or surpassing the base of the beak of the perigynium, the center greenish and 3-nerved, the margins hyaline; perigynia up to 20 per spike, 2.5–3.5 mm long, 1.5–2.2 mm wide, oval to orbicular, widest near the base, spreading or ascending, plano-convex, green to stramineous, subcoriaceous, strongly nerved on the outer face, with 2–4 fine nerves on the inner face, winged to the base, substipitate, the beak 1.0–1.3 mm long, serrulate, bidentate; achenes lenticular, 1.0–1.6 mm long, 1.0–1.3 mm wide, apiculate, stipitate, weakly jointed with the deciduous style; stigmas 2, short, slender, brownish.

Common Name: Sedge.
Habitat: Moist prairies, moist savannas, low woods.
Range: Massachusetts to southern Minnesota, south to southeastern Nebraska, southeastern Kansas, eastern Texas, and Georgia.
Illinois Distribution: Scattered in Illinois; apparently more frequent in the northern half of the state.

This species is similar to and sometimes confused with *C. tenera, C. longii,* and *C. albolutescens.* Specimens with separated spikes that resemble *C. tenera* may be distinguished by the strongly clavate base of the spikes and by the broadly ovate to suborbicular rather than narrowly ovate

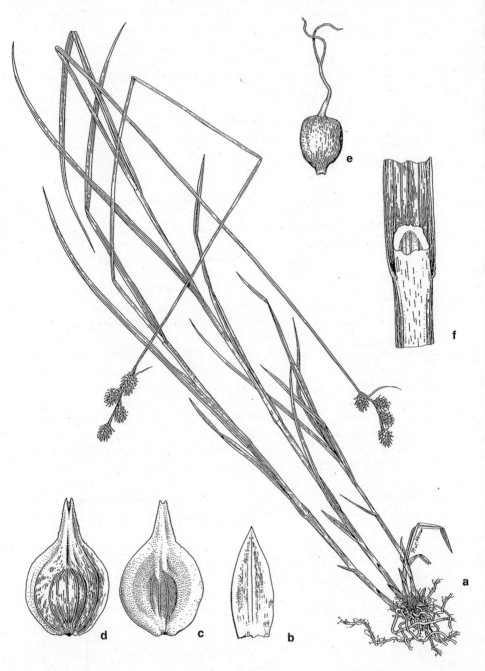

51. Carex festucacea.
a. Habit.
b. Pistillate scale.

c. Perigynium, dorsal view.
d. Perigynium, ventral view.

e. Achene.
f. Sheath with ligule.

perigynia. Specimens with more crowded spikes tend to resemble *C. longii* and *C. albolutescens* but differ in having the perigynia broadest near the base rather than nearest the middle. *Carex festucacea* flowers from late April to early June.

52. Carex albolutescens Schw. Ann. Lyc. N. Y. 1:66. 1824. Fig. 52.

Plants perennial, densely cespitose, from short, black, fibrillose rootstocks; culms to 1 m tall, triangular, slightly scabrous on the angles, at least beneath the inflorescence, longer than the leaves, light brown to brownish black at the base, with the old leaves often remaining as stubble; leaves 3–5 per culm, ascending, 2.0–3.5 mm wide, flat, pale green, the margins scabrous toward the apex; sheaths somewhat loose, green-nerved throughout or with a hyaline ventral band, prolonged at the summit; spikes 2–8 per culm, 5–13 mm long, not strongly clavate at base, greenish to stramineous, approximate or somewhat separated but rarely moniliform, the inflorescence 1.5–4.0 cm long, gynecandrous; lowest bract setaceous, the upper bracts scalelike; pistillate scales lanceolate, acute to acuminate, flat, shorter than the perigynia, the center greenish and 3-nerved, the nerves reaching the tip of the scale, the margins silver-hyaline; perigynia many per spike, 2.6–4.5 mm long, 1.5–2.7 mm wide, obovate, widest near the middle, spreading, plano-convex, green to stramineous, papery, strongly nerved on the outer face, with 4–7 raised nerves on the inner face, winged to the base but not quite to the tip, substipitate, the beak 0.5–1.0 mm long, serrulate, bidentate; achenes lenticular, 1.3–1.7 mm long, 0.7–1.0 mm wide, apiculate, stipitate, jointed with the deciduous style; stigmas 2, short, reddish.

Common Name: Sedge.
Habitat: Moist woods.
Range: Rhode Island to southeastern Michigan, south to southeastern Missouri, eastern Texas, and central Georgia.
Illinois Distribution: Known only from Pope and Union counties.

This species is similar and often confused with *C. longii*. It differs from *C. longii* in its spreading rather than ascending perigynia, in the wing not quite reaching the tip of the perigynium, and in the midvein of the pistillate scale reaching the tip.

After studying *C. albolutescens* and *C. longii* in detail, Rothrock (1991) has determined that nearly all collections from Illinois previously called *C. albolutescens* are actually *C. longii.*

Mackenzie (1931, 1940) erroneously called this species *C. straminea,* but *C. straminea* is a different species. *Carex albolutescens* flowers from mid-April through May.

53. Carex longii Mack. Bull. Torrey Club 49:372. 1923. Fig. 53.

Plants perennial, densely cespitose, from short, black, fibrillose rootstocks; culms to 1.2 m tall, wiry, triangular, scabrous on the angles beneath the inflorescence or smooth, longer than the leaves, pale brown to brownish black at the base;

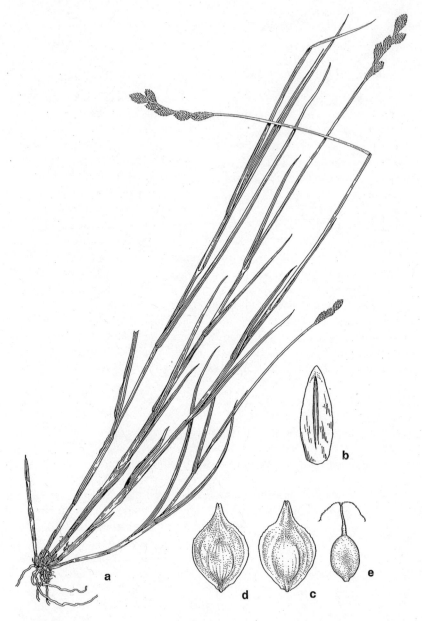

52. Carex albolutescens.
a. Habit.

b. Pistillate scale.
c. Perigynium, dorsal view.

d. Perigynium, ventral view.
e. Achene.

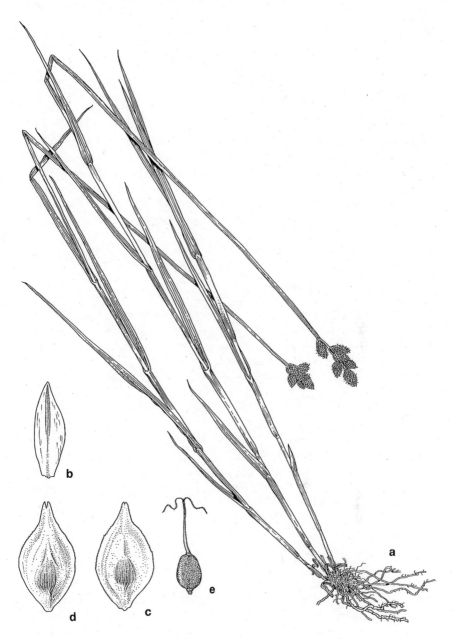

53. Carex longii.
a. Habit.

b. Pistillate scale.
c. Perigynium, dorsal view.

d. Perigynium, ventral view.
e. Achene.

leaves 2–4 per culm, ascending, 2–5 mm wide, flat, the upper surface minutely
papillose, the veins on the lower surface and the margins usually scabrous; sheaths
somewhat loose, green-nerved throughout, hyaline ventrally, prolonged at the
summit, the lowest ones stramineous; spikes 3–10 per culm, 6–13 mm long, not
strongly clavate at base, green to brownish, approximate or somewhat separated
but rarely moniliform, the inflorescence 1.0–4.5 cm long, gynecandrous; lowest
bract usually setaceous, scabrous, the upper bracts scalelike; pistillate scales lanceolate, obtuse to acute, concave, shorter than the perigynia, the center greenish
and 3-nerved, the nerves not quite reaching the tip of the scale, the margins silver-
hyaline; perigynia many per spike, 3.0–4.5 mm long, 1.6–2.6 mm wide, obovate,
widest near the middle, appressed-ascending, plano-convex, green to brownish,
papery, strongly nerved on the outer face, with 4–7 raised nerves on the inner face,
winged all the way to the tip as well as to the base, substipitate, the beak 0.5–1.0
mm long, serrulate, bidentate; achenes lenticular, 1.3–1.7 mm long, 0.7–1.0 mm
wide, apiculate, stipitate, weakly jointed with the deciduous style; stigmas 2, short,
reddish.

Common Name: Long's Sedge.
Habitat: Flatwoods, mesic sand prairies, wet woods.
Range: Nova Scotia to southern Michigan and Illinois, south to eastern
Texas and Florida; Mexico; Central America; western South America;
West Indies; introduced in Hawaii.
Illinois Distribution: Scattered throughout the state.

Until 1991, when Rothrock clearly distinguished *C. longii* from *C. albolutescens*, all Illinois collections of this species were called *C. albolutescens*.

Carex longii differs from *C. albolutescens* in its appressed-ascending perigynia that
are winged from the base all the way to the tip of the beak and in its convex pistillate scales whose midrib fails to reach the tip. This species flowers from late April
through most of June.

54. Carex cumulata (Bailey) Mack. Bull. Torrey Club 49:366. 1923. Fig. 54.
Carex albolutescens Schw. var. *cumulata* Bailey, Bull. Torrey Club 20:422. 1893.

Plants perennial, densely cespitose, from short, black, fibrillose rootstocks;
culms to 90 cm tall, triangular, the angles scabrous just below the inflorescence,
brownish at the base, with the remains of previous year's leaves usually present;
leaves 2–4 per culm, ascending, 3–6 mm wide, flat, firm, light green, scabrous
along the margins; sheaths loose, green-nerved ventrally nearly to the prolonged
summit; spikes 5–30 per culm, 6–10 mm long, not clavate at base, green to
brownish, crowded into an inflorescence up to 4.5 cm long, gynecandrous; lowest
bract cuspidate, the upper bracts scalelike; pistillate scales lanceolate, acute, usually reaching the base of the beak of the perigynium, the center greenish and 1- to
3-nerved, the margins white-hyaline; perigynia numerous per spike, 3–4 mm long,
2.0–3.5 mm wide, rhombic-orbicular, widest near the middle, ascending, plano-

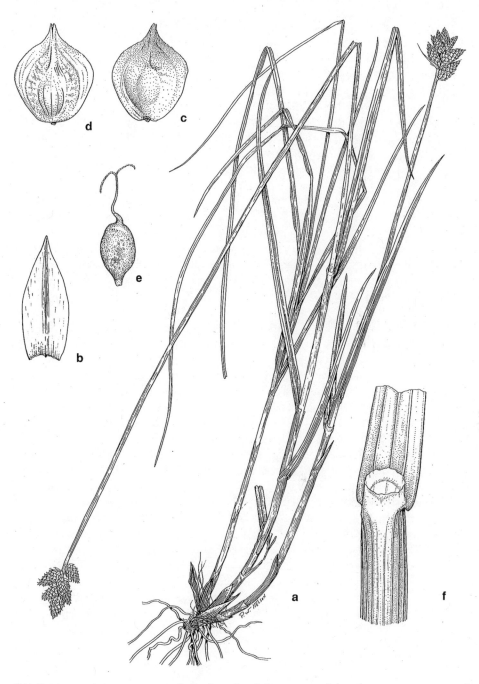

54. Carex cumulata.
a. Habit.
b. Pistillate scale.

c. Perigynium, dorsal view.
d. Perigynium, ventral view.

e. Achene.
f. Sheath with ligule.

convex, green to brownish, finely nerved on the outer face, nerveless on the inner face, winged all the way to the base, substipitate, the beak 0.5–1.0 mm long, serrulate, bidentate; achenes lenticular, about 2 mm long, 1.2–1.5 mm wide, yellow-brown, apiculate, stipitate; stigmas 2, short, reddish.

Common Name: Sedge.
Habitat: Mesophytic depression in black oak savanna (in Illinois).
Range: Prince Edward Island to Saskatchewan, south to northeastern Illinois, Ohio, and New Jersey.
Illinois Distribution: Known only from Kankakee County (east of St. Anne).

This species is distinguished by its crowded spikes and by its perigynia that are widest near the middle, usually more than 2 mm wide, and nerveless on the inner face. In Illinois, this species is known only from the vicinity of St. Anne in Kankakee County where it was first collected more than fifty years ago by Schneider and rediscovered in 1987 by Ken Dritz. *Carex cumulata* flowers during May.

55. Carex suberecta (Olney) Britt. Man. Fl. N. States, ed. 2, 1057. 1905. Fig. 55.
Carex tenera Dewey var. *suberecta* Olney, Caric. Bor. Am. 3. 1871.

Plants perennial, densely cespitose, from short, black, fibrillose rootstocks; culms to 1 m tall, sharply triangular, the angles scabrous, at least beneath the inflorescence, light brown at the base, with the remains of previous year's leaves usually present; leaves 3–5 per culm, ascending, 1.5–3.3 mm wide, firm, light green, scabrous along the margins; sheaths more or less tight, green-nerved nearly throughout, prolonged at the summit, the lowermost stramineous to brown; spikes 2–5 per culm, 7–12 mm long, tapering to each end, yellow-brown, approximate but distinct in an inflorescence 1.5–3.0 cm long, gynecandrous; lowest bract cuspidate, upper bracts scalelike; pistillate scales lanceolate, long-acuminate to subaristate, usually reaching the base of the beak of the perigynium, the center green and 3-nerved, the margins hyaline; perigynia numerous per spike, 4–5 mm long, 2.0–2.8 mm wide, rhombic, widest above the middle, appressed, plano-convex, yellow-brown, faintly nerved on the outer face, nerveless or with 1–5 faint nerves

on the inner face, the wings diminishing toward the base, substipitate, the beak 0.5–1.5 mm long, serrulate, bidentate, usually greenish; achenes lenticular, 1.5–1.8 mm long, about 1 mm wide, apiculate, substipitate; stigmas 2, short, reddish brown.

Common Name: Sedge.
Habitat: Marshes, fens, wet prairies, wet meadows.
Range: Ontario to Minnesota, south to Missouri, Illinois, Ohio, and Virginia.
Illinois Distribution: Apparently confined to the northern two-thirds of the state.

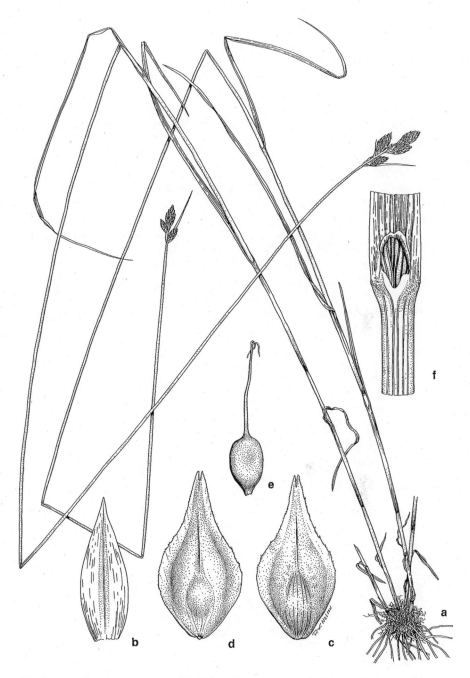

55. Carex suberecta.
a. Habit.

b. Pistillate scale.
c. Perigynium, dorsal view.

d. Perigynium, ventral view.
e. Achene.

Carex suberecta is distinguished by its rather crowded spikes and its moderately large perigynia that are widest above the middle and more or less cuneate at the base. The similar *C. cumulata* has round-based perigynia.

This species is fairly common in northeastern Illinois but is much less common elsewhere. *Carex suberecta* flowers from early May to early June.

56. Carex brevior (Dewey) Mack. Am. Midl. Nat. 4:235. 1915. Fig. 56.
Carex straminea Willd. var. *brevior* Dewey, Am. Journ. Sci. 11:158. 1826.
Carex festucacea Schk. var. *brevior* (Dewey) Fern. Proc. Am. Acad. 37:477. 1902.

Plants perennial, cespitose, from short, black, fibrillose rootstocks; culms to 1 m tall, rather stiff, triangular, the angles scabrous beneath the inflorescence or smooth, with old leaf bases often persisting; sterile shoots usually present; leaves 3–6 per culm, ascending, 1.5–4.0 mm wide, flat, firm, light green, scabrous along the margins; sheaths tight, with a narrow hyaline ventral band, prolonged at summit; spikes 3–10 per culm, 7–15 mm long, ovoid to subglobose, clavate at base, green, crowded into an inflorescence or separated but rarely moniliform in an inflorescence up to 5 cm long; lowest bract setaceous, scabrous, the upper bracts usually scalelike; pistillate scales lanceolate, acute to acuminate, the tip surpassing the base of the beak of the perigynium, the center green and 3-nerved, the margins hyaline; perigynia 8–20 per spike, 4.0–5.5 mm long, 2.5–3.5 mm wide, orbicular, widest near the middle, coriaceous, appressed-ascending, plano-convex, green, strongly nerved on the outer face, nerveless on the inner face, winged to the base, substipitate, the beak 0.8–1.2 mm long, serrulate, bidentate, sometimes red-tipped; achenes lenticular, 1.7–2.0 mm long, 1.5–1.8 mm wide, yellow-brown, apiculate, substipitate; stigmas 2, long, reddish brown.

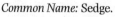

Common Name: Sedge.
Habitat: Sandy prairies, dry woods, along railroads, often in disturbed areas.
Range: Maine to British Columbia, south to Oregon, New Mexico, Texas, Tennessee, and Delaware.
Illinois Distribution: Scattered to common in the northern three-fourths of the state, very uncommon elsewhere.

Carex brevior is distinguished by its large, orbicular perigynia that are essentially nerveless on their inner face. It differs from *C. suberecta* in its spreading perigynia. It is sometimes difficult to separate *C. brevior* from *C. molesta* with which it seems to intergrade. In general, the inner face of the perigynium of *C. brevior* is nerveless, while the inner face of the perigynium of *C. molesta* is finely nerved.

This species is found most often in disturbed, dry habitats. *Carex brevior* flowers from late April to early June.

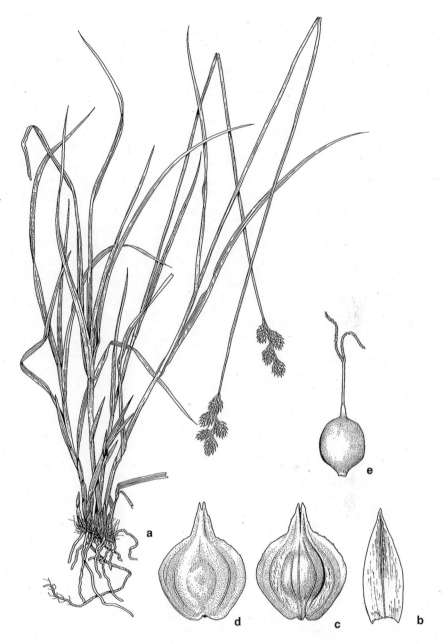

56. Carex brevior.
a. Habit.

b. Pistillate scale.
c. Perigynium, dorsal view.

d. Perigynium, ventral view.
e. Achene.

57. Carex molesta Mack. ex Bright, Trillia 9:30. 1930. Fig. 57.

Plants perennial, cespitose, from short, black, fibrillose rootstocks; culms to 1 m tall, wiry, triangular, the angles scabrous beneath the inflorescence, with old leaf bases often persisting; sterile shoots usually not present; leaves 4–7 per culm, ascending, 2.0–3.5 mm wide, flat, rather thin, light green, scabrous along the margins; sheaths tight, the ventral band narrowed quickly below the summit, pale green, prolonged at the summit, the lowermost brownish; spikes 2–5 per culm, 6–10 mm long, nearly globose, not clavate at base, green to greenish white, crowded into a dense inflorescence up to 3 cm long, or sometimes with the lower spikes distant; bracts mostly scalelike; pistillate scales lanceolate, obtuse to acute, barely reaching the base of the beak of the perigynium, the center green and 3-nerved, the margins hyaline; perigynia 15–30 per spike, 4.0–5.5 mm long, 2–3 mm wide, broadly ovate to suborbicular, widest near the middle, submembranous, ascending, plano-convex, green or greenish white, faintly nerved on the outer face, finely nerved on the inner face, winged to the base, substipitate, the beak 1.0–1.3 mm long, serrulate, bidentate, tipped with yellow-brown; achenes lenticular, 1.6–1.9 mm long, 1.2–1.3 mm wide, yellow-brown, apiculate, substipitate; stigmas 2, short, reddish brown.

Common Name: Sedge.
Habitat: Old fields, moist prairies, swamps, wet depressions, ditches.
Range: Vermont to Saskatchewan, south to Colorado, Kansas, Arkansas, Tennessee, and Delaware.
Illinois Distribution: Scattered throughout the state.

This is one of the more confusing species of section Ovales. Its globose spikelets resemble those of *C. cristatella*, but all of the perigynia of *C. molesta* are ascending. *Carex molesta* is also similar to *C. brevior*, but *C. molesta* generally does not have quite orbicular perigynia, and the inner face of each perigynium is finely nerved.

Carex molesta apparently is one of the most widespread species of section Ovales in Illinois, where it grows in a variety of dry or moist areas. The flowers appear from late April to early June.

58. Carex reniformis (Bailey) Small, Fl. S. E. U. S. 220. 1903. Fig. 58.
Carex straminea Willd. var. *reniformis* Bailey, Mem. Torrey Club 1:73. 1889.

Plants perennial, cespitose, from short, black, fibrillose rootstocks; culms to 1 m tall, triangular, scabrous on the angles beneath the inflorescence, light brown near the base, with old leaf bases often persisting; sterile shoots usually not present; leaves 4–5 per culm, ascending, 2.0–2.5 mm wide, flat, firm, light green, scabrous along the margins; sheaths tight, narrowly hyaline, green, prolonged at the summit; spikes 3–6 per culm, 6–10 mm long, ellipsoid to obovoid, not clavate at the base, silvery brown to green, gynecandrous, approximate but separated in a

57. Carex molesta.
a. Habit.
b. Pistillate scale.
c. Perigynium, dorsal view.
d. Perigynium, ventral view.
e. Achene.

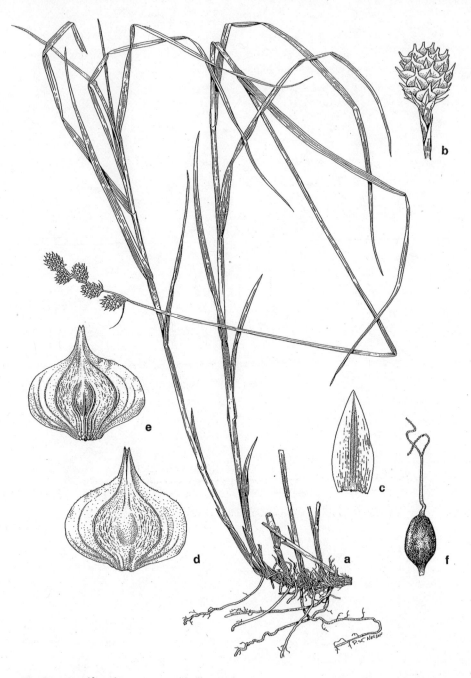

58. Carex reniformis.
a. Habit.
b. Spike.

c. Pistillate scale.
d. Perigynium, dorsal view.

e. Perigynium, ventral view.
f. Achene.

flexuous inflorescence up to 4.5 cm long, or the lower spikes distant; lowest bract setaceous, scabrous, the upper bracts scalelike; pistillate scales lanceolate, obtuse to acute, usually reaching above the base of the beak of the perigynium, the center green and 3-nerved, the margins hyaline; perigynia numerous per spike, 4–5 mm long, 3.5–5.0 mm wide, orbicular to reniform, widest at or below the middle, subcoriaceous, appressed-ascending, flat, stramineous, finely nerved on the outer face, usually nerveless on the inner face, winged nearly to base, substipitate, the beak 1.0–1.5 mm long, serrulate, bidentate, green; achenes lenticular, 1.9–2.2 mm long, about 1.5 mm wide, apiculate, stipitate; stigmas 2, reddish brown.

Common Name: Sedge.
Habitat: Wet ground.
Range: Southeast Virginia to southeast Illinois, south to Texas and Florida.
Illinois Distribution: Originally found at Mermet Conservation Area, Massac County and now known from a second locality in Massac County.

This is one of the better marked species of section Ovales by virtue of its very broad perigynia, which are often wider than they are high, bordered by a very broad wing. The flowers appear in May.

59. Carex bicknellii Britt. in Britt. & Brown, Ill. Fl. 1:360. 1896. Fig. 59.
Carex straminea Willd. var. *meadii* Boott, Ill. Gen. Carex 121. 1862.

Plants perennial, cespitose, from stout, black rootstocks; culms to 1 m tall, triangular, the angles scabrous beneath the inflorescence, light brown near the base, with old leaf bases often persisting; sterile shoots often present; leaves 3–6 per culm, ascending, 2.5–4.5 mm wide, flat, firm, scabrous-papillate on the upper surface, scabrous along the margins, smooth on the lower surface; sheaths tight, white-hyaline ventrally, not particularly prolonged at summit; spikes 3–7 per culm, 10–18 mm long, ovoid to obovoid, sometimes clavate at base, silvery brown to green, gynecandrous, crowded into an inflorescence 2–4 cm long or with the spikes distant; lowest bract setaceous, scabrous, the upper bracts scalelike; pistillate scales lanceolate, obtuse to acute, not reaching the base of the beak of the perigynium, the center green and 3-nerved, the margins hyaline; perigynia numerous per spike, 4.5–7.5 mm long, 2.8–5.0 mm wide, broadly ovate, widest at or a little below the middle, membranous, appressed-ascending, flat, greenish to stramineous, the margins translucent, becoming strongly tinged with golden brown and sometimes ciliate, strongly nerved on both faces, broadly winged to base, substipitate, the beak 1.0–1.5 mm long, serrulate, bidentate, often red-tipped; achenes lenticular, 1.7–2.0 mm long, 1.5–1.7 mm wide, apiculate, substipitate; stigmas 2, short, reddish.

Common Name: Bicknell's Sedge.
Habitat: Dry prairies, old fields, dry slopes.

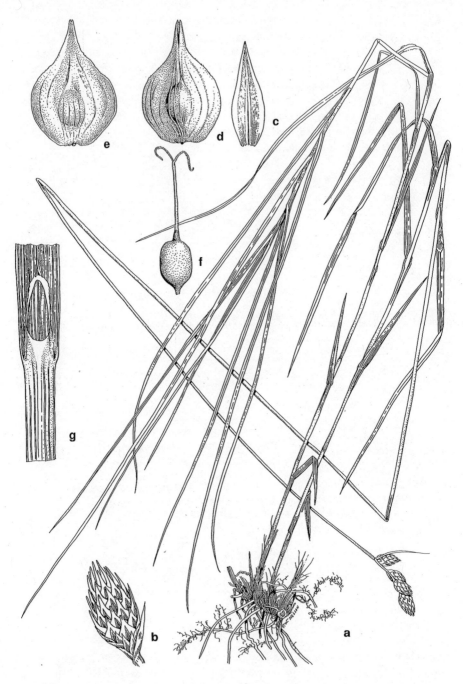

59. Carex bicknellii.
a. Habit.
b. Spike.

c. Pistillate scale.
d. Perigynium, dorsal view.

e. Perigynium, ventral view.
f. Achene.
g. Sheath with ligule.

Range: Maine to Saskatchewan, south to New Mexico, Oklahoma, Arkansas, and Delaware.

Illinois Distribution: Occasional to common in the northern three-fourths of the state, apparently absent elsewhere.

This species has the largest perigynia of any member of section Ovales, except *C. muskingumensis.* The perigynia are widest below the middle and strongly nerved on both faces. The first collection of this species from Illinois was by S. B. Mead from Hancock County in the early 1840s. Boott described Mead's plant as *Carex straminea* var. *meadii.* It was not until 1896 that this plant was elevated to the rank of species when it became known as *C. bicknellii.*

Carex bicknellii occurs in a variety of dry habitats in the northern three-fourths of Illinois. This species flowers from early May to mid-June.

60. **Carex alata** Torr. & Gray, Ann. Lyc. N. Y. 3:396. 1836. Fig. 60.

Plants perennial, cespitose, from short, black, fibrillose rootstocks; culms to 1.2 m tall, triangular, scabrous on the angles beneath the inflorescence, light brown near the base, with old leaf bases persisting; sterile shoots sometimes present; leaves 3–7 per culm, ascending, 2.0–5.5 mm wide, flat, firm, deep green, scabrous along the margins; sheaths tight, green-nerved ventrally, prolonged at the yellow-brown summit; spikes 3–8 per culm, 8–16 mm long, ovoid to ellipsoid, usually somewhat clavate at base, silvery brown to green, gynecandrous, usually crowded in an inflorescence up to 4 cm long; lower bracts setaceous, scabrous, the upper bracts scalelike or even absent; pistillate scales narrowly lanceolate, acuminate or more commonly awn-tipped and scabrous, usually as long as the perigynia, the center greenish and 1- to 3-nerved, the margins white-hyaline; perigynia numerous per spike, 4–5 mm long, 2.5–4.0 mm wide, obovate to suborbicular, broadest above the middle, membranous, appressed-ascending, flat, light brown to greenish, usually nerveless on the outer face, faintly 3-nerved on the inner face, broadly winged nearly to the base, substipitate, the beak 0.7–0.9 mm long, serrulate, bidentate; achenes lenticular, 1.6–1.9 mm long, about 1 mm wide, yellow-brown, apiculate, stipitate; stigmas 2, short, light reddish.

Common Name: Broad-winged Sedge.

Habitat: Wet ground.

Range: Massachusetts to Pennsylvania, southwest across southeastern Illinois and southern Missouri to Texas, east to Florida; also in northern Indiana and southern Michigan.

Illinois Distribution: Known from Jackson, Massac, Pope, and Wabash counties.

This species is distinguished by its relatively large, broadly winged perigynia that are broadest just above the middle and are only faintly nerved or nerveless. Rothrock has annotated a specimen from the Oakwood Bot-

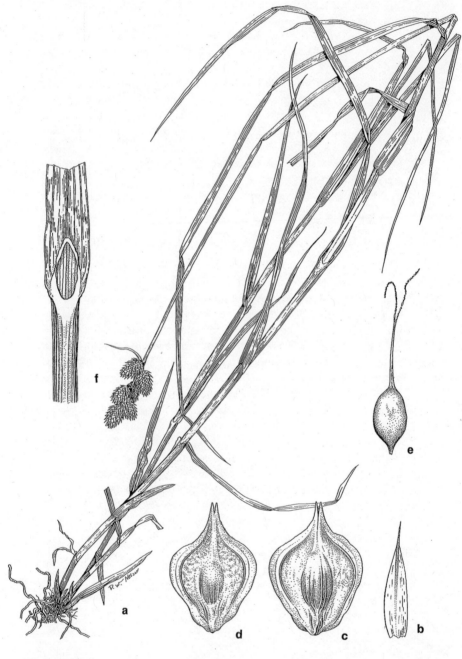

60. Carex alata.
a. Habit.
b. Pistillate scale.

c. Perigynium, dorsal view.
d. Perigynium, ventral view.

e. Achene.
f. Sheath with ligule.

toms in Jackson County as a possible hybrid between *C. alata* and *C. tribuloides*. This species flowers during May.

61. Carex praticola Rydb. Mem. N. Y. Bot. Gard. 1:84. 1900. Fig. 61.

Plants perennial, cespitose, from short, black, fibrillose rootstocks; culms to 70 cm tall, very slender, weak, triangular, scabrous on the angles, at least beneath the inflorescence, brownish at the base, with old leaf bases persisting; sterile culms usually not present; leaves 2–4 per culm, ascending, 1.0–3.5 mm wide, flat, weak, light green, scabrous along the margins, papillate on the lower surface; sheaths tight, pale, with a broad, hyaline ventral band, prolonged at the summit; spikes 2–7 per culm, 1.0–1.3 cm long, contracted at base, gynecandrous, more or less separated in a flexuous, moniliform inflorescence up to 5 cm long, or the uppermost spikes approximate; bracts usually scalelike; pistillate scales broadly lanceolate, obtuse to acute, longer and broader than the perigynium, the center green and 1- to 3-nerved, the margins hyaline; perigynia 6–20 per spike, 4.5–6.5 mm long, 1.5–2.0 mm wide, lance-ovate, widest at or below the middle, membranous, appressed-ascending, flat, green to chestnut-brown, faintly nerved on the outer face, usually nerveless on the inner face, winged to the base, substipitate, the beak 1.2–2.0 mm long, serrulate, bidentate, green; achenes lenticular, 1.5–2.0 mm long, 1.3–1.6 mm wide, apiculate, substipitate; stigmas 2, reddish brown.

Common Name: Sedge.
Habitat: Adventive in a cemetery in Illinois.
Range: Greenland to Alaska, south to California, Colorado, North Dakota, northern Michigan, and northern Maine; adventive in Illinois.
Illinois Distribution: Known only from a cemetery in Thornton in Cook County, where it was first discovered by Ken Dritz in the 1980s.

The distinguishing features of this *Carex* are its long but narrow perigynia that are completely concealed by the pistillate scales.
 Carex praticola grows natively far to the north of Illinois. The Illinois collections are undoubtedly from adventive populations in a cemetery that is regularly mowed. This species flowers during May.

62. Carex leptalea Wahl. Sv. Vet. Akad. 24:139. 1803. Fig. 62.
Carex polytrichoides Willd. Sp. Pl. 4:213. 1805.

Plants perennial, densely cespitose, with slender, scaly stolons; culms to 60 cm tall, capillary, weak, sharply triangular, longer than the leaves, brown at base, with old leaf bases persisting; leaves usually 2 per culm, filiform, up to 1.3 mm wide, flat or canaliculate, deep green, smooth or scabrous on the margins, minutely papillose on the upper surface; sheaths tight, the ventral band hyaline, sometimes strongly nerved and becoming fibrillose, truncate or concave at summit, the lowest pale or brown; spike 1 per culm, terminal, up to 15 cm long, up to 3 mm wide, ovoid to oblongoid, androgynous with a few appressed staminate scales at the tip; bracts ab-

61. Carex praticola.
a. Habit.

b. Pistillate scale.
c. Perigynium, dorsal view.

d. Perigynium, ventral view.
e. Achene.

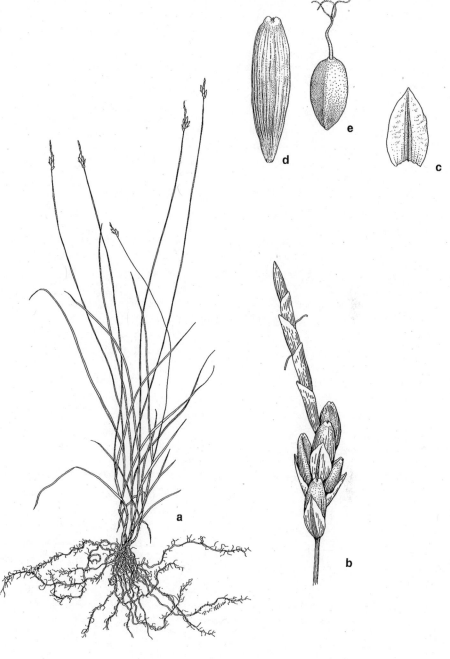

62. Carex leptalea.
a. Habit.
b. Inflorescence.
c. Scale.
d. Perigynium.
e. Achene.

sent; pistillate scales ovate, obtuse to acute to short-awned, about half as long as the perigynia, with a green center and hyaline margins; perigynia 1–10 per spike, 2.5–4.0 mm long, ellipsoid, strongly ascending, flattened and 2-edged, pale green to yellowish, several-nerved, more or less spongy at base, substipitate, beakless; achenes trigonous, 2.0–2.5 mm long, 1.5–2.0 mm wide, shiny; stigmas 3, short, reddish brown.

Common Name: Slender Sedge.
Habitat: Bogs, fens, wet meadows.
Range: Labrador to Alaska, south to California, Colorado, North Dakota, Missouri, Tennessee, and North Carolina.
Illinois Distribution: Known from several northeastern counties as well as Fayette, Peoria, Tazewell, Vermilion, Washington, and Woodford counties.

This slender, delicate species is recognized by its capillary culms, filiform leaves, and solitary spikes with a few male flowers at the tip. The beakless perigynia are several-nerved and about twice as long as the subtending scales.

This species, which forms stoloniferous mats, is characteristic of bogs and fens. The flowering time for *Carex leptalea* is during May.

63. Carex jamesii Schw. Ann. Lyc. N. Y. 1:67. 1824. Fig. 63.
Carex steudelii Kunth, Enum. Pl. 2:480. 1837.

Plants perennial, densely cespitose, from short, black rootstocks; culms to 30 cm tall, triangular, capillary but slightly winged above, weak, scabrous on the angles, at least beneath the inflorescence, red-brown near the base, usually shorter than the leaves; leaves 4–6 per culm, ascending, 2.0–3.7 mm wide, flat, lax, deep to pale green, scabrous along the margins, usually minutely papillate; sheaths tight, nerved, hyaline, the lower dark brown; spikes 2–3 per culm, 2–10 mm long, greenish, androgynous, the uppermost sessile, the lowest on elongated capillary peduncles, the staminate part 3–10 mm long, about 0.5 mm thick and composed of 4–several tiny and broadly obtuse to truncate scales; lowest pistillate scales leaflike, ovate, long-acuminate, up to 5 cm long, the margins hyaline; perigynia

2–3 per spike, 5–6 mm long, 2.3–2.5 mm wide, turgid, orbicular, erect, green, nerveless but with 2 keels decurrent from the base of the beak, spongy at base, with a thick stipe up to 1 mm long, the beak 2–3 mm long, flat, serrulate; achenes trigonous, about 2.5 mm long, nearly as broad, broadly stipitate, umbonate at tip; stigmas 3, short, red-brown.

Common Name: James' Sedge.
Habitat: Mesic woods.
Range: Ontario and southern New York to Michigan, south to Iowa, Kansas, Missouri, Tennessee, and Virginia.
Illinois Distribution: Common throughout the state.

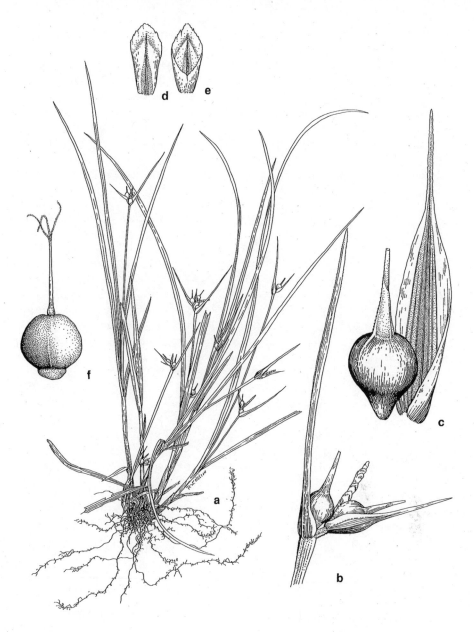

63. Carex jamesii.
a. Habit.
b. Inflorescence.
c. Pistillate scale and perigynium.
d. Staminate scale, dorsal view.
e. Staminate scale, ventral view.
f. Achene.

This is one of the most easily recognized species of *Carex* by virtue of its spikes, which have a slender series of obtuse to truncate staminate scales above two or three orbicular, 2-keeled perigynia. Each pistillate scale subtending the perigynia is leaflike. The only similar species is *C. willdenowii*, which differs in its more ellipsoid perigynia, each with a broadly triangular beak. The tips of the staminate scales of *C. willdenowii* are tapering rather than obtuse to truncate.

This is a common species in mesic woods in Illinois where it grows in dense, short clumps. *Carex jamesii* flowers from mid-April through May.

64. Carex willdenowii Schk. in Willd. Sp. Pl. 4:211. 1805. Fig. 64.

Plants perennial, densely cespitose, from short, black rootstocks; culms to 30 cm long, triangular, slightly winged above, weak, scabrous on the angles, red-brown near the base, usually shorter than the leaves, with old leaf bases persisting; leaves 3–5 per culm, ascending, 2–4 mm wide, flat to canaliculate, lax, deep to pale green, scabrous along the margins, papillate; sheaths tight, nerved, hyaline, the lower brown; spikes 1–3 per culm, 4–20 mm long, greenish, androgynous, the upper-most sessile, the lowest on elongated capillary peduncles, the staminate part 2–8 mm long, about 0.5 mm thick and composed of 6–several scales tapering at the tip; pistillate scales acute to cuspidate, sometimes scalelike, green-striate, with hyaline margins; perigynia 3–10 per spike, 4.5–5.5 mm long, about 1.5 mm wide, turgid, ellipsoid, erect, green to yellow-green, nerveless but with 2 keels decurrent from the base of the beak, spongy at base, with a slender stipe, the beak 2.0–2.5 mm long, tri-angular, serrulate; achenes trigonous, about 2.5 mm long, about 1.5 mm broad, stipitate, umbonate at tip; stigmas 3, short, red-brown.

Common Name: Willdenow's Sedge.
Habitat: Rocky woods.
Range: Vermont to southern Ontario to Minnesota, south to Texas and Florida.
Illinois Distribution: Confined to the southern eighth of the state; also Iroquois County.

This species differs from the similar and much more common *C. jamesii* in its narrower perigynia and its staminate scales that taper at the tip.

Carex willdenowii was first discovered in 1984 by Annette Parker in a rocky woods at Pounds Hollow, Gallatin County. It has subsequently been found in similar habitats in the southern tip of Illinois. *Carex willdenowii* flowers from mid-April to mid-May.

65. Carex pensylvanica Lam. Encycl. 3:388. 1792. Fig. 65.

Plants perennial, densely cespitose, from slender, scaly, fibrillose stolons; culms to 40 cm tall, wiry, triangular, longer than to equalling the leaves, usually smooth on the angles beneath the inflorescence, red-purple near the base, with persistent tufts of fibers at the base; sterile shoots often present; leaves 2–8 per culm, 1.5–3.0 mm wide, flat or becoming canaliculate at base, firm to lax, scabrous along the

64. Carex willdenowii.
a. Habit.

b. Pistillate scale.

c. Perigynium.
d. Achene.

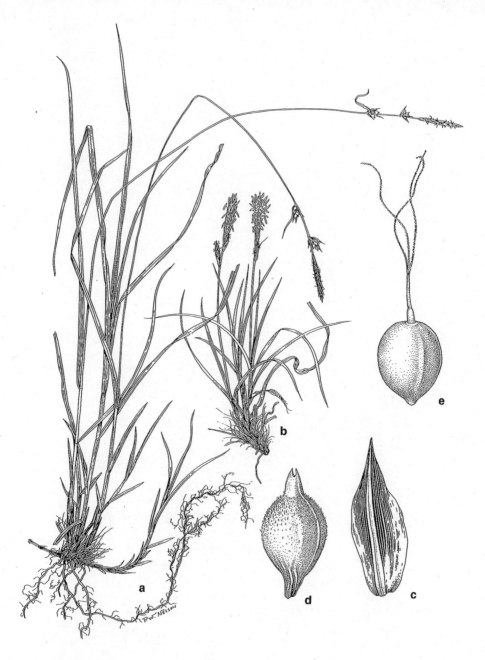

65. Carex pensylvanica.
a. Habit.

b. Young plant at anthesis.
c. Pistillate scale.

d. Perigynium.
e. Achene.

margins; sheaths tight, green, with a hyaline ventral band truncate to deeply cleft at the summit, sometimes red-dotted, the lower sheaths red, becoming fibrous; terminal spike staminate, nearly sessile, 8–20 mm long, 2–3 mm wide, the scales obtuse to acuminate, usually reddish brown; pistillate spikes 1–4, lateral, approximate or less commonly slightly separated, 3–12 mm long, globose to ovoid, barely longer than wide, sessile; bracts reduced, reddish brown, with hyaline margins; pistillate scales ovate to lanceolate, obtuse to acute to acuminate, equalling or surpassing the perigynia, reddish purple to brownish; perigynia up to 20 per spike, 1.5–3.0 mm long, ovoid to orbicular, trigonous, 2-keeled, membranous, pale green, puberulent, stipitate, abruptly contracted to a beak 0.2–0.9 mm long, the beak rarely more than fourth as long as the body, bidentate; achenes trigonous, 1.5–2.0 mm long, brownish, substipitate; stigmas 3, red-brown.

Common Name: Pennsylvania Sedge.
Habitat: Savannas, open woods.
Range: Quebec to Ontario and North Dakota, south to Missouri, Tennessee, and South Carolina.
Illinois Distribution: Occasional to common in the northern half of the state, much rarer southward.

Carex pensylvanica is distinguished from other similar species except *C. lucorum*, *C. heliophila*, and *C. communis* by its ovoid to orbicular rather than ellipsoid to obovoid perigynia. It differs from *C. communis* in that it possesses slender stolons. *Carex pensylvanica* has angular perigynia up to 3 mm long, while *C. heliophila*, a species more common in the western prairies and plains of the United States, has nearly terete perigynia, 3.0–3.5 mm long. *Carex pensylvanica* differs from *C. lucorum* in its shorter perigynial beak (a fourth to an eighth as long as the body) and shorter perigynia (up to 3 mm long), while the perigynial beak in *C. lucorum* is at least two-thirds as long as the body, and the perigynia are 3–4 mm long.

Some botanists combine *Carex pensylvanica*, *C. heliophila*, and *C. lucorum* as one, while others consider *C. pensylvanica* to consist of three varieties or three subspecies. Without convincing evidence to the contrary, I am recognizing all three taxa at the rank of species. During a conversation in 1957 with the eminent late botanist Julian A. Steyermark when I was wrestling with whether to combine two species in a different group, Steyermark advised me not to combine hastily and without direct proof, because the botanists who named the species had seen characteristics that had convinced them of their differences.

Carex pensylvanica is a prominent member of black oak savannas in northern Illinois. Elsewhere, it is often found in dry woods. Pennsylvania sedge flowers from mid-April to late May.

66. Carex lucorum Willd. in Link, Enum. Pl. 2:380. 1822. Fig. 66.
Carex pensylvanica Lam. var. *distans* Peck, N. Y. State Mus. Rep. 48:174. 1896.
Carex pensylvanica Lam. var. *lucorum* (Willd.) Fern. Proc. Am. Acad. 37:505. 1902.

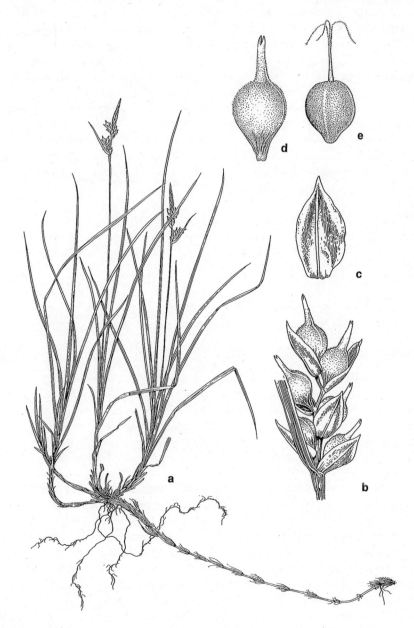

66. Carex lucorum.
a. Habit.

b. Pistillate inflorescence.
c. Pistillate scale.

d. Perigynium.
e. Achene.

Plants perennial, densely cespitose, from slender, scaly, fibrillose stolons; culms to 30 cm tall, longer than to equalling the culms, triangular, strongly scabrous on the angles beneath the inflorescence, red-purple near the base, with persistent, fibrous old leaves at the base; sterile shoots usually present; leaves up to 8 per culm, 1.5–2.5 mm wide, flat or canaliculate, usually rather lax, scabrous along the margins; sheaths tight, green, the hyaline ventral band deeply concave or truncate at the summit; terminal spike staminate, nearly sessile, 12–20 mm long, 2.5–3.0 mm wide, the scales obtuse to acute, reddish purple; pistillate spikes 2–3, lateral, approximate or sometimes the lowest one somewhat distant, up to 12 mm long, a little longer than wide, sessile; bracts reduced, reddish purple with hyaline margins; pistillate scales broadly lanceolate, obtuse to acute to acuminate, usually surpassing the perigynia, reddish purple with white-hyaline margins; perigynia up to 10 per spike, 3.2–4.0 mm long, orbicular, trigonous, 2-keeled, membranous, yellow-green, puberulent, stipitate, abruptly contracted to a beak, 0.9–2.0 mm long, the beak usually at least two-thirds as long as the body, bidentate; achenes trigonous, 1.7–2.0 mm long, brown, substipitate; stigmas 3, brown.

Common Name: Sedge.
Habitat: Woods.
Range: Quebec to Ontario, south to Illinois, Tennessee, and South Carolina.
Illinois Distribution: Known only from Hayes Creek Canyon, Pope County, where it was collected by Lawrence Stritch in 1981.

This species differs from *C. pensylvanica* in its longer perigynium with a much longer beak at least two-thirds as long as the body of the perigynium. Some botanists consider this taxon merely a variety of *C. pensylvanica*, while some others combine it with *C. pensylvanica*. It differs from *C. heliophila* in its more lax leaves, its trigonous achenes, and its longer beaks of the perigynia.

Because the range of this species is predominantly northern, the collection from southeastern Illinois was unexpected. There is a collection from Lake County that Mohlenbrock and Ladd (1978) considered *C. lucorum*, but further study of the plant indicates that the specimen may be closer to *C. pensylvanica*. In the Chicago area, Swink and Wilhelm (1994) state that it would be a practical impossibility to recognize two species. This species flowers during April in southern Illinois.

67. Carex heliophila Mack. Torreya 13:15. 1913. Fig. 67.
Carex pensylvanica Lam. var. *digyna* Boeckeler, Linnaea 41:220. 1877.
Carex pensylvanica Lam. ssp. *heliophila* (Mack.) W. A. Weber, Brittonia 33:375. 1981.
Carex inops Bailey ssp. *heliophila* (Mack.) Crins, Can. Journ. Bot. 61:1709. 1983.

Plants perennial, densely cespitose, from slender, scaly, fibrillose stolons; culms to 40 cm tall, stiff, triangular, scabrous on the angles beneath the inflorescence, usually longer than the leaves, red-brown near the base, with last year's leaf bases persisting; sterile shoots usually present; leaves 5–10 per culm, to 2.5 mm wide, rather stiff, usually revolute, harshly scabrous along the margins, canaliculate at base; sheaths tight, green, with a hyaline ventral band, shallow concave at the summit, the lower sheaths becoming fibrillose; terminal spike staminate, nearly sessile, 8–20 mm long, 3–5 mm wide, the scales obtuse to acuminate, purple or brown with scarious margins; pistillate spikes 1–3, lateral, approximate, 3–6 mm long, often as wide as long, sessile; bracts reduced, reddish brown; pistillate scales ovate, obtuse to acuminate, about equalling the perigynia, reddish brown, with white-hyaline margins; perigynia up to 15 per spike, usually much fewer, 3–4 mm long, suborbicular, more or less terete and not trigonous, 2-keeled, membranous, green,

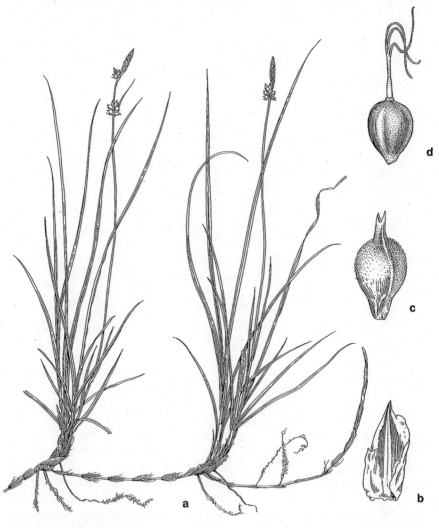

67. Carex heliophila.
a. Habit.

b. Pistillate scale.

c. Perigynium.
d. Achene.

puberulent, strongly stipitate, abruptly contracted to a beak 0.7–0.8 mm long, about a third as long as the body, bidentate; achenes trigonous, 1.8–2.0 mm long, nearly as broad, brownish, substipitate; stigmas 3, red-brown.

Common Name: Plains Sedge.
Habitat: Prairies, plains, savannas.
Range: Ontario to Alberta, south to New Mexico, Kansas, Illinois, and New York; also British Columbia.
Illinois Distribution: Known from Hardin, Jo Daviess, and Rock Island counties.

This species is distinguished by its orbicular, nearly terete perigynia that are 3–4 mm long; its perigynial beak about a third as long as the body; and its generally stiff stems and firm, harshly scabrous leaves.

Some botanists consider this plant to be a variety of *C. inops. Carex heliophila* is a common sedge in the prairies and plains to the west of Illinois. The flowers appear in May.

68. Carex communis L. H. Bailey, Mem. Torrey Club 1:41. 1889. Fig. 68.

Plants perennial, densely cespitose, without stolons; culms to 60 cm tall, triangular, slender but firm, scabrous on the angles beneath the inflorescence, reddish purple near the base, usually longer than the leaves; sterile shoots usually many; leaves 2–6 per culm, up to 5 mm wide, rarely wider, flat, thin, pale green, scabrous along the margins; sheaths tight, deeply tinged with purple, concave at the summit; terminal spike staminate, sessile to short-stalked, up to 18 mm long, up to 2 mm wide, the scales obtuse to acuminate; pistillate spikes 2–5, lateral, usually somewhat separated to even remote, 4–8 mm long, longer than wide, sessile; bracts reduced, purplish; pistillate scales ovate, acute to cuspidate, about as long as the perigynia, usually purplish; perigynia up to 10 per spike, 2.5–4.0 mm long, subglobose to ellipsoid, trigonous, 2-keeled, membranous, pale green, puberulent, stipitate, abruptly contracted to a beak 0.5–0.7 mm long, bidentate; achenes trigonous, 1.5–2.0 mm long, 1.2–1.3 mm wide, light brown, substipitate; stigmas 3, reddish brown.

Common Name: Sedge.
Habitat: Mesic woods, dry woods.
Range: Quebec to Ontario, south to Minnesota, Illinois, Arkansas, and Georgia.
Illinois Distribution: Scattered in Illinois, but not particularly common.

The distinguishing features of this sedge are its usually suborbicular perigynia and its absence of stolons. *Carex pensylvanica, C. lucorum,* and *C. heliophila* have similarly shaped perigynia, but all three of these species have stolons. *Carex communis* may occur in either moist or dry woods. This species flowers during May.

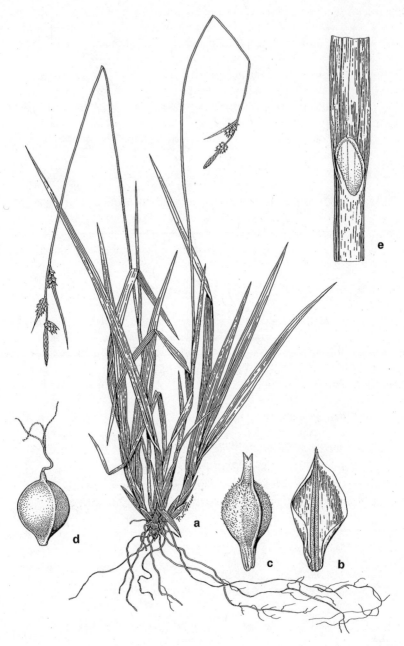

68. Carex communis.
a. Habit.

b. Pistillate scale.
c. Perigynium.

d. Achene.
e. Sheath with ligule.

69. Carex emmonsii Dewey, Ann. Lyc. N. Y. 3:411. 1836. Fig. 69.
Carex lucorum Willd. var. *emmonsii* (Dewey) Chapm. Fl. South. U.S. 539. 1860.
Carex albicans Willd. var. *emmonsii* (Dewey) Rettig, Sida 14:133. 1990.

Plants perennial, densely cespitose, from brownish, scaly rootstocks; culms to
40 cm tall, filiform, weak, arching to reclining, triangular, scabrous on the angles
beneath the inflorescence, red-purple near the base, with overwintering leaves and
leaves of the season present; sterile shoots few or even absent; leaves 2–6 per culm,
up to 1.5 mm wide, spreading, canaliculate, green, scabrous along the margins;
sheaths tight, reddish, concave at the summit; terminal spike staminate, nearly sessile, up to 8 mm long, 1.5–2.0 mm wide, the scales cucullate and long-acuminate,
reddish; pistillate spikes 2–3, lateral, approximate or the lowermost remote, up to
5 mm long, globose to ovoid to ellipsoid, sessile; bracts purplish, reduced; pistillate
scales oblong to obovate, acute to cuspidate, about equalling the perigynia, purplish with a green center and hyaline margins; perigynia 4–10 per spike, 2.0–3.3
mm long, ellipsoid, trigonous, 2-keeled, membranous, olive green to yellow-green,
puberulent, long-stipitate, tapering to a distinct beak 0.5–1.0 mm long, the beak
bidentate, sometimes hyaline; achenes trigonous, 1.4–1.5 mm long, about 1 mm
wide, dark brown, apiculate; stigmas 3, pale red-brown.

Common Name: Sedge.
Habitat: Dry woods.
Range: Nova Scotia to Michigan and Wisconsin, south to Illinois, Ohio,
New York, and Vermont.
Illinois Distribution: Known from Alexander, Gallatin, Hardin, and Un-
ion counties in southern Illinois. Wilhelm and Swink (1994) report it
from Cook and Kankakee counties.

The distinguishing features of *C. emmonsii* are its ellipsoid perigynia;
its weak, arching or reclining stems; and its absence of stolons. *Carex
albicans* is similar but has merely obtuse to acute rather than long-acuminate
staminate scales and less reclining or arching stems. *Carex physorhyncha* is also
similar but possesses slender stolons.

Some botanists consider this taxon to be a variety of *C. albicans*, but the differences seem to me to be enough to merit species status.

The southern Illinois collections, all southern disjuncts in the overall range of
this species, came from dry woods where *Carex emmonsii* blooms during April.

70. Carex albicans Willd. in Spreng. Syst. Veg. 3:818. 1826. Fig. 70.
Carex varia Muhl. Sv. Vet. Akad. 14:159. 1803, non Lamn. (1791).
Carex pensylvanica Lam. var. *muhlenbergii* Gray, N. Am. Gram. 2:163. 1835.
Carex artitecta Mack. N. Am. Fl. 18:189. 1935.
Carex albicans Willd. var. *muhlenbergii* (Gray) Rettig, Sida 14:132. 1990.

Plants perennial, densely cespitose, from brownish, scaly rootstocks; culms
to 50 cm tall, slender, rather stiff, erect or ascending, longer than the leaves, trian-

gular, scabrous on the angles beneath the inflorescence, red-purple near the base, with overwintering leaves and leaves of the season present; sterile shoots numerous; leaves 3–8 per culm, up to 2.5 mm wide, ascending, canaliculate, green, scabrous along the margins; sheaths tight, red-purple, concave at the summit; terminal spike staminate, nearly sessile, up to 1.5 cm long, 1.2–2.0 mm wide, the scales obtuse to acute, never long-acuminate, reddish to reddish brown; pistillate spikes 1–4, lateral, approximate, or the lowest not overlapping the one above, up to 7 mm long, sessile; bracts reddish brown, at least at the base, reduced; pistillate scales lanceolate, acute to cuspidate, about equalling the perigynia, purple-red with a green center and hyaline margins; perigynia 4–12 per spike, 2.8–3.5 mm long, ellipsoid, trigonous, 2-keeled, membranous, yellow-green, puberulent, long-stipitate, tapering to a distinct beak about 1 mm long, the beak about a third the length of the perigynia, bidentate; achenes trigonous, 1.3–1.5 mm long, about 1 mm wide, dark brown, apiculate; stigmas 3, pale red-brown.

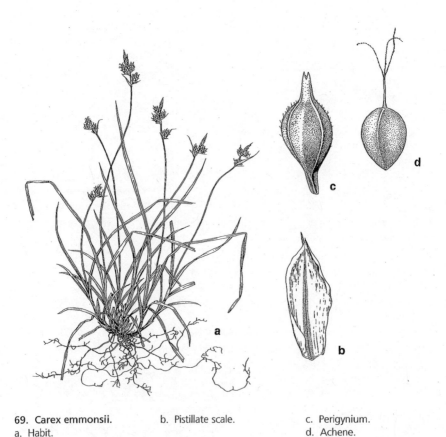

69. Carex emmonsii.   b. Pistillate scale.   c. Perigynium.
a. Habit.   d. Achene.

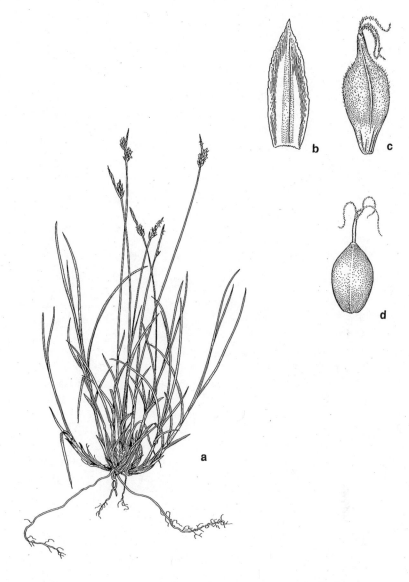

70. Carex albicans.
a. Habit.

b. Pistillate scale.

c. Perigynium.
d. Achene.

Common Name: Sedge.
Habitat: Dry woods.
Range: Quebec to Ontario, south to eastern Texas and South Carolina.
Illinois Distribution: Common throughout the state.

This species is similar to *C. emmonsii* from which it differs in its obtuse to acute staminate scales and its stiffer, more erect stems. It differs from *C. physorhyncha* in its lack of stolons.

This is the species that has universally been called *Carex artitecta* in the past. Rettig (1989, 1990) discusses the nomenclature of the *Carex albicans* complex. This is one of the most common species of *Carex* in Illinois in dry woods. *Carex albicans* flowers from mid-April to mid-May.

71. Carex physorhyncha Liebm. ex Steud. Syn. Pl. Gl. 2:219. 1855. Fig. 71.
Carex varia Muhl. var. *australis* Bailey, Bot. Gaz. 16:153. 1892.
Carex albicans Willd. var. *australis* (Bailey) Rettig, Sida 14:133. 1990.

Plants perennial; culms loosely cespitose, from brown, scaly stolons; culms to 40 cm tall, about as long as the leaves, slender, ascending, triangular, scabrous on the angles beneath the inflorescence, reddish near the base, with old leaves usually persisting; sterile shoots few to many; leaves 4–8 per culm, 1.5–3.0 mm wide, ascending, usually flat, pale green, scabrous along the margins; sheaths tight, reddish, concave at the summit; terminal spike staminate, sessile or more often short-stalked, up to 8 mm long, up to 2 mm wide, the scales acute; pistillate spikes 3–4, lateral, sessile or short-stalked, usually not overlapping; bracts scalelike, or the uppermost reduced leaflike; pistillate scales lanceolate, cuspidate, slightly shorter to equalling the perigynia, purple-red or yellow-brown with a 3-nerved green center and hyaline margins; perigynia 6–12 per spike, 2.5–3.0 mm long, ellipsoid, trigonous, 2-keeled, membranous, pale green, puberulent, stipitate, tapering to a distinct beak 0.75–1.00 mm long, about a third as long as the body, the beak bidentate; achenes trigonous, 1.3–1.5 mm long, about 1 mm wide, dark brown, apiculate; stigmas 3, pale red-brown.

Common Name: Sedge.
Habitat: Rocky woods, particularly in chert in Illinois.
Range: Southeastern Virginia to southern Illinois, southeastern Missouri, and Oklahoma, south to Texas and Florida; Mexico.
Illinois Distribution: Known from Jackson, Randolph, Union, and Williamson counties; also Effingham County.

This species differs from the very similar-appearing *C. albicans* and *C. emmonsii* in that it possesses slender stolons.

Some botanists consider this plant to be a variety of *C. albicans*, but I believe it deserves species status.

At two of the five locations for this species in Illinois, it occurs in dry, rocky,

71. Carex physorhyncha.
a. Habit.

b. Pistillate scale.

c. Perigynium.
d. Achene.

cherty woods associated with *Pinus echinata*. *Carex physorhyncha* flowers during late April and early May.

72. Carex nigromarginata Schw. Ann. Lyc. N. Y. 1:68. 1824. Fig. 72.
Carex lucorum Willd. var. *nigromarginata* (Schw.) Chapm. Fl. South. U.S. 539. 1860.

Plants perennial, densely to loosely cespitose, from very short, slender, scaly stolons; culms to 30 cm tall, much shorter than the leaves, the shorter ones crowded among the leaf bases, slender, scabrous on the angles beneath the inflorescence, red-purple near the base, fibrillose; overwintering leaves much longer than the culms, new leaves very short, 0.8–4.0 mm wide, flat or canaliculate near the base, rather stiff, green, scabrous along the margins; sheaths tight, reddish, truncate at the summit; terminal spike staminate, sessile or nearly so, to 15 mm long, purple-brown to green; pistillate spikes 2–3, lateral, ovoid to ellipsoid, usually approximate to the staminate spike, sessile or nearly so; lowest bract setaceous, up to 2.5 cm long, the upper one reduced; pistillate scales broadly lanceolate, acute to cuspidate, purple to purple-brown, with a pale center, about as long as or longer than the perigynia; perigynia 6–15 per spike, 3–4 mm long, 1.2–1.5 mm thick, broadly ellipsoid, trigonous, 2-keeled, membranous, yellow-green, puberulent, long-stipitate, tapering to a distinct beak about 1 mm long, the beak bidentate; achenes trigonous, 1.3–1.5 mm long, about 1 mm wide, brown, apiculate; stigmas 3, red-brown.

Common Name: Sedge.
Habitat: Woods.
Range: Connecticut across New York and Pennsylvania to southern Indiana, southern Illinois, and southeastern Missouri, south to Louisiana and Florida.
Illinois Distribution: Known only from Jackson, Hardin, Montgomery, Pope, Union, and Wabash counties.

This is one of four species of *Carex* in this section that has all or most of its spikes hidden among the leaf bases. It differs from the other three in its broadly ellipsoid perigynia that are only a half to two-thirds as wide as long as opposed to being nearly as wide as long.

This species has been found only a few times in Illinois where it occurs in rather dry woods. *Carex nigromarginata* flowers in late April and early May.

73. Carex umbellata Schk. in Willd. Sp. Pl. 4:290. 1805. Fig. 73.

Plants perennial, densely cespitose, from fibrous rootstocks; culms up to 15 cm long, usually much shorter and hidden among the leaf bases, slender, lax to stiff, scabrous on the angles beneath the inflorescence, red-brown near the base, fibrillose; leaves soft or firm, 1.5–2.5 mm wide, flat but canaliculate at the base, light

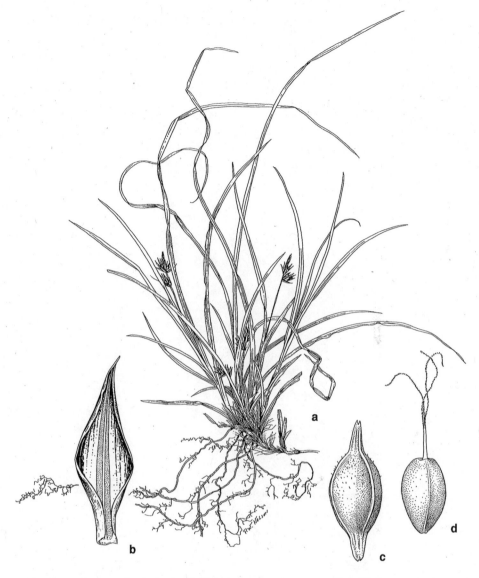

72. Carex nigromarginata.
a. Habit.

b. Pistillate scale.

c. Perigynium.
d. Achene.

green, scabrous along the margins; sheaths tight, elongated at the summit; terminal spike staminate, sessile or nearly so, up to 12 mm long, red-brown; pistillate spikes 2–4, lateral, ellipsoid, usually approximate to the staminate spike, sessile or nearly so; bracts scalelike or reduced; pistillate scales broadly lanceolate, usually acuminate, reddish brown with a green center, about as long as the perigynia; perigynia up to 20 per spike, 2.2–3.3 mm long, 1.0–1.5 mm thick, the body nearly orbicular, trigonous, 2-keeled, membranous, dull green, puberulent, long-stipitate, abruptly tapering to a distinct beak 0.9–1.7 mm long, the beak often as long as to about three-fourths the length of the body, bidentate; achenes trigonous, 1.5–1.7 mm long, dark brown, apiculate; stigmas 3, red-brown.

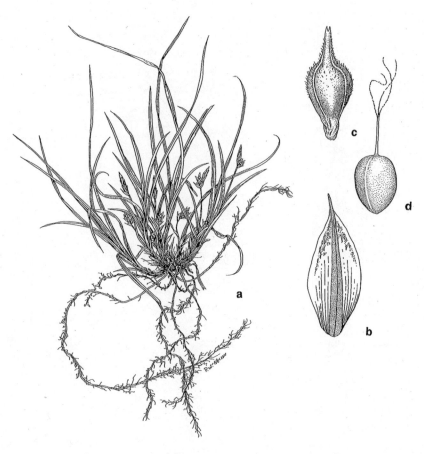

73. Carex umbellata.
a. Habit.
b. Pistillate scale.
c. Perigynium.
d. Achene.

Common Name: Sedge.
Habitat: Dry, rocky woods.
Range: Newfoundland to Saskatchewan, south to Minnesota, Illinois, Tennessee, and Virginia; British Columbia.
Illinois Distribution: This species is apparently known only from Jackson and Randolph counties.

This species is similar to *C. abdita* by virtue of its hidden spikes, but it differs in its longer proportioned beak of the perigynium and its acuminate rather than acute scales. All northern Illinois specimens with beaks up to 1 mm long and with acute pistillate scales are referred to *C. abdita*. *Carex umbellata* flowers during April.

74. Carex abdita Bickn. Bull. Torrey Club 35:492. 1908. Fig. 74.

Plants perennial, densely cespitose, from fibrous rootstocks; culms up to 20 cm tall but usually much shorter and hidden among the leaf bases, slender, lax to stiff, scabrous on the angles beneath the inflorescence, red-brown near the base, fibrillose; leaves soft or firm, to 2.5 mm wide, flat but canaliculate at the base, light green, scabrous along the margins; sheaths tight, elongated at the summit; terminal spike staminate, sessile or nearly so, up to 10 mm long, red-brown; pistillate spikes 2–4, lateral, ellipsoid, usually approximate to the staminate spike, sessile or nearly so; bracts scalelike or reduced; pistillate scales broadly lanceolate, acute, reddish brown with a green center, about as long as the perigynia; perigynia up to 20 per spike, 3.2–4.5 mm long, 1.2–1.8 mm thick, the body nearly orbicular, trigonous, 2-keeled, membranous, dull green, puberulent, long-stipitate, abruptly tapering to the distinct beak 0.8–1.0 mm long, the beak half to two-thirds the length of the body of the perigynium, bidentate; achenes trigonous, 1.5–1.8 mm long, gray-black, apiculate; stigmas 3, red-brown.

Common Name: Sedge.
Habitat: Dry prairies, dry woods, sandy soil.
Range: Nova Scotia to Minnesota, south to Missouri, Illinois, Indiana, and Virginia.
Illinois Distribution: Throughout the state but more common in the northern counties.

This species differs from the very similar *C. umbellata* in its larger perigynia, its shorter proportioned perigynial beak, and its merely acute rather than long-acuminate pistillate scales.

This species was referred previously by me (1975, 1985) as *C. umbellata*. In northern Illinois, this species is common in sandy woods and sandy prairies. At the relatively few stations in southern Illinois, *C. abdita* is found in woods with scattered sandstone boulders. *Carex abdita* flowers in April and early May.

75. Carex tonsa (Fern.) Bickn. Bull. Torrey Club 35:492. 1908. Fig. 75.
Carex umbellata Schk. var. *tonsa* Fern. Proc. Am. Acad. 37:507. 1902.

Plants perennial, densely cespitose, from fibrous rootstocks; culms up to 15 cm tall but usually much shorter and hidden among the leaf bases, slender, lax to stiff, scabrous on the angles beneath the inflorescence, reddish near the base, fibrillose; leaves up to 4.5 mm wide, stiff, canaliculate, dark green, scabrous along the margins; sheaths tight, reddish; terminal spike staminate, sessile or nearly so, 6–12 mm long, reddish brown; pistillate spikes 2–3, crowded at the base of the plant; bracts setaceous or scalelike; pistillate scales broadly lanceolate, acuminate, purple-brown to stramineous with a green center, as long as or longer than the perigynia; perigynia up to 20 per spike, 3–5 mm long, 1.3–1.5 mm thick, the body more or less orbicular, trigonous, 2-keeled, membranous, light green, glabrous or very slightly puberulent, long-stipitate, abruptly tapering to a distinct beak 1–3 mm long, bidentate; achenes trigonous, 1.7–2.0 mm long, brown, apiculate; stigmas 3, red-brown.

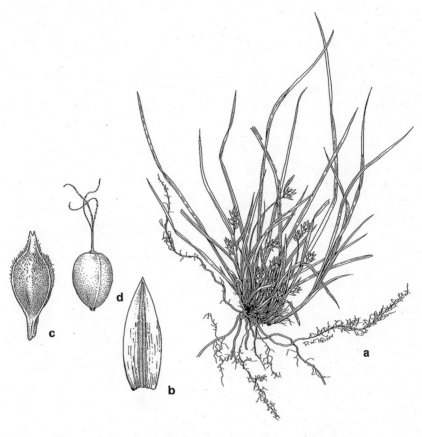

74. Carex abdita.
a. Habit.

b. Pistillate scale.

c. Perigynium.
d. Achene.

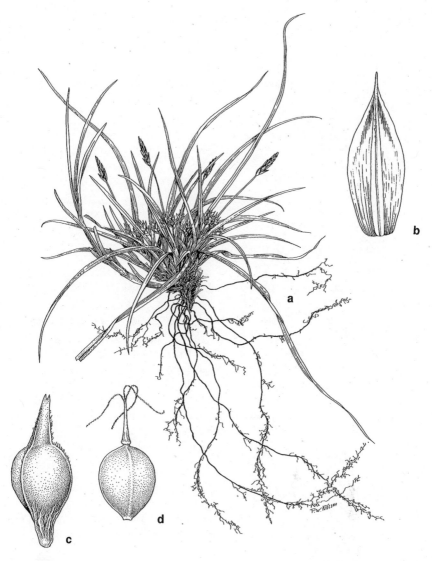

75. **Carex tonsa.**
a. Habit.

b. Pistillate scale.

c. Perigynium.
d. Achene.

Common Name: Shaved Sedge.
Habitat: Sand dunes, sand prairies, rocky woods.
Range: Quebec to Alberta, south to Minnesota, southern Illinois, and Virginia.
Illinois Distribution: Known from counties in extreme northwestern Illinois; also Lake, Mason, and Pope counties.

This species is characterized by its hidden spikes and its glabrous or only sparsely puberulent perigynia. The first collection of *Carex tonsa* in Illinois was made by William Hopkins along the banks of Lusk Creek, Pope County in the 1960s. It has been found most recently in sand prairies at Illinois Beach State Park in Lake County. It is extremely common in sandy areas in northwestern Illinois. *Carex tonsa* flowers in April and early May.

76. Carex pedunculata Muhl. ex Willd. Sp. Pl. 4:222. 1805. Fig. 76.

Plants perennial, densely cespitose, from stout, woody rootstocks; culms to 30 cm tall, capillary, shorter than the leaves, purple at the base; leaves 2–3 mm wide, firm, strap-shaped, acute, pale green to glaucous, brownish toward the tip, often with a purple line between the brown and green portions, scabrous on the margins and on the veins; sheaths loose, deep red or purple and abruptly green at the ligule; terminal spike staminate or androgynous, up to 15 mm long; lateral spikes 3–4, androgynous or pistillate, on capillary peduncles, the tiny, long-sheathing bracts red-purple; pistillate scales broadly oblanceolate, short-acuminate to aristate, green or purple, with a 3-nerved green center; perigynia 1–8 per spike, 3.8–4.5 mm long, trigonous, obovoid, 2-keeled, membranous, dark green, sparsely puberulent, spongy at the cuneate base, with a minute beak entire, usually bent; achenes 2.5–3.5 mm long, trigonous, apiculate, stipitate; stigmas 3.

Common Name: Long-stalked Hummock Sedge.
Habitat: Dry, calcareous slopes, moist ravines.
Range: Newfoundland to Saskatchewan, south to South Dakota, Iowa, Illinois, Ohio, and Delaware; northwestern Georgia.
Illinois Distribution: Known from Cook, Jo Daviess, Kane, Lake, McHenry, and Winnebago counties.

This species is distinguished by its lateral spikes borne on capillary peduncles and its puberulent, trigonous perigynia. It differs from the similar *C. richardsonii* in its cuspidate pistillate scales; in its usually androgynous terminal spike; and in the presence of lower spikes on long, capillary peduncles.

The few locations for this species in Illinois are from both moist and dry areas. George Vasey was apparently the first to collect this species in Illinois when he found it near Elgin, Kane County in the nineteenth century. *Carex pedunculata* flowers during March and April.

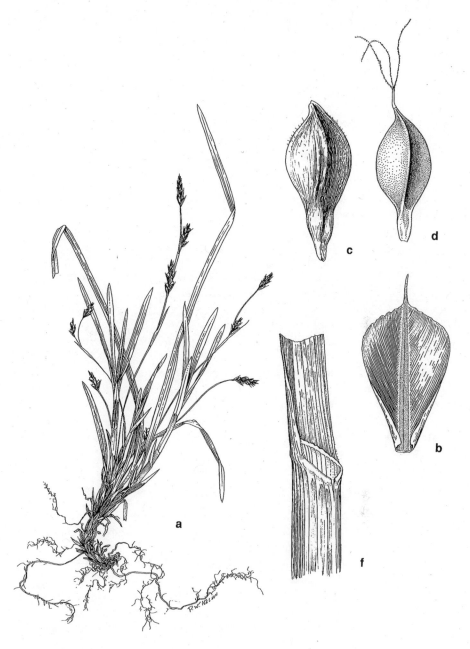

76. Carex pedunculata.
a. Habit.

b. Pistillate scale.
c. Perigynium.

d. Achene.
e. Sheath with ligule.

77. Carex richardsonii R. Br. Frankl. Journey 751. 1823. Fig. 77.
Carex richardsonii R. Br. f. *exserta* Fern. Rhodora 44:290. 1942.

Plants perennial, loosely cespitose, from stout, dark brown, scaly, slender rhizomes; culms to 30 cm tall, flexuous, scabrous on the narrow ridges, not noticeably triangular, brown-fibrillose at the base; leaves 2.0–3.5 mm wide, coriaceous, canaliculate, pale green, scabrous on the margins; sheaths tight, deep purple, becoming fibrous; terminal spike entirely staminate, up to 2.5 cm long, up to 3.5 mm thick, short-pedunculate, with obtuse, purple scales with silver margins; lateral spikes usually 2, pistillate, 1–2 cm long, appressed-ascending; bracts purplish, the lowest sheathing; pistillate scales broadly lanceolate, obtuse to acute, longer than the perigynia, purple with hyaline margins; perigynia 10–25 per spike, 2.2–3.0 mm long, obovoid, subterete to trigonous, pubescent, 2-keeled, stramineous to light brown, spongy-stipitate at the base, with a short, entire beak about 0.5 mm long; achenes trigonous, 1.5–2.0 mm long, brown; stigmas 3, dark brown to black.

Common Name: Hummock Sedge.
Habitat: Rocky areas, hill prairies, sand dunes.
Range: Vermont to Alberta, south to South Dakota, Minnesota, central Illinois, northern Ohio, and western New York.
Illinois Distribution: Confined to the northern half of the state.

This species differs from *C. pedunculata* in having an entirely staminate terminal spike, in lacking capillary peduncles on the lowest pistillate spikes, and in having obtuse to acute pistillate scales.

The first Illinois collection of this species was made by Samuel B. Mead on May 22, 1843 from Augusta, Hancock County. *Carex richardsonii* flowers from mid-April through May.

78. Carex eburnea Boott ex. Hook. Fl. Bor. Am. 2:226. 1839. Fig. 78.

Plants perennial, cespitose, from slender, pale brown rhizomes, forming dense mats; culms to 30 cm tall, capillary, wiry, smooth, obscurely triangular, longer than the leaves, brownish near the base; leaves 3–6 per culm, filiform, involute, up to 0.5 (–1.0) mm wide, often recurved, green, scabrous; sheaths tight, yellow-brown with hyaline margins; terminal spike staminate, up to 10 mm long, 0.5–1.5 mm thick, sessile or nearly so, surpassed and partially hidden by the uppermost pistillate spikes; lateral spikes 2–4, pistillate, crowded except sometimes for a remote lower one, 2–7 mm long, 1.5–6.0 mm thick, ovoid to ellipsoid, borne on slender, erect peduncles; bracts tubular, truncate at the tip, greenish and often with hyaline margins; staminate scales elliptic, obtuse to subacute, 2.5–4.0 mm long, with hyaline margins; pistillate scales broadly ovate, obtuse to acute, 1–2 mm long, as long as to nearly equalling the perigynia, whitish or yellow-brown with a green center; perigynia 2–6 per spike, 1.5–2.2 mm long, trigonous, ellipsoid, yellow-green to black, shiny, membranous, 2-keeled, finely nerved, glabrous, with a short beak

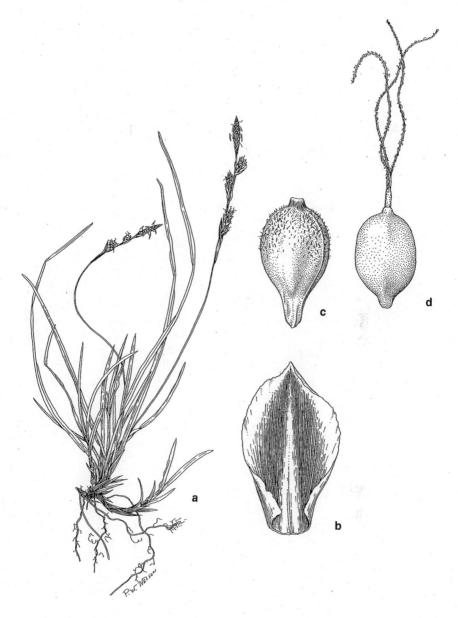

77. Carex richardsonii.
a. Habit.

b. Pistillate scale.

c. Perigynium.
d. Achene.

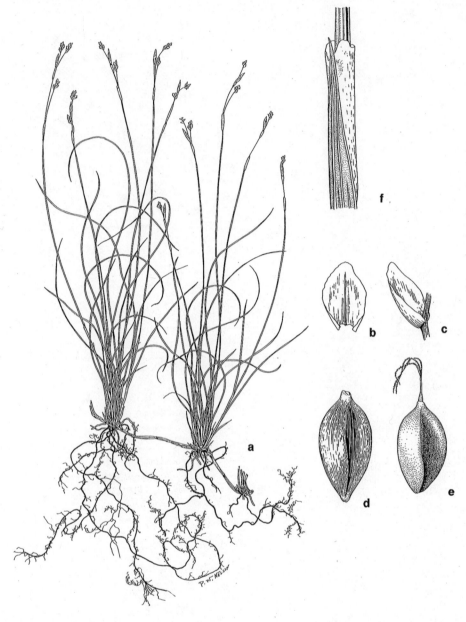

78. Carex eburnea.
a. Habit.

b, c. Pistillate scales.
d. Perigynium.

e. Achene.
f. Sheath.

0.2–0.5 mm long; achenes trigonous, 1.4–1.8 mm long, up to 1 mm wide, brown, apiculate; stigmas 3, brown.

Common Name: Sedge.
Habitat: Calcareous ledges, wooded ravines.
Range: Newfoundland to Alaska, south to South Dakota, Arkansas, Alabama, and Virginia.
Illinois Distribution: Occasional in the northern half of Illinois; also Monroe and Union counties.

This species is recognized by its mat-forming habit, its capillary stems and filiform leaves, and its ellipsoid perigynia, 1.5–2.2 mm long.

Most Illinois collections have come from calcareous rocky areas, although some Lake County populations are in a wooded ravine near Lake Michigan. This species flowers from mid-April to mid-May.

79. Carex hirtifolia Mack. Bull. Torrey Club 37:244. 1910. Fig. 79.
Carex pubescens Muhl. ex Willd. Sp. Pl. 4:281. 1805, non Poir. (1789).

Plants perennial, loosely cespitose, from slender rootstocks; culms to 80 cm tall, slender, firm, longer than the leaves, pubescent, triangular, scabrous on the upper angles; leaves to 1 cm wide, lax, flat, hirsute, pale green; sheaths tight, green to reddish brown, except for the hyaline ventral band; terminal spike staminate, 1–2 cm long, to 2.5 mm thick, sessile or nearly so; lateral spikes 2–4, pistillate, appressed-ascending, to 2 cm long, to 5 mm thick, approximate or separate, the lowest usually short-pedunculate; lowest bract setaceous, up to 7 cm long, the upper bracts smaller; pistillate scales 3–5 mm long, obovate to nearly orbicular, cuspidate to awned, whitish with a green midvein, ciliate, about as long as the perigynia; perigynia up to 25 per spike, 3.5–5.0 mm long, broadly ellipsoid, trigonous, densely pubescent, firm, nerveless, stipitate, tapering to a slender beak, the beak 0.8–1.2 mm long, bidentate; achenes trigonous, 2.5–2.8 mm long, 1.2–1.5 mm broad, apiculate; stigmas 3, reddish brown.

Common Name: Hairy Sedge.
Habitat: Dry or mesic woods.
Range: New Brunswick and Nova Scotia to Ontario, south to eastern Kansas, Missouri, Kentucky, and Maryland.
Illinois Distribution: Common throughout the state.

This is the most pubescent native species of *Carex* in Illinois, with the culms, leaves, and perigynia densely pubescent. *Carex hirtifolia* flowers from late March to mid-May.

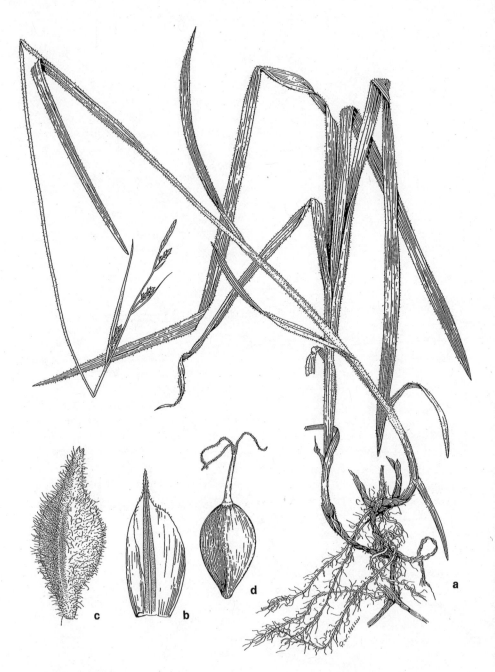

79. Carex hirtifolia.
a. Habit.

b. Pistillate scale.

c. Perigynium.
d. Achene.

80. Carex garberi Fern. Rhodora 37:253. 1935. Fig. 80.

Plants perennial, loosely cespitose, usually with slender rhizomes; culms to 40 cm tall, stiff, scabrous on the angles beneath the inflorescence; leaves 5–7 per culm, 1.0–3.5 mm wide, rather stiff, flat, smooth or scabrous along the margins, with pellucid, oblong cells between the often papillate nerves beneath, at least the uppermost leaves longer than the culms; lowermost bracts long-sheathing, the sheaths concave at summit; terminal spike gynecandrous, up to 20 mm long, 1.5–2.5 mm thick, sessile or nearly so; lateral spikes 2–6, pistillate, up to 20 mm long, up to 4.5 mm thick, thick-cylindric, the upper usually crowded, the lowest remote and borne on a scabrous peduncle; pistillate scales oblong to obovate, obtuse to acute, 2–3 mm long, membranous, brown to purple, with a hyaline margin; perigynia up to 30 per spike, 2–3 mm long, biconvex, plump, ellipsoid to obovoid, rounded and beakless at the tip, white-papillate at least in the upper half, stipitate; achenes lenticular, ellipsoid, 1.3–2.0 mm long, 1.2–1.5 mm broad, brown, apiculate; stigmas 2, blackish.

Common Name: Garber's Sedge
Habitat: Sandy beaches near Lake Michigan.
Range: Quebec to Alaska, south to California, northern Illinois, northern Ohio, and Maine.
Illinois Distribution: Known only from Cook and Lake counties.

This species is similar to *C. aurea* but differs in its white-papillate perigynia and its gynecandrous terminal spike. This species flowers in June in Illinois.

81. Carex aurea Nutt. Gen. 2:205. 1818. Fig. 81.

Plants perennial, loosely cespitose, often with slender rhizomes; culms to 50 cm tall, slender, scabrous on the angles beneath the inflorescence; leaves 4–7 per culm, 1–3 mm wide, rather stiff, flat, smooth or scabrous along the margins, with pellucid, oblong cells between the pale, sometimes papillate nerves beneath, the uppermost leaves longer than the culms; lowermost bracts long-sheathing, the sheaths concave at summit; terminal spike staminate, rarely with a few perigynia at the tip, up to 10 mm long, sessile or short-pedunculate; lateral spikes 3–5, pistillate, up to 20 mm long, up to 5 mm thick, loosely flowered, the uppermost usually crowded, the lowest remote and borne on a scabrous peduncle; pistillate scales broadly ovate, acute to cuspidate, yellow-brown with a hyaline margin and green midvein, shorter than the perigynia; perigynia up to 20 per spike, 2.0–3.2 mm long, biconvex, plump, obovoid, rounded and beakless at the tip, yellow or orange at maturity, coarsely nerved, stipitate; achenes broadly lenticular, 1.3–1.8 mm long, 1.2–1.5 mm broad, brown, apiculate; stigmas 2, blackish.

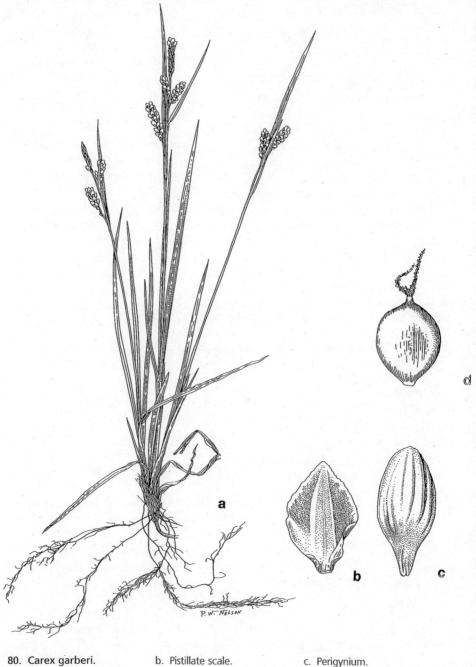

P. W. NELSON

80. Carex garberi.
a. Habit.

b. Pistillate scale.

c. Perigynium.
d. Achene.

Common Name: Golden Sedge.
Habitat: Calcareous swales and sandy prairies.
Range: Newfoundland to Alaska, south to California, New Mexico, Texas, Nebraska, northern Illinois, northern Ohio, and New England.
Illinois Distribution: Known from Cook, Kane, Lake, and Menard counties.

This species is similar to *C. garberi* but may be distinguished by its usually staminate terminal spike and by its yellow or orange mature perigynia.

Apparently, the first Illinois collection was made by F. C. Gates on

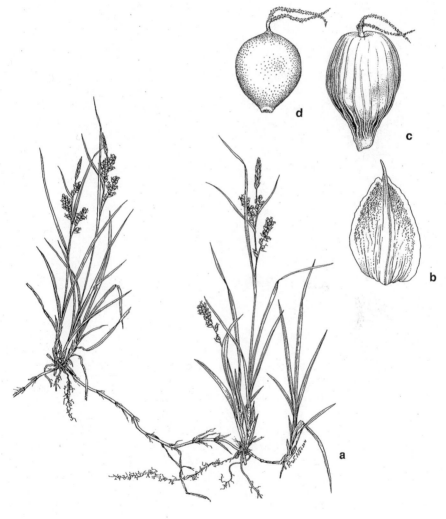

81. Carex aurea.
a. Habit.

b. Pistillate scale.

c. Perigynium.
d. Achene.

June 8, 1908 from Waukegan, Lake County. *Carex aurea* flowers from mid-May to mid-June.

82. Carex crinita Lam. Encycl. 3:393. 1791.

Plants perennial, cespitose, from stout rootstocks and slender stolons; culms to 1.5 m tall, stout, scabrous on the angles, often reddish near the base; leaves 3–5 per culm, up to 1.4 cm wide, somewhat scabrous along the slightly revolute margins, septate between the nerves, the uppermost leaves longer than the culms; lowest sheaths red-brown to purple, strongly nerved and becoming fibrillose at maturity, the hyaline ventral bands thin; uppermost 1–3 spikes staminate, or with some pistillate flowers at the tip, up to 6 cm long, at maturity pendulous on slender peduncles; staminate scales acuminate to awned, conspicuous; lower 2–6 spikes pistillate, or with a few staminate flowers at the tip, narrowly cylindric, up to 12 cm long, pendulous on slender, smooth peduncles; lowest bracts foliaceous; pistillate scales ovate, retuse, abruptly rough-awned, red or yellow-brown with a green center, much longer than the perigynia; perigynia numerous, 2–4 mm long, ellipsoid to ovoid to suborbicular, green or pale brown or stramineous, lustrous, inflated, nerveless, beakless or with a beak 0.25 mm long, substipitate; achenes lenticular, 1.5–1.7 mm long, brownish, often slightly crimped on one side at the middle; stigmas 2.

Two varieties that tend to intergrade may be recognized in Illinois.

1. Lowest 2–6 spikes entirely pistillate; perigynia 2.0–3.5 mm long, often crimped on one side . . . . 82a. *C. crinita* var. *crinita*
1. Lowest 2–6 spikes pistillate but some of them with a few staminate flowers at apex; perigynia 3–4 mm long, not crimped. . . . . 82b. *C. crinita* var. *brevicrinis*

82a. Carex crinita Lam. var. crinita Fig. 82e.

Lowest 2–6 spikes entirely pistillate; awns of the lowest pistillate scales usually at least four times as long as the perigynia; perigynia 2.0–3.5 mm long, usually crimped on one side.

Common Name: Fringed Sedge.
Habitat: Swampy woods, marshes.
Range: Newfoundland to Minnesota, south to Louisiana, northern Georgia, and South Carolina.
Illinois Distribution: Throughout the state.

The fringed sedge derives its common name from the fringed appearance of the pistillate spikes due to the elongated awn on each of the pistillate scales.

In addition to swampy woods and marshes, this plant is characteristic of marshy areas behind high dunes in the Chicago region, according to Swink and Wilhelm (1994). This is the more common variety of *Carex crinita* in Illinois. The flowers bloom from May to August.

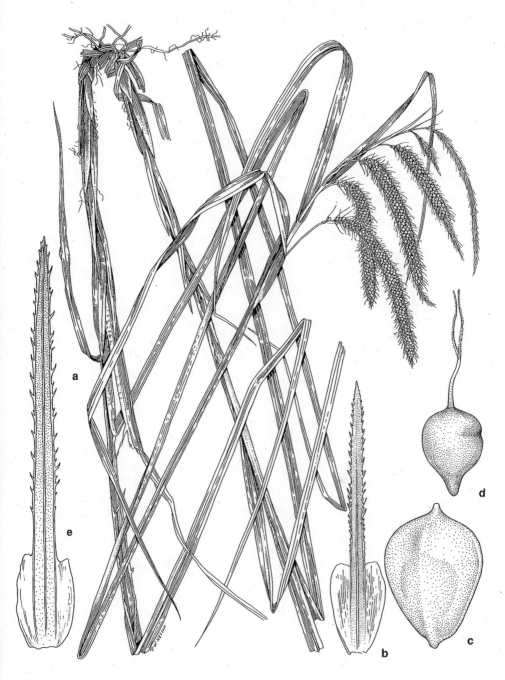

82. **Carex crinita var.
brevicrinis.**
a. Habit.

b. Pistillate scale.
c. Perigynium.

d. Achene. var.
crinita.
e. Pistillate scale.

82b. Carex crinita Lam. var. **brevicrinis** Fern. Rhodora 48:54. 1946. Fig. 82 a–d.

Lowest 2–6 spikes pistillate but some of them with a few staminate flowers at apex; awns of the lowest pistillate scales not more than twice as long as the perigynia; perigynia 3–4 mm long, not crimped.

Common Name: Fringed Sedge.
Habitat: Swampy woods, marshes.
Range: Coastal Plain from North Carolina to eastern Texas, up the Mississippi Embayment to Indiana, Illinois, and Missouri.
Illinois Distribution: Scattered in Illinois.

This variety is recognized by the presence of staminate flowers at the tip of some or all of the lowest two to six spikes. Since the variety and typical *C. crinita* are sometimes found growing together, it may not be proper to give it varietal status. There may be some correlation between the presence of androgynous spikes and the size of the perigynia. Specimens with androgynous spikes on the average have slightly longer perigynia.

Although Fernald (1950) remarks that this is a plant of the Coastal Plain and the interior up to Kentucky and Missouri, several specimens in the Chicago area have the characteristics of var. *brevicrinis.* This variety flowers from May to August.

83. Carex aquatilis Wahlenb. var. **substricta** Kukenth. in Engl. Pflanzenr. 4 (20):309. 1909. Fig. 83.
Carex variabilis Bailey var. *altior* Rydb. Mem. N. Y. Bot. Gard. 1:76. 1900.
Carex substricta (Kukenth.) Mack. in Rydb. Fl. Rocky Mts. 139. 1938.
Carex aquatilis Wahlenb. var. *altior* (Rydb.) Fern. Rhodora 44:295. 1942.

Plants perennial, cespitose, from elongated, scaly stolons; sterile shoots common; culms to 1 m tall, smooth or scabrous on the angles, reddish brown at the base, often with last year's leaves persistent; leaves up to 15 per culm, 6.5–8.0 mm wide, flat, scabrous along the margins, papillate on both surfaces, septate between the nerves, the uppermost longer than the culms; sheaths glabrous, the hyaline ventral band usually red-speckled; uppermost 1–3 spikes staminate, up to 5 cm long, erect, densely flowered; lower 3–6 spikes pistillate, often with a few staminate flowers at the tip, erect, up to 6 cm long, densely flowered; lowest bract foliaceous; pistillate scales oblanceolate to obovate, rounded or mucronate at the tip, red-brown with a greenish midvein, a little longer to a little shorter than the perigynia; perigynia numerous, flat, 2.7–3.3 mm long, obovate, broadest above the middle, pale brown, spreading to appressed-ascending, papillate, 2-ribbed, otherwise nerveless, with a minute, entire beak, stipitate; achenes lenticular, 1.3–1.5 mm long, dark brown, short-apiculate, substipitate; stigmas 2.

Common Name: Aquatic Sedge.
Habitat: Marshes, wet meadows, wet ditches, along streams.

83. Carex aquatilis var. substricta.

a. Habit.
b. Pistillate scale.
c. Perigynium.
d. Achene.

Range: Newfoundland to Alberta, south to Nebraska, Illinois, northern Ohio, and New Jersey.

Illinois Distribution: Occasional in the northernmost counties, much rarer elsewhere.

This taxon is often considered to be a distinct species, where it is called *C. substricta.*

Apparently, the best way to treat this taxon is as a variety of *C. aquatilis,* calling it var. *substricta.* This variety differs from typical var. *aquatilis* in its more sharply triangular stems and its more rounded and slightly larger perigynia.

This plant is somewhat similar in appearance to *Carex emoryi, C. haydenii,* and *C. stricta.* It differs from all three in the well-developed blades of the lowermost sheath and in its perigynia, which are broadest above the middle.

Typical var. *aquatilis* ranges far north and west of Illinois. *Carex aquatilis* var. *substricta* flowers from mid-April to early June.

84. Carex stricta Lam. Encycl. 3:387. 1792. Fig. 84.
Carex strictior Dewey in Wood, Class-Book 582. 1845.
Carex stricta Lam. var. *strictior* (Dewey) Carey in Gray, Man. Bot. 548. 1848.

Plants perennial, cespitose, producing scaly, brown stolons; sterile shoots common; culms to 1.5 m tall, firm, sharply triangular, scabrous on the angles, red-brown at the base, often with last year's leaves persistent; leaves up to 6 mm wide, stiff, keeled, often revolute, usually scabrous along the margins, often glaucous when young, usually shorter than the culms; lowest sheaths bladeless, becoming fibrous, brown or nearly black to reddish or green, smooth or scabrous, often glaucous when young, the ligule longer than the width of the blade, forming a sharp V; upper 2–3 spikes staminate, up to 6.5 cm long, erect, on short peduncles, densely flowered; staminate scales brown, often tinged with purple; lower 3–4 spikes pistillate but with a few staminate flowers at the tip, up to 10 cm long, erect, subsessile or on short peduncles up to 1.5 cm long; pistillate scales ovate, obtuse to acute, 1.5–3.5 mm long, red-brown or hyaline and brown-dotted, shorter than the perigynia; perigynia numerous, more or less flat, 1.7–3.4 mm long, pale brown, ovate to elliptic, broadest at or below the middle, appressed-ascending, 2-ribbed,

otherwise nerveless, often granular-papillate, with a minute beak 0.1–0.2 mm long, substipitate; achenes lenticular, 1.0–1.8 mm long, dull brown, short-apiculate, substipitate; stigmas 2.

Common Name: Tussock Sedge.
Habitat: Sedge meadows, fens, marshes.
Range: Quebec to North Dakota, south to northern Nebraska, eastern Texas, Mississippi, and South Carolina; also in New Mexico.
Illinois Distribution: Occasional to common in the northern three-fourths of Illinois, absent elsewhere.

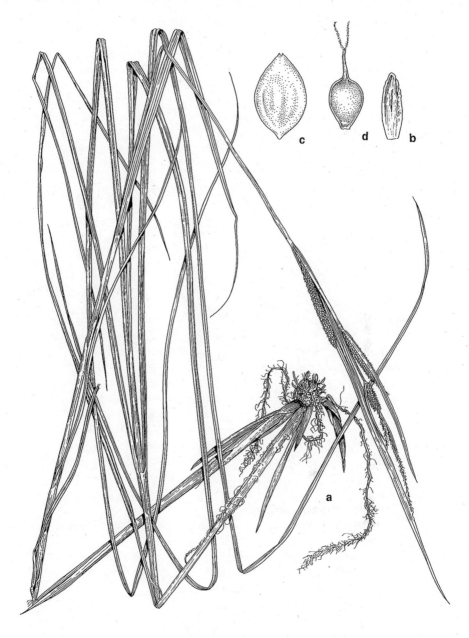

84. Carex stricta.
a. Habit.

b. Pistillate scale.

c. Perigynium.
d. Achene.

The common name for this species is derived from the dense, huge clumps, or tussocks, that it forms.

There have been two taxa created for *C. stricta* by many botanists. They are either recognized as *C. stricta* var. *stricta* and *C. stricta* var. *strictior* or as two distinct species—*C. stricta* and *C. strictior*. Although differences between the two are sometimes detectable, I am unable to convince myself that two taxa are involved. Therefore, I am including all material under *C. stricta*. Variety *strictior* is said to differ in its glaucous younger leaves and scabrous sheaths.

In northern Illinois, *Carex stricta* is a common inhabitant of sedge meadows and fens.

Carex stricta differs from the similar *C. emoryi* in its V-shaped ligule and from *C. haydenii* in its shorter pistillate scales and its purple-tinged staminate scales. *Carex stricta* flowers from mid-April to early June.

85. Carex nebrascensis Dewey, Am. Journ. Sci. (II) 18:102. 1854. Fig. 85.

Plants perennial, cespitose, producing stout, scaly rhizomes; sterile shoots common; culms to 1 m tall, stout, usually scabrous on the angles, brown or purple-brown at the base, often with last year's leaves persistent; leaves up to 12 mm wide, thick, flat or somewhat keeled, usually septate between the veins, mostly shorter than the leaves; lowest sheaths bladeless, pale, with the ligule as wide as long; upper 1–3 spikes staminate, erect, up to 4 cm long, subsessile to short-pedunculate; staminate scales narrowly oblong, purple-brown with a green midvein; lower 2–5 spikes pistillate, erect, up to 6 cm long, ascending, subsessile or short-pedunculate; pistillate scales lanceolate, up to 5 mm long, purple-brown with a green midvein, usually excurrent at the tip, about as long as the perigynia; perigynia numerous, flat to biconvex, 2.5–4.0 mm long, broadly elliptic to obovate, dull brown, strongly few-nerved on both faces, the beak 0.3–0.6 mm long, bidentate, usually dark-tipped, substipitate; achenes lenticular, 1.3–1.5 mm long, blackish, dull, stipitate; stigmas 2.

Common Name: Nebraska Sedge.
Habitat: Swale along railroad and alkaline edge of highway pavement (in Illinois).
Range: North Dakota to British Columbia, south to California, northern Arizona, northern New Mexico, and Kansas; adventive in Illinois and Missouri.
Illinois Distribution: Du Page County: swale along railroad, West Chicago Prairie; Kane County: along Interstate 74, three miles south of Woodhull.

This is one of the few species of *Carex* in Illinois that is not native.

This species differs from the rather similar *C. aquatilis* var. *substricta*, *C. stricta*, *C. emoryi*, and *C. haydenii* in its distinct perigynial beak and its strongly nerved perigynia. *Carex nebrascensis* flowers during May and June.

85. Carex nebrascensis.
a. Habit.

b. Pistillate scale.

c. Perigynium.
d. Achene.

86. Carex emoryi Dewey in Torr. Bot. Mex. Bound. Survey 230. 1859. Fig. 86.
Carex stricta Lam. var. *emoryi* (Dewey) Bailey, Proc. Am. Acad. Sci. 22:85. 1886.

Plants perennial, cespitose, with rather stout, brown, scaly rhizomes; sterile shoots common; culms to 1.1 m tall, stout, usually scabrous on the angles, red-purple at the base, often with last year's leaves persistent; leaves up to 6 mm wide, flat to revolute, scabrous along the margins, septate between the veins, usually longer than the culms; lowest sheaths bladeless, red to purple-brown to blackish, usually disintegrating into fibrillose filaments, convex at the mouth, the ligule shorter than the width of the leaf; upper 1–3 spikes staminate, up to 5 cm long, more or less erect, pedunculate; staminate scales obovate, rounded at the tip, brown with a tawny midvein; lowest 3–5 spikes pistillate, often with a few staminate flowers at the tip, up to 10 cm long, erect, subsessile to short-pedunculate; pistillate scales 2.0–3.5 mm long, ovate to elliptic, subacute to acuminate, hyaline to reddish brown with a tawny midvein, about as long as to surpassing the perigynia; perigynia numerous, flat or biconvex, elliptic to ovate, 1.7–3.2 mm long, rounded at the tip, dull green to stramineous, 3- to 5-nerved, usually granular-papillate, with a minute beak up to 0.3 mm long; achenes lenticular, 1–2 mm long, dull brown; stigmas 2.

Common Name: Emory's Sedge.
Habitat: Along streams, sedge meadows.
Range: Connecticut to North Dakota and eastern Wyoming, south to New Mexico, Texas, northwestern Arkansas, southern Illinois, and southeastern Virginia.
Illinois Distribution: Occasional in the northern two-thirds of the state; also in Johnson, Randolph, and Union counties.

Emory's sedge very closely resembles *C. stricta* but differs in its very short ligules and the fewer granular papillae on its perigynia. It differs from *C. aquatilis* var. *substricta* in its lack of leafy lower sheaths. This species flowers from mid-April to early June.

87. Carex haydenii Dewey, Am. Journ. Sci. II, 18:103. 1854. Fig. 87.
Carex stricta Lam. var. *haydenii* (Dewey) Kukenth. Pflanzenr. IV, 20:330. 1909.

Plants perennial, cespitose, with short, erect rhizomes; sterile shoots common; culms to 1.2 m tall, stout, scabrous on the angles, red-brown at the base, with last year's leaves persistent; leaves 2.5–5.0 mm wide, keeled, usually revolute, firm, scabrous on the margins, shorter than the culms; lowest sheaths bladeless, brown, sometimes fibrillose, the uppermost smooth with a hyaline ventral band, the ligule longer than the width of the leaf; upper 1–2 spikes staminate, erect, up to 4 cm long, subsessile to pedunculate; staminate scales 3–4 mm long, obovate, subacute, brown to red-brown; lower 2–3 spikes pistillate, erect, sometimes with a few staminate flowers at the tip, up to 5 cm long, sessile or short-pedunculate; pistillate scales 2.0–3.5 mm long, oblong to ovate, acute to acuminate, brown or red-brown

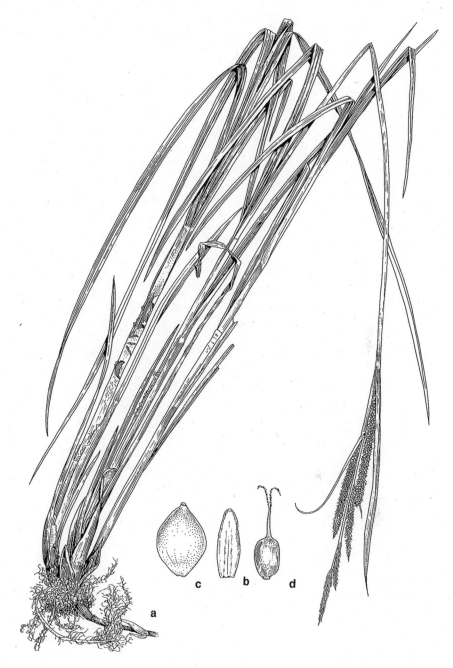

86. Carex emoryi.
a. Habit.

b. Pistillate scale.

c. Perigynium.
d. Achene.

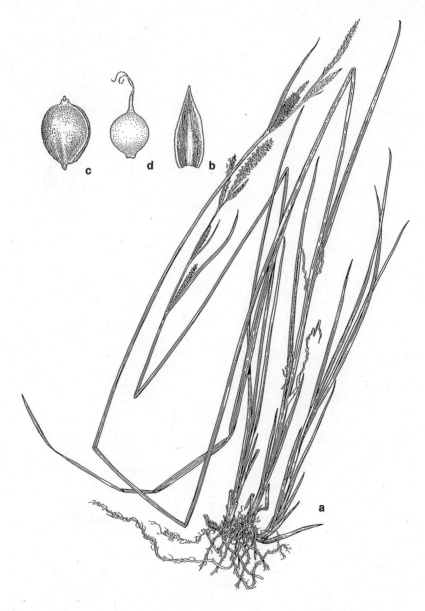

87. Carex haydenii.
a. Habit.

b. Pistillate scale.

c. Perigynium.
d. Achene.

with a green or tawny midvein, longer than the perigynia; perigynia numerous, biconvex but plump, 1.5–2.8 mm long, ellipsoid to obovoid, olive green to brown, papillate, red-dotted, inflated, nerveless, substipitate, minutely beaked; achenes lenticular, 1.0–1.5 mm long, pale brown, iridescent; stigmas 2.

Common Name: Hayden's Sedge.
Habitat: Sedge meadows, sandy wetlands.
Range: Quebec and Nova Scotia to Ontario and Minnesota, south to northern Missouri, northern Illinois, central Indiana, Pennsylvania, and New Jersey; also in South Dakota.
Illinois Distribution: Occasional to rare in the northern half of the state; also Wabash County.

This species differs from *C. aquatilis* var. *substricta* in its bladeless lower sheaths and from *C. emoryi* and *C. stricta* in its nearly orbicular, nerveless perigynia that are exceeded by the pistillate scales. *Carex haydenii* flowers from mid-May to mid-June.

88. Carex torta Boott ex Tuckerm. Enum. Meth. 11. 1843. Fig. 88.

Plants perennial, cespitose, usually with slender rhizomes; sterile shoots common; culms to 75 cm tall, usually smooth on the angles, red-purple at the base, with last year's leaves persistent; leaves 3–5 mm wide, flat, scabrous along the margins and on the upper surface, shorter than the culms; lowest sheaths bladeless, red-tinged, the upper ones with sheaths green-nerved, with a broad, usually russet-spotted hyaline ventral band; terminal 1–2 spikes staminate, erect, up to 4.5 cm long; staminate scales red-brown with a hyaline margin and paler midvein; lateral spikes 3–6, pistillate or with a few staminate flowers at the tip, up to 9 cm long, ascending to pendulous, sessile or subsessile or on capillary peduncles; pistillate scales elliptic, obtuse, purple-black with a hyaline margin and a green midvein, barely reaching the base of the perigynium; perigynia numerous, 2.5–4.5 mm long, flat to plano-convex, ovate to obovate, deep green, 2-ribbed, otherwise nerveless, contracted into a short, twisted beak, substipitate; achenes lenticular, dull brown, 1.8–2.0 mm long, substipitate; stigmas 2.

Common Name: Twisted Sedge.
Habitat: Along and in streams.
Range: Quebec to Minnesota, south to Arkansas and Georgia.
Illinois Distribution: Occasional in the southern fourth of the state; also in Whiteside County.

This species is readily distinguished by its nerveless perigynia, each with a minute, twisted beak and by its purple-black pistillate scales with hyaline margins and green midveins.

Carex torta is found often in rocky streambeds where it is subjected to continuous flowing water. This species flowers from mid-April to early June.

88. Carex torta.
a. Habit.

b. Pistillate scale.

c. Perigynium.
d. Achene.

89. Carex buxbaumii Wahlenb. Sv. Vet. Akad. Handl. 24:163. 1803. Fig. 89.

Plants perennial, cespitose, with long, horizontal rhizomes; culms up to 1 m tall, firm, slender, scabrous on the angles, at least beneath the inflorescence, red-purple at the base; leaves up to 4 mm wide, flat but with revolute margins, the lower surface glaucous and papillate, scabrous along the margins; lowest sheaths red-brown, fibrillose, upper sheaths with a hyaline ventral band russet-spotted, the ligule longer than wide; terminal spike gynecandrous, less commonly entirely staminate, up to 4 cm long, the lateral spikes 1–4, pistillate, erect to ascending, sessile or sub-sessile, the upper ones overlapping, up to 2 cm long; pistillate scales ovate, aristate, dark red-purple with a green midvein, a little longer than the perigynia; perigynia up to 40 per spike, ovoid, biconvex, 2.5–4.0 mm long, glaucous green, papillate, 2-ribbed and several-nerved, substipitate, minutely beaked; achenes trigonous, 1.5–1.8 mm long, brown, puncticulate, apiculate, stipitate; stigmas 3.

Common Name: Buxbaum's Sedge.
Habitat: Marshes, wet prairies, swales, usually in calcareous areas.
Range: Newfoundland to Alaska, south to California, Colorado, Arkansas, Kentucky, and North Carolina.
Illinois Distribution: Occasional in the northern half of the state; also in Montgomery, Richland, Shelby, St. Clair, and Washington counties.

This distinctive species is recognized by its dark red-purple, aristate pistillate scales and by its minutely beaked, glaucous perigynia.

Occasional specimens with the terminal spike entirely staminate may occur. This species flowers during May.

90. Carex limosa L. Sp. Pl. 977. 1753. Fig. 90.

Plants perennial, with long-running, slender, brown, scaly stolons; culms to 60 cm tall, slender, scabrous on the sharp angles, purple-red at the base; leaves 1–3 near the base of the plant, up to 2.5 mm wide, conduplicate to involute, keeled, glaucous, scabrous along the margins; sheaths reddish, the hyaline ventral band russet-spotted and pale-nerved near the apex; terminal spike staminate, to 3.5 cm long, on long, slender peduncles; lateral spikes 1–3, pistillate, sometimes with a few staminate flowers at the tip, up to 2 cm long, 4–8 mm thick, on slender, ascending peduncles; pistillate scales ovate, obtuse to acuminate, often mucronate, yellow-brown to purple-brown, with a green midvein, about as long as or slightly longer than the perigynia; perigynia up to 30 per spike, 3–4 mm long, ovoid, trigonous but compressed, glaucous green, papillose, several-nerved, substipitate, minutely beaked; achenes trigonous, 2.0–2.3 mm long, dark brown, apiculate; stigmas 3.

Common Name: Sedge.
Habitat: Sphagnum bogs.
Range: Labrador to Alaska, south to California and Nevada, Montana,

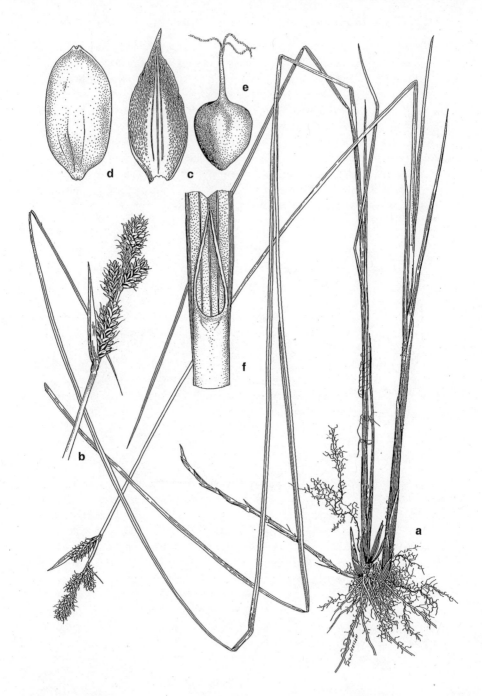

89. Carex buxbaumii.
a. Habit.
b. Inflorescence.
c. Pistillate scale.
d. Perigynium.
e. Achene.
f. Sheath with ligule.

90. Carex limosa.
a. Habit.

b. Pistillate scale.

c. Perigynium.
d. Achene.

northeastern Nebraska, Iowa, northern Illinois, northern Ohio, and Delaware.

Illinois Distribution: Found in sphagnum bogs in northeastern Illinois; also in Peoria and Tazewell counties.

This slender species of sphagnum bogs is further distinguished by its very narrow, often involute leaves; its long-pedunculate staminate spike; and its ovoid, glaucous green perigynia. *Carex limosa* flowers from mid-June to mid-July.

91. Carex shortiana Dewey, Am. Journ. Sci. 30:60. 1836. Fig. 91.
Carex shortii Torr. Ann. Lyc. N. Y. 3:407. 1836.

Plants perennial, cespitose, with dark, thick, fibrillose rhizomes; culms to 90 cm tall, rather stout, smooth or scabrous on the angles, brownish at the base; leaves 4–8 mm wide, flat, white-punctate, the margins and often the midrib scabrous; sheaths greenish white, septate, with a broad, hyaline ventral band, sometimes red-tinged; terminal spike gynecandrous, up to 3.5 cm long, 4–5 mm thick, erect; lateral spikes 4–5, entirely pistillate or sometimes with a few staminate flowers at the base, narrowly cylindrical, erect, subsessile or short-pedunculate; pistillate scales ovate, acuminate to cuspidate, reddish brown with a green midvein, slightly shorter than to about as long as the perigynia; perigynia 20–60 per spike, crowded, squarrose, 2–3 mm long, compressed-obovoid, olive green to dark brown, rugulose, 2-ribbed, nerveless, stipitate, with a beak up to 0.2 mm long; achenes trigonous, 1.7–2.0 mm long, granular, apiculate, substipitate; stigmas 3.

Common Name: Short's Sedge.
Habitat: Mesic woods, bottomlands, at the head of ravines, wet meadows.
Range: Ontario to Iowa, south to Oklahoma, Tennessee, and Virginia.
Illinois Distribution: Common in the southern half of the state, occasional to rare elsewhere.

This distinctive species is recognized by its cylindrical, squarrose spikes that are gynecandrous or pistillate and by its compressed-obovoid perigynia that turn dark brown at maturity.

The earliest Illinois botanists called this species *C. shortii*, but that binomial is preceded by *C. shortiana* by a few months. *Carex shortiana* flowers from May to late June.

92. Carex X deamii F. J. Herm. Rhodora 40:81. 1938. Fig. 92.

Plants perennial, cespitose, with fibrillose rhizomes; culms to 90 cm tall, triangular, smooth or sometimes scabrous on the angles, brownish at the base; leaves up to 9 mm wide, flat, the margins often scabrous; sheaths septate, with a broad, hyaline ventral band, sometimes red-tinged; terminal spike gynecandrous, up to

91. Carex shortiana.
a. Habit.
b. Inflorescence.

c. Staminate scale.
d. Pistillate scale.

e. Perigynium, dorsal view.
f. Perigynium, side view.
g. Achene.

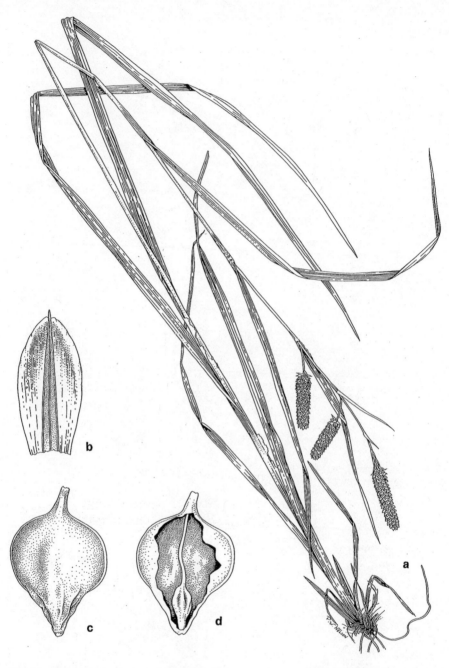

92. Carex X deamii.
a. Habit.

b. Pistillate scale.
c. Perigynium.

d. Perigynium cut away to
show aborted achene.

3.5 cm long, 7–8 mm thick, erect; lateral spikes 4–5, usually entirely pistillate, narrowly cylindrical, erect, subsessile or short-pedunculate; pistillate scales ovate, acute to acuminate, slightly shorter than to about as long as the perigynia; perigynia up to 70 per spike, crowded, squarrose, 2–3 mm long, obovoid, olive green to brown, 2-ribbed, nerveless, stipitate, with a beak 1.0–1.5 mm long; achenes trigonous, seldom developed; stigmas 3.

Common Name: Deam's Sedge.
Habitat: Wet woods, alluvial woods.
Range: Indiana, Illinois, Missouri.
Illinois Distribution: Known only from Fayette, Macon, Pike, and Shelby counties.

This plant is reputed to be a hybrid between *Carex shortiana* and *C. typhina*. At most sites, including the Illinois stations, both *C. shortiana* and *C. typhina* are usually present, as is *C. squarrosa*.

    Carex X deamii more nearly resembles *C. shortiana* than it does either *C. typhina* or *C. squarrosa*. It differs from *C. shortiana* primarily in its thicker spikes and the longer beak of the perigynium. Mature achenes are rarely formed. *Carex X deamii* flowers during May.

93. Carex lasiocarpa Ehrh. var. **americana** Fern. Rhodora 44:304. 1942. Fig. 93.

Plants perennial, solitary or in small tufts, with long, scaly stolons; sterile shoots common; culms to 1.2 m tall, wiry, usually smooth, purple-red at the base; leaves up to 2 mm wide, convolute, septate between the veins, but with no midvein, usually roughened near the tip; lowest sheaths reddish, becoming fibrillose; upper two spikes usually staminate, erect, to 6 cm long, on slender peduncles; staminate scales light red-brown; lowest 1–3 spikes pistillate, sometimes with a few staminate flowers at the tip, to 5 cm long, erect, sessile or subsessile; pistillate scales lanceolate, acute to awned, purple-brown with a green center, usually slightly shorter than the perigynia; perigynia up to 50 per spike, broadly ovoid to ellipsoid, 3–5 mm long, densely hairy, obscurely nerved, with a short, bidentate beak, the teeth of the beak up to 0.6 mm long; achenes trigonous, 1.7–1.9 mm long, yellow-brown, punctate, jointed with the style; stigmas 3.

Common Name: Hairy-fruited Sedge.
Habitat: Sphagnum bogs, sedge meadows, sometimes in shallow water.
Range: Newfoundland to British Columbia, south to Washington, Idaho, northern Illinois, Pennsylvania, and New Jersey.
Illinois Distribution: Confined to the northern third of the state; also Jefferson County.

Fernald considers the United States plants different from the typical variety found in Europe and Asia. The Eurasian plants have perigynia 4–6 mm long, with the teeth of the beak 0.7–1.0 mm long.

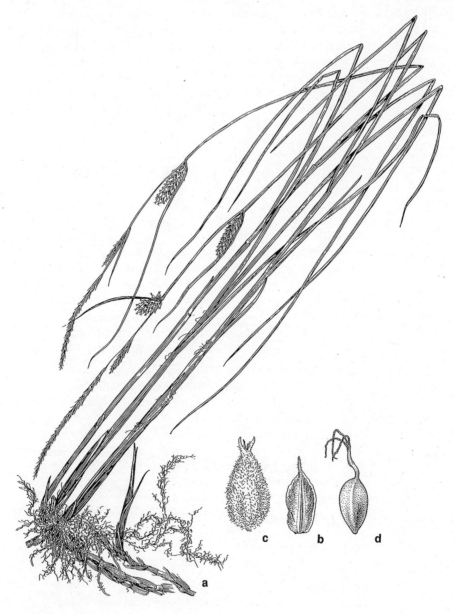

93. Carex lasiocarpa var. americana.

a. Habit.
b. Pistillate scale.

c. Perigynium.
d. Achene.

Carex lasiocarpa var. *americana* is readily distinguished by its very narrow, convolute leaves without a midvein; its smooth culms; and its hairy perigynia.

Some early botanists called this plant *C. filiformis*, but that Linnaean binomial belongs to a different species. This plant flowers from mid-May to mid-June.

94. Carex pellita Willd. in Schk. Riedgr. 84. 1801. Fig. 94.
Carex lanuginosa Michx. Fl. Bor. Am. 2:175. 1803.
Carex filiformis L. var. *latifolia* Boeck. Linnaea 41:309. 1877.

Plants perennial, loosely cespitose, with long, scaly stolons; sterile shoots common; culms to 1 m tall, stiff, usually scabrous on the angles, sometimes smooth, purple-red at the base; leaves up to 5 mm wide, flat except for the revolute margins, septate, with a conspicuous midvein, scabrous toward the tip; lowest sheaths dark red, becoming fibrillose; upper 1–3 spikes staminate, erect, up to 4 cm long, long-pedunculate; staminate scales light reddish brown; lowest 2–3 spikes pistillate, occasionally with a few staminate flowers at the tip, up to 5 cm long, erect, sessile or subsessile; pistillate scales broadly lanceolate, acuminate to awned, red-brown with a green center, as long as or longer than the perigynia; perigynia up to 75 per spike, broadly ovoid, 2.0–3.5 mm long, densely pubescent, many-nerved, with a bidentate beak about 1 mm long, the teeth of the beak 0.4–0.8 mm long; achenes trigonous, 1.7–1.9 mm long, yellow-brown, punctate, jointed with the style; stigmas 3.

Common Name: Woolly Sedge.
Habitat: Wet prairies, fens, marshes, sedge meadows, swamps.
Range: Quebec to British Columbia, south to California, Arizona, New Mexico, Texas, Tennessee, and Virginia.
Illinois Distribution: Occasional to common in Illinois.

This species is readily distinguished by its flat blades up to 5 mm wide, its usually multiple staminate spikes, and its densely hairy perigynia. It differs from the very similar *C. lasiocarpa* var. *americana* by its broader leaves with a distinct midvein.

Although this species is known throughout the country as *C. lanuginosa*, the binomial *C. pellita* predates it and must be used. *Carex pellita* flowers from early April to mid-June.

95. Carex hirta L. Sp. Pl. 975. 1753. Fig. 95.

Plants perennial, loosely cespitose, with long, scaly stolons; sterile shoots common; culms to 1 m tall, rather slender, usually scabrous on the angles to nearly smooth, brown or purplish at the base; leaves up to 6 mm wide, hairy, flat, obscurely septate, with a conspicuous midvein, scabrous toward the tip; sheaths tight, densely hairy, concave at the mouth; upper 1–3 spikes staminate, erect, up to 3 cm long, long-pedunculate; staminate scales pubescent; lowest 2–3 spikes pistillate, up to 5 cm long, erect, sessile or subsessile; pistillate scales broadly lanceo-

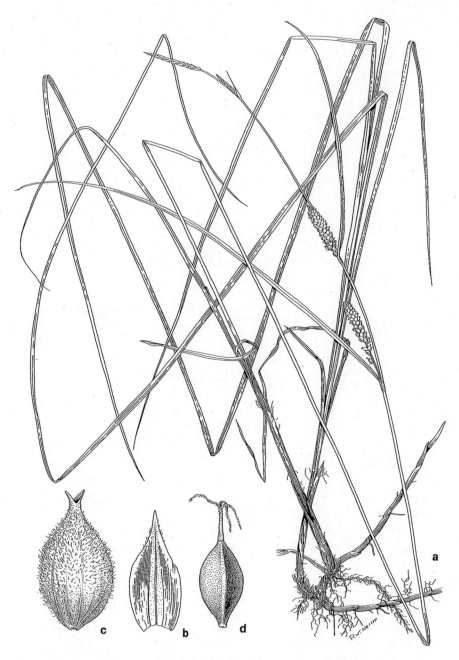

94. Carex pellita.
a. Habit.

b. Pistillate scale.

c. Perigynium.
d. Achene.

95. Carex hirta.
a. Habit.

b, Pistillate scale.

c. Perigynium.
d. Achene.

late, acuminate to awned, hairy, purple-brown with a green center, usually longer than the perigynia; perigynia up to 35 per spike, narrowly ovoid, 5–9 mm long, densely hairy but not obscuring the 15–20 strong nerves, with a bidentate beak 1.5–2.5 mm long, the teeth of the beak 1–2 mm long; achenes trigonous, 2.7–3.0 mm long, yellowish, jointed with the style; stigmas 3.

Common Name: Hairy Sedge.
Habitat: In a prairie restoration in Illinois.
Range: Native of Europe; adventive from Prince Edward Island to Michigan, south to central Illinois, Pennsylvania, and New Jersey.
Illinois Distribution: Known only from a prairie restoration area in Ford County, where it was collected by Don Gardner on June 21, 1991.

This native of Europe has been introduced into a prairie restoration site in Illinois. The seeds were probably mixed with other seeds used in the restoration. The Illinois population is vigorous and produces seeds.

Carex hirta is similar to *C. pellita* in appearance, but the nerves of the perigynia are more pronounced and not obscured, the teeth of the beak of the perigynium are 1–2 mm long, and the leaves and the sheaths are pubescent. *Carex hirta* flowers in June in Illinois.

96. Carex pallescens L. Sp. Pl. 977. 1753. Fig. 96.
Carex pallescens L. var. *neogaea* Fern. Rhodora 44:306. 1942.

Plants perennial, cespitose, with short stolons; culms to 75 cm tall, slender, scabrous on the angles, pubescent, reddish tinged at the base; leaves 2–3 per culm, up to 3 mm wide, flat except for the revolute margins, softly pubescent on the lower surface; sheaths softly pubescent, brown, concave at the summit, the ligules longer than wide; terminal spike staminate, erect, up to 3 cm long, on a stalk up to 1.5 cm long; staminate scales acute, green to yellow-brown; lateral spikes 2–4, pistillate, erect to spreading, cylindrical to nearly globose, up to 2 cm long, densely flowered, on stalks up to 1.5 cm long, or the uppermost nearly sessile; lowest bract leaflike; pistillate scales narrowly ovate, acute to short-awned, yellow-brown with a green center, about as long as the perigynia; perigynia up to 40 per spike, ellipsoid-triangular, up to 3 mm long, glabrous, faintly nerved, beakless; achenes trigonous,

1.7–2.0 mm long, short-apiculate, substipitate; stigmas 3.

Common Name: Pale Sedge.
Habitat: Rocky barrens (in southern Illinois).
Range: Newfoundland to Ontario, south to northeastern Minnesota, northern Wisconsin, northern Michigan, northeastern Ohio, Pennsylvania, and New Jersey; also in Illinois.
Illinois Distribution: There are old records from Fulton, Hancock, and McHenry counties in central Illinois as well as recent collections from Johnson and Saline counties in southern Illinois.

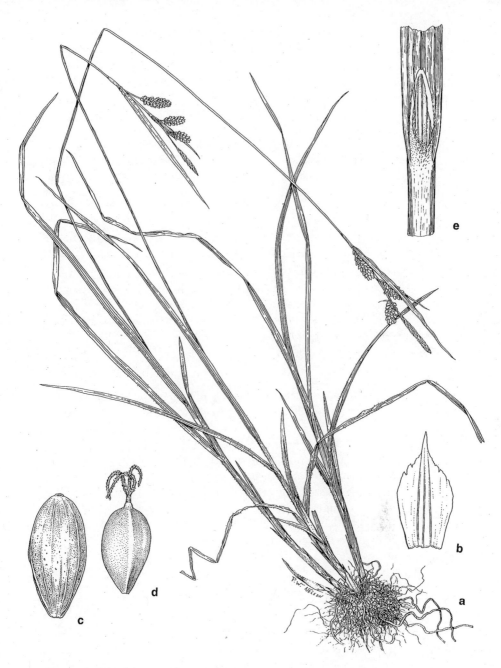

96. Carex pallescens.
a. Habit.

b. Pistillate scale.
c. Perigynium.

d. Achene.
e. Sheath with ligule.

This species strongly resembles *C. hirsutella*, but the presence of a terminal stami-nate spike readily separates the two. There is a collection made by George Vasey in the latter part of the nineteenth century in which the label reads merely "Ring-wood." Many of Vasey's early collections have only "Ringwood" as their locality, and although Vasey lived there for a while, some botanists question that any or all of these collections actually came from McHenry County.

The Johnson County records are based on collections made at the edge of Simpson Township Barrens in rocky soil that harbors a number of prairie species. The area was forested until the late 1980s when Forest Service personnel became aware of prairie species growing at the periphery of the woods. By cutting a num-ber of trees and periodically burning the area, the woods became transformed into a prairielike community known as a barrens. It is here that *C. pallescens* grows. Fer-nald (1950) believes that the North American plants assignable to *C. pallescens* dif-fer from European plants of the same species, because the European plants have a short beak on the perigynium. He calls the North American plants var. *neogaea*. This species flowers in Johnson County during May and June.

97. Carex hirsutella Mack. Bull. Torrey Club 50:349. 1923. Fig. 97.
Carex hirsuta Willd. Sp. Pl. 4:252. 1805, non Suter (1802).
Carex triceps var. *hirsuta* L. H. Bailey, Mem. Torrey Club 1:35. 1899.
Carex complanata Torr. & Hook. var. *hirsuta* (L. H. Bailey) Gl. Phytologia 4:22. 1952.

Plants perennial, loosely cespitose, with short stolons; culms slender, wiry, sharply triangular, to 90 cm tall, pubescent, scabrous on the angles, reddish at the base; leaves up to 6 per culm, up to 4 mm wide, flat except for the revolute margins, softly pubescent on both surfaces, shorter than the inflorescence; sheaths softly pubescent, reddish brown to green, with a densely hairy hyaline ventral band, the ligules shorter than wide; terminal spike gynecandrous, erect, up to 2 cm long, up to 7 mm thick, thick-cylindric with a clavate base, usually on a short peduncle; staminate scales acuminate; lateral spikes 2–4, pistillate, erect, up to 1.8 cm long, up to 5 mm thick, sessile or on very short peduncles; pistillate scales ovate, acute to cuspidate, with a green center, shorter than the perigynia; perigynia 15–30 per spike, obovoid but with a flat inner face, pale but usually green above the middle, glabrous, 2–3 mm long, weakly nerved, usually short-pointed at the tip; achenes trigonous, 1.6–2.0 mm long, apiculate, rounded to substipitate at the base; stigmas 3.

Common Name: Hairy-leaved Sedge.
Habitat: Dry woods, fields, dry meadows.
Range: Quebec and Maine to Ontario, south to eastern Nebraska, north-eastern Texas, Mississippi, Alabama, and South Carolina.
Illinois Distribution: Throughout the state.

This species is similar to *C. bushii*, *C. caroliniana*, and *C. complanata* by virtue of its terminal spike that is gynecandrous and clavate at the

97. Carex hirsutella.
a. Habit.

b. Inflorescence.
c. Perigynium with scale.

d. Perigynium.
e. Achene.

base and by its pubescent leaves. It differs from *C. bushii* and *C. complanata* in having the inner face of the perigynium flattish, while the inner face of the perigynium of *C. bushii* and *C. caroliniana* is rounded. It differs from *C. complanata*, which is as yet unknown from Illinois, in its softer, hairier leaves that lack a keel. Gleason (1952) considers *C. hirsutella* to be a variety of *C. complanata*, calling it var. *hirsuta*.

Until 1900, this species was referred to by Illinois botanists as *C. triceps* Michx., but Michaux's binomial was preceded by Schrank's *C. triceps* for another species in 1789.

Carex hirsutella is a species of dry woods, old fields, and dry meadows. This species flowers in May and June.

98. Carex caroliniana Schw. Ann. Lyc. N. Y. 1:67. 1824. Fig. 98.
Carex triceps var. *smithii* Porter ex L. H. Bailey, Bot. Gaz. 13:88. 1888.

Plants perennial, loosely cespitose, with short stolons; culms slender, rather stiff, sharply triangular, up to 1 m tall, scabrous at least near the inflorescence, glabrous or nearly so, reddish at the base; leaves up to 6 per culm, up to 4 mm broad, flat except for the revolute margins, soft, glabrous or nearly so, shorter than the inflorescence; sheaths pilose, reddish brown to green, the ligules about as wide as long; terminal spike gynecandrous, erect, up to 2 cm long, up to 7 mm thick, thick-cylindric with a clavate base, sessile or on a short peduncle; staminate scales acuminate, stramineous; lateral spikes usually 2, entirely pistillate, erect, up to 1.6 cm long, up to 6 mm thick, sessile or nearly so; pistillate scales ovate, acute to acuminate, mucronate, reddish brown with a green center, shorter than the perigynia; perigynia 15–30 per spike, obovoid, turgid, rounded on all faces, olive green to brownish, glabrous or sparsely pilose, 2–3 mm long, strongly nerved, pointed at the tip; achenes trigonous, 1.5–2.2 mm long, apiculate, usually tapering to the base; stigmas 3.

Common Name: Sedge.
Habitat: Wet meadows, wet woods.
Range: New Jersey to southern Ohio, southern Illinois, southern Missouri, and southeastern Kansas, south to eastern Texas, northern Georgia, and North Carolina.
Illinois Distribution: Occasional in the southern half of the state; also Henry and McDonough counties.

This essentially southern species is similar in overall appearance to *C. hirsutella* and *C. bushii*, but it lacks the hairy leaves and culms of both of these species. Its round-angled perigynia are very similar to those of *C. bushii*, but the perigynia differ readily from the flat inner face of the perigynium of *C. hirsutella*.

This species tends to grow in much wetter habitats than either *C. hirsutella* or *C. bushii*.

Although *C. caroliniana* is distributed throughout most of the southern half of

98. Carex caroliniana.
a. Habit.

b. Pistillate scale.

c. Perigynium.
d. Achene.

Illinois, it was not discovered in Illinois until May 17, 1950, when Julius R. Swayne found it in Jackson County. *Carex caroliniana* flowers in May and June.

99. Carex bushii Mack. Bull. Torrey Club 37:241. 1910. Fig. 99.

Plants perennial, loosely cespitose, with short stolons; culms slender, rather stiff, sharply triangular, up to 80 cm tall, scabrous on the angles, usually sparsely pubescent, reddish at the base; leaves up to 5 per culm, up to 4 mm wide, flat except for the revolute margins, firm, softly pubescent, shorter than the inflorescence; sheaths more or less pubescent, at least the lowest ones reddish, the ligules as long as or longer than wide; terminal spike gynecandrous, erect, up to 2 cm long, up to 8 mm thick, thick-cylindric, with a clavate base, sessile or on a short peduncle; staminate scales lanceolate, acuminate; lateral spikes 1–2, entirely pistillate, erect, up to 1.6 cm long, up to 8 mm thick, sessile or nearly so; pistillate scales lanceolate, acuminate to awned, usually pilose, reddish brown with a green center, longer than the perigynia; perigynia 15–30 per spike, obovoid, turgid, rounded on all faces, olive green to brownish, glabrous or nearly so, 2.5–4.0 mm long, strongly nerved, without a beak; achenes trigonous, 2.0–2.5 mm long, apiculate, substipitate; stigmas 3.

Common Name: Bush's Sedge.
Habitat: Usually dry woods, dry meadows, old fields.
Range: Massachusetts to Pennsylvania and southeastern Nebraska, south to northeastern Texas, northern Mississippi, northern Georgia, and Delaware; also in Michigan.
Illinois Distribution: Common in the southern half of Illinois, becoming less common northward; apparently adventive in Cook, Du Page, and Lake counties.

This species is very similar in appearance to *C. caroliniana* and *C. hirsutella*. It differs from *C. caroliniana* in its hairy leaves and from *C. hirsutella* in its perigynia, which are rounded on all faces instead of being flat on the inner face.

Swink and Wilhelm (1994) suggest that the collections of this species from Cook and Du Page counties are from recently adventive plants. *Carex bushii* flowers during May and June.

100. Carex virescens Muhl. ex Willd. Sp. Pl. 4:251. 1805. Fig. 100.

Plants perennial, densely cespitose, with short stolons; culms slender, triangular, up to 1 m tall, scabrous on the angles, pubescent, reddish at the base; sterile shoots often present; leaves up to 5 per culm, up to 4 mm wide, flat, soft, pubescent, shorter than the inflorescence; sheaths pubescent, at least the lowest sometimes reddish brown, the ligules longer than wide; terminal spike gynecandrous, erect, up to 4 cm long, up to 4 mm thick, narrowly cylindric, sessile or nearly so; staminate scales ovate, obtuse to acuminate; lateral spikes pistillate, up to 3.5 cm long, up to 4 mm thick; pistillate scales ovate, usually awned, whitish with a green center, about as long as the perigynia; perigynia up to 60 per spike, ellipsoid to nar-

99. Carex bushii.
a. Habit.

b. Pistillate scale.

c. Perigynium.
d. Achene.

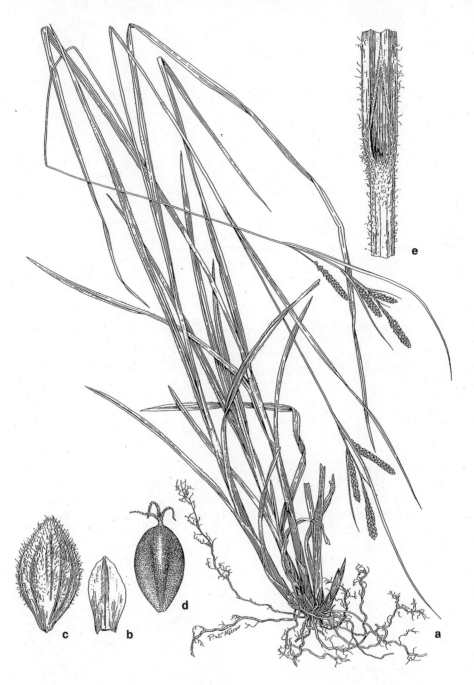

100. Carex virescens.
a. Habit.

b. Pistillate scale.
c. Perigynium.

d. Achene.
e. Sheath with ligule.

rowly ovoid, somewhat compressed, triangular, green, pubescent, 2.0–2.5 mm long, strongly nerved, beakless, substipitate; achenes trigonous, 1.5–2.0 mm long, dark brown, apiculate, tapering to the base; stigmas 3.

Common Name: Hairy-fruited Sedge.
Habitat: Dry woods.
Range: Quebec to southern Michigan, south to southern Illinois and southern Missouri, northern Mississippi, northern Georgia, and Virginia.
Illinois Distribution: Apparently nearly confined to the southern half of the state; also Vermilion County.

Carex virescens is distinguished by its slenderly cylindric spikes in which the terminal one is gynecandrous and by its hairy leaves, culms, and perigynia. The similar *C. swanii* has shorter, thicker spikes and finely nerved rather than strongly nerved perigynia.

Carex virescens is a species found primarily in dry woods. Although there are several reports of this species from the northern half of Illinois, all of these reports are based apparently on specimens of *C. swanii*. This species flowers during May and June.

101. **Carex swanii** (Fern.) Mack. Bull. Torrey Club 37:246. 1910. Fig. 101.
Carex virescens Muhl. var. *swanii* Fern. Rhodora 8:183. 1906.

Plants perennial, densely cespitose, with short rhizomes; culms slender, triangular, up to 80 cm tall, scabrous on the angles, glabrous or sparsely pubescent, reddish at the base; sterile shoots often present; leaves up to 6 per culm, up to 3 mm wide, flat, softly pubescent, about as long as or longer than the inflorescence; sheaths pubescent, the lowest reddish brown, the ligules longer than wide; terminal spike gynecandrous, erect, up to 2 cm long, up to 5 mm thick, short-cylindric, sessile or nearly so; staminate scales ovate, obtuse to acuminate; lateral spikes 1–4, pistillate, up to 2 cm long, up to 4 mm thick; pistillate scales ovate, aristate, hyaline with a green center, about as long as the perigynia; perigynia up to 30 per spike, obovoid to broadly ovoid, 1.7–2.4 mm long, somewhat compressed, triangular, green, densely pubescent, obscurely finely nerved, beakless, substipitate; achenes trigonous, 1.3–1.8 mm long, yellow-brown, apiculate, substipitate; stigmas 3.

Common Name: Swan's Sedge.
Habitat: Savannas, dry woods, swamp forests.
Range: Nova Scotia to southern Wisconsin, south to northeastern Arkansas, Tennessee, northern Georgia, and North Carolina.
Illinois Distribution: Scattered throughout the state.

Carex swanii is distinguished by its terminal, gynecandrous spike, which is up to 2.5 cm long and up to 5 mm thick. It also has pubescent

101. Carex swanii.
a. Habit.

b. Pistillate scale.
c. Perigynium, dorsal view.

d. Perigynium, ventral view.
e. Achene.

leaves and perigynia. The short, thick spikes resemble those of *C. hirsutella*, *C. bushii*, and *C. caroliniana*, but these latter species do not have hairy perigynia.

Carex virescens is similar to *C. swanii* but has longer and more slender spikes.

Although this species is regularly found in savannas and dry woods, it occurs in swamp forests in the dune region near Lake Michigan. *Carex swanii* flowers during May and June.

102. **Carex prasina** Wahl. Kongl. Vet. Acad. Handl. II 24:161. 1803. Fig. 102. *Carex miliacea* Muhl. ex Willd. Sp. Pl. 4:290. 1805.

Plants perennial, cespitose, with short rhizomes; culms slender, sharply triangular, up to 60 cm tall, scabrous, glabrous, the base often purplish; sterile shoots often present; leaves up to 5 per culm, 3–6 mm wide, flat, pale green, glabrous, scabrous on the margins, usually shorter than the inflorescence; sheaths sometimes pink or purplish; terminal spike entirely staminate or with pistillate flowers above and staminate flowers below, up to 4 cm long, up to 4 mm thick, more or less erect, on short peduncles; staminate scales lanceolate, acute to awned; lateral spikes 2–4, pistillate, up to 6 cm long, up to 5 mm thick, narrowly cylindric, ascending or pendulous, on peduncles up to 4 cm long; pistillate scales ovate, cuspidate to awned, whitish with a green center, shorter than to as long as the perigynia; perigynia up to 50 per spike, ovoid, 2.5–4.0 mm long, triangular, pale, glabrous, finely nerved, stipitate, with a green, sometimes bent beak up to 2 mm long; achenes trigonous, 1.3–1.6 mm long, pale brown, apiculate, stipitate; stigmas 3.

Common Name: Sedge.
Habitat: Rich woods.
Range: Quebec to Ontario, south to Illinois, Tennessee, and South Carolina.
Illinois Distribution: Scattered in the northwestern quadrant of Illinois; also Johnson and Lawrence counties.

This species is recognized by its narrowly cylindric spikes and the long, curved beak of the perigynium. Its nearest relatives are *C. gracillima*, which has beakless perigynia, and *C. oxylepis* and *C. davisii*, both of which have pubescent leaves. *Carex prasina* flowers during May and June.

103. **Carex gracillima** Schwein. Ann. Lyc. N. Y. 1:66. 1824. Fig. 103.

Plants perennial, cespitose, with slender rhizomes; culms slender, triangular, up to 1 m tall, rarely scabrous, glabrous, the base purple-red; sterile shoots common; leaves up to 4 per culm, up to 8 mm wide, flat, dark green, glabrous, shorter than the inflorescence; lowest sheaths dark red, the upper pale and often purple-spotted on the hyaline ventral bands; terminal spike mostly staminate, but usually with a few pistillate flowers at the tip, up to 6 cm long, up to 3.5 mm thick, narrowly cylindric, on an arching, capillary peduncle often nearly as long as the spike; staminate scales narrowly ovate, acute, whitish with a green center; lateral spikes 3–4, entirely pistillate, up to 5 cm long, up to 3 mm thick, on pendulous, capillary

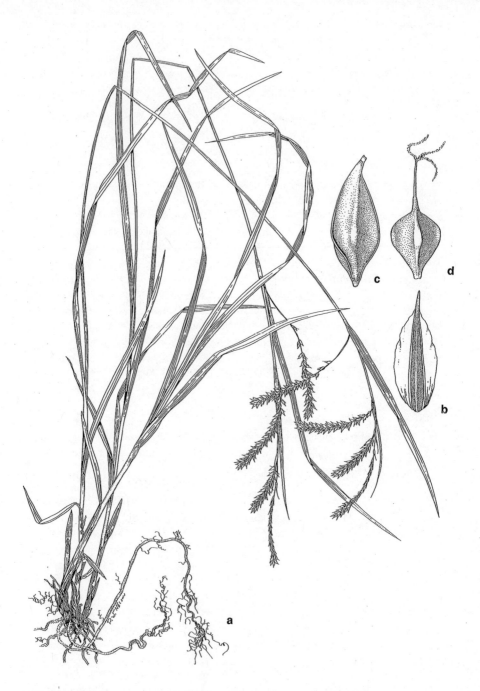

102. Carex prasina.
a. Habit.

b. Pistillate scale.

c. Perigynium.
d. Achene.

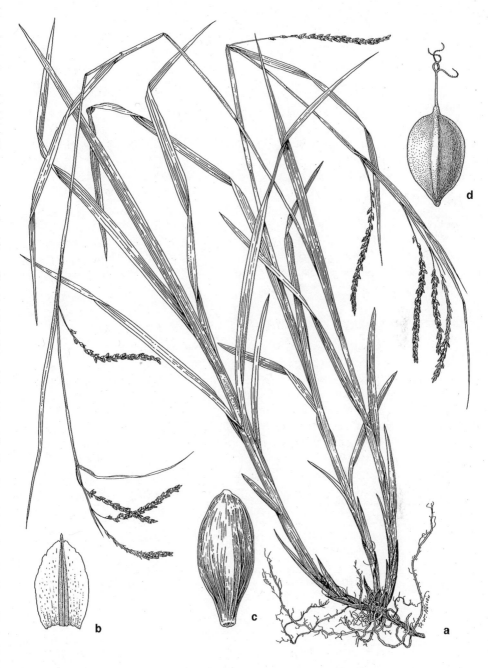

103. Carex gracillima.
a. Habit.

b. Pistillate scale.

c. Perigynium.
d. Achene.

peduncles often as long as the spikes; pistillate scales narrowly ovate, obtuse to acute, yellow-brown with a green center, shorter than the perigynia; perigynia up to 50 per spike, crowded except near the base, narrowly ellipsoid, 2.0–3.5 mm long, triangular, with a few strong nerves, glabrous, beakless; achenes trigonous, 1.3–2.6 mm long, apiculate, more or less rounded at the base; stigmas 3.

Common Name: Graceful Sedge.
Habitat: Moist or dry woods.
Range: Newfoundland to Manitoba, south to Missouri and Virginia, and in the mountains of North Carolina and Tennessee.
Illinois Distribution: Occasional in the northern fourth of Illinois; also Alexander, Jackson, and McDonough counties.

This species is readily recognized by its pendulous, narrowly cylindric spikes. It differs from the similar *C. formosa* in its beakless perigynia.

One of its locations in the southern half of Illinois is near a spring in the woods near Kinkaid Lake in Jackson County. *Carex gracillima* flowers from April to June.

104. **Carex oxylepis** Torr. & Hook. Ann. Lyc. N. Y. 3:409. 1836.

Plants perennial, cespitose, with slender rhizomes; culms slender, triangular, to 1 m tall, glabrous or slightly pubescent, purplish at the base; sterile shoots sometimes present; leaves up to 5 per culm, 3–6 mm wide, soft, pubescent, at least on the lower surface, usually shorter than the inflorescence; sheaths pubescent; terminal spike gynecandrous, up to 4.5 cm long, up to 3 mm thick, on a short, slender peduncle; staminate scales acute to acuminate; lateral spikes 2–3, pistillate, up to 4.5 cm long, up to 4 mm thick, narrowly cylindrical, on curving or pendulous slender hairy peduncles; pistillate scales lanceolate, awned, pale with a green center, shorter than the perigynia; perigynia up to 40 per spike, narrowly ovoid, 3.5–5.0 mm long, somewhat triangular, strongly nerved, glabrous or pubescent, narrowed to the base, with a short, bidentate beak; achenes trigonous, 1.8–2.0 mm long, yellowish, apiculate, substipitate; stigmas 3.

Two varieties occur in Illinois, separated by the following key:

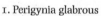

1. Perigynia glabrous .... 104a. *Carex oxylepis* var. *oxylepis*
1. Perigynia pubescent .... 104b. *Carex oxylepis* var. *pubescens*

104a. **Carex oxylepis** Torr. & Hook. var. **oxylepis** Fig. 104.

Perigynia glabrous.

Common Name: Sedge.
Habitat: Swampy woods.
Range: Virginia to Missouri, south to Texas and Florida.
Illinois Distribution: Confined to the extreme southern tip of Illinois.

104. Carex oxylepis.
a. Habit.

b. Pistillate scale.

c. Perigynium.
d. Achene.

Most plants of this rare southern species in Illinois are the typical variety with glabrous perigynia. *Carex oxylepis* is recognized by its pubescent leaves and sheaths, its slender terminal spike that is gynecandrous, and its pistillate scales that are shorter than the perigynia. This species flowers during April and May.

104b. Carex oxylepis Torr. & Hook. var. **pubescens** J. K. Underw. Am. Midl. Nat. 33:635. 1945. Not illustrated.

Perigynia pubescent.

Common Name: Sedge.
Habitat: Swampy woods.
Range: Arkansas; southern Illinois; Tennessee.
Illinois Distribution: Known only from Hardin County.

105. Carex formosa Dewey, Am. Journ. Sci. 8:98. 1824. Fig. 105.

Plants perennial, cespitose, with short rhizomes; culms to 1 m tall, triangular, slender, slightly pubescent on the angles, purplish at the base; sheaths pubescent, with the hyaline band red-spotted near the summit; leaves 3–4 per culm, up to 1.8 mm wide, lax, membranous, septate-nodulose, dark green, at least the upper ones pubescent along the margins; terminal spike gynecandrous, to 3 cm long, to 2 mm thick, on a long, slender, flexuous peduncle; staminate scales obtuse to awned, white-hyaline with a green center; lateral spikes 3–4, pistillate, oblong-cylindric, to 2.5 cm long, to 4 mm thick, on long, slender, flexuous peduncles; lower bracts leaflike; pistillate scales broadly lanceolate, obtuse to acute, greenish white with a green center, shorter than the perigynia, some of the basal ones empty; perigynia up to 20 per spike, ellipsoid, tapering to both ends, somewhat inflated, 3.5–5.0 mm long, 1.5–2.0 mm wide, glabrous, membranous, dark green, few-nerved, substipitate, with a usually short-bidentate beak 0.2–0.3 mm long; achenes triangular, 1.8–2.0 mm long, 1.3–1.5 mm wide, substipitate, jointed with the style; stigmas 3, short.

Common Name: Sedge.
Habitat: Moist savannas.
Range: Quebec to Ontario and Minnesota, south to northeastern Iowa, northeastern Illinois, southern Michigan, New York, and Connecticut.
Illinois Distribution: Known only from Cook and Lake counties, where it was first found in 1992.

This species differs from the similar *C. prasina* and *C. gracillima* in its glabrous sheaths and leaves; from *C. davisii* in its shorter, awnless pistillate scales and its generally shorter perigynia; and from *C. oxylepis* in its longer peduncles and its empty basal pistillate scales. This species flowers during June.

105. Carex formosa.
a. Habit.

b. Pistillate scale.

c. Perigynium.
d. Achene.

106. Carex davisii Schwein. & Torr. Ann. Lyc. N. Y. 1:326. 1825. Fig. 106.
Carex torreyana Dewey, Am. Journ. Sci. 10:47. 1825.

Plants perennial, cespitose, with slender rhizomes; culms slender, triangular, to 1 m tall, usually pubescent, purplish at the base; leaves up to 5 per culm, 3–8 mm wide, flat, soft, pubescent, at least on the lower surface, sometimes longer than the inflorescence; sheaths pubescent, the ventral band often speckled near the summit; terminal spike gynecandrous, up to 3.5 cm long, sessile or with a short peduncle; staminate scales acute to acuminate; lateral spikes 2–3, entirely pistillate, up to 4 cm long, up to 6 mm thick, curving to pendulous on slender peduncles; pistillate scales hyaline with a green center, awned, as long as or longer than the perigynia; perigynia up to 40 per spike, oblong-ovoid, turgid, 4–5 mm long, glabrous, with several strong nerves, stipitate, with a bidentate beak up to 0.5 mm long; achenes trigonous, 2.2–2.7 mm long, yellowish, apiculate, substipitate; stigmas 3.

Common Name: Davis's Sedge.
Habitat: Moist woods, wet ditches.
Range: Maine to Minnesota, south to Texas, Tennessee, and Maryland.
Illinois Distribution: Throughout the state.

Carex davisii is distinguished by its hairy leaves and sheaths, its spikes with the awned pistillate scales exserted from the spikes, and its terminal spike that is gynecandrous.

Very early botanists such as Engelmann (1843) and Mead (1846) called this species *C. torreyana*, but the binomial *C. davisii* precedes *C. torreyana* by a few months. *Carex davisii* flowers during May and June.

107. Carex debilis Michx. var. **rudgei** Bailey, Mem. Torrey Club 1:34. 1889. Fig. 107.
Carex flexuosa Muhl. ex Willd. Sp. Pl. 4:297. 1895.

Plants perennial, cespitose, with short rhizomes; culms slender, triangular, arching, up to 1 m tall, scabrous on the angles, glabrous, purplish at the base; overwintering basal leaves up to 7 mm wide; other leaves up to 3 per culm, 2–4 mm wide, flat, soft, glabrous, scabrous along the margins, usually longer than the inflorescence; sheaths glabrous, yellow-brown to red, speckled near the summit ventrally; terminal spike up to 5 cm long, 1–2 mm thick, very slender, either entirely staminate or with some pistillate flowers at the top, usually on a scabrous peduncle; staminate scales obtuse to acute; lateral spikes 3–4, entirely pistillate, to 8 cm long, to 4 mm thick, rather loosely flowered, on curving or pendulous scabrous peduncles; pistillate scales oblong, obtuse, stramineous or brownish with a green center, shorter than the perigynia; perigynia up to 25 per spike, fusiform to narrowly lanceoloid, 4–7 mm long, 1.5–2.0 mm wide, ascending, green, glabrous, strongly 2-ribbed and with several faint nerves, with a stout bidentate beak up to 1 mm long; achenes trigonous, 2.0–2.2 mm long, apiculate, stipitate; stigmas 3.

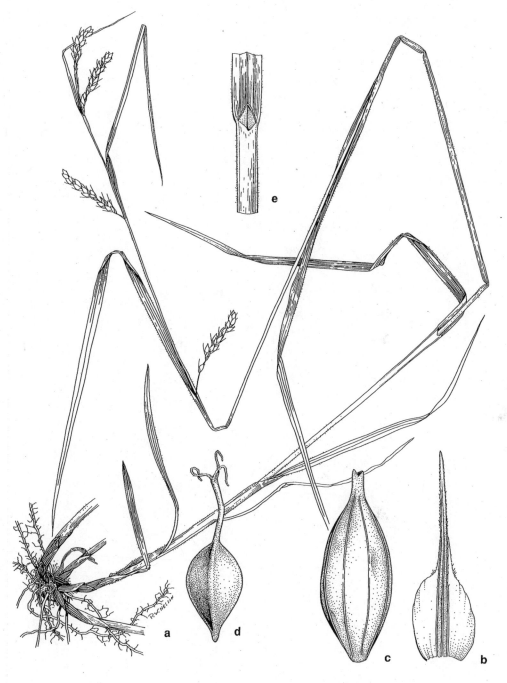

106. Carex davisii.
a. Habit.

b. Pistillate scale.
c. Perigynium.

d. Achene.
e. Sheath with ligule.

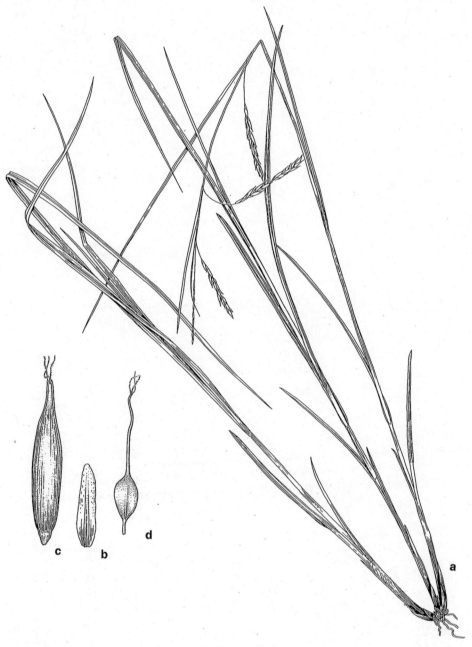

107. Carex debilis var. rudgei.

a. Habit.
b. Pistillate scale.
c. Perigynium.
d. Achene.

Common Name: Sedge.
Habitat: Dry woods.
Range: Newfoundland to Ontario, south to Missouri, Tennessee, and North Carolina.
Illinois Distribution: Known only from Hardin County in extreme southeastern Illinois and Carroll County in northwestern Illinois.

This readily distinguishable species has fusiform perigynia that are 4–7 mm long and only about 1.5–2.0 mm wide. The terminal spike is either entirely staminate or with some pistillate flowers at the tip.

Typical var. *debilis*, which apparently ranges slightly to the east and south of Illinois, has longer perigynia and whitish pistillate scales.

Earlier botanists considered var. *rudgei* a distinct species, calling it *C. flexuosa*. *Carex debilis* var. *rudgei* flowers during May and June.

108. **Carex sprengelii** Dewey in Spreng. Syst. 3:827. 1826. Fig. 108.
Carex longirostris Torr. ex Schw. Ann. Lyc. N. Y. 1:71. 1824, non Krock. (1814).

Plants perennial, cespitose, from a coarse, fibrous, matted rhizome; culms slender, triangular, up to 1 m long, erect to decumbent, usually scabrous on the angles, glabrous, brownish at the base; leaves up to 8 per culm, 2.5–4.0 mm wide, flat, scabrous along the margins, shorter than the inflorescence; sheaths glabrous, usually septate, the lowermost brown, uppermost 1–3 spikes entirely staminate, rarely with a few perigynia at the base, up to 2 cm long, up to 2 mm thick, stramineous, on scabrous peduncles often as long as the spike; staminate scales acute to cuspidate, hyaline with a greenish center; lower spikes 2–5, entirely pistillate, up to 3.5 cm long, up to 1 cm thick, loosely flowered, borne on scabrous, pendulous peduncles often longer than the spikes; pistillate scales lance-ovate, acuminate to awned, stramineous to greenish, equalling or longer than the perigynia; perigynia up to 40 per spike, 5.0–6.5 mm long, 1.8–2.0 mm wide, spreading to ascending, the body globose, glabrous, 2-ribbed, otherwise nerveless, abruptly tapering to a bidentate beak at least as long as the body of the achene; achenes trigonous, up to 2.5 mm long, yellowish, apiculate, substipitate; stigmas 3.

Common Name: Long-beaked Sedge.
Habitat: Moist woods, wooded terraces.
Range: Quebec to Alberta, south to Colorado, Nebraska, northern Illinois, Pennsylvania, and Delaware.
Illinois Distribution: Occasional to common in the northern half of the state; also Washington County.

The major distinction of this species is its long bidentate beak of the perigynium that is usually at least as long as the body. There are usually 1–3 staminate spikes and 2–5 pendulous pistillate spikes.

Early Illinois botanists called this plant *C. longirostris* Torr., but this binomial

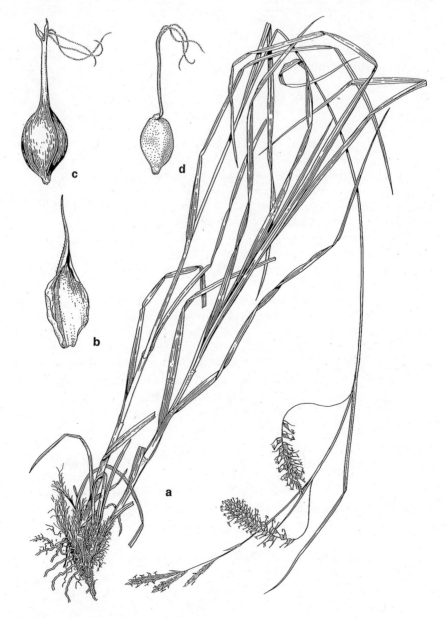

108. Carex sprengelii.
a. Habit.

b. Pistillate scale.

c. Perigynium.
d. Achene.

had been used previously for a different species and is not valid for *C. sprengelii*. This species flowers in May.

109. Carex granularis Muhl. ex Willd. Sp. Pl. 4:179. 1805.

Plants perennial, cespitose, from short rhizomes; culms slender, spreading to ascending, up to 80 cm tall, triangular, smooth or only slightly scabrous, brownish at the base; overwintering leaves 5–12 mm wide; other leaves up to 5 per culm, 5–10 mm wide, flat, soft, more or less glaucous, usually scabrous along the margins, sometimes longer than the inflorescence; sheaths yellow-brown, red-spotted on the hyaline ventral band; terminal spike 1, entirely staminate, up to 2.5 cm long, up to 2.5 mm thick, sessile or subsessile; staminate scales acuminate, reddish brown with a green center; lateral spikes 2–5, entirely pistillate, up to 3 cm long, up to 6 mm thick, the uppermost sessile or subsessile, the lower ones on peduncles of varying lengths; bracts leaflike, often 10 cm long or longer; pistillate scales white to reddish brown, red-spotted, acuminate to awned, shorter than to equalling the perigynia; perigynia up to 50 per spike, ellipsoid to ovoid to substipitate, 2.3–4.0 mm long, 1.0–2.5 mm wide, usually inflated, olive green to brownish, glabrous, with many strong nerves, with a short, entire beak up to 0.5 mm long, or sometimes beakless; achenes trigonous, 1.6–1.8 mm long, yellow-brown, apiculate, stipitate; stigmas 3.

Two varieties that tend to intergrade occur in Illinois.

1. Pistillate spikes 5–6 mm thick; perigynia strongly inflated, ovoid to subglobose, 2.5–4.0 mm long, 1.5–2.5 mm wide .... 109a. *C. granularis* var. *granularis*
1. Pistillate spikes 3–5 mm thick; perigynia less inflated, ellipsoid, 2.3–3.0 mm long, 1.0–1.5 mm wide.. 109b. *C. granularis* var. *haleana*

109a. Carex granularis Muhl. var. granularis Fig. 109 a–g.
Carex granularis Muhl. var. *recta* Dewey in Wood, Class-book 763. 1860.

Pistillate spikes 5–6 mm thick; perigynia strongly inflated, ovoid to subglobose, 2.5–4.0 mm long, 1.5–2.5 mm wide.

Common Name: Sedge.
Habitat: Woods, old fields, fens, wet meadows.
Range: Vermont to Ontario, south to eastern Kansas, Louisiana, and Florida.
Illinois Distribution: Common throughout Illinois.

Carex granularis is recognized by its single, sessile, staminate spike; its short-cylindric pistillate spikes with densely crowded perigynia; and its somewhat glaucous appearance. It differs from var. *haleana* in its slightly thicker pistillate spikes and its larger and broader perigynia. This variety is common in a variety of habitats in Illinois. *Carex granularis* var. *granularis* flowers from late April until early June.

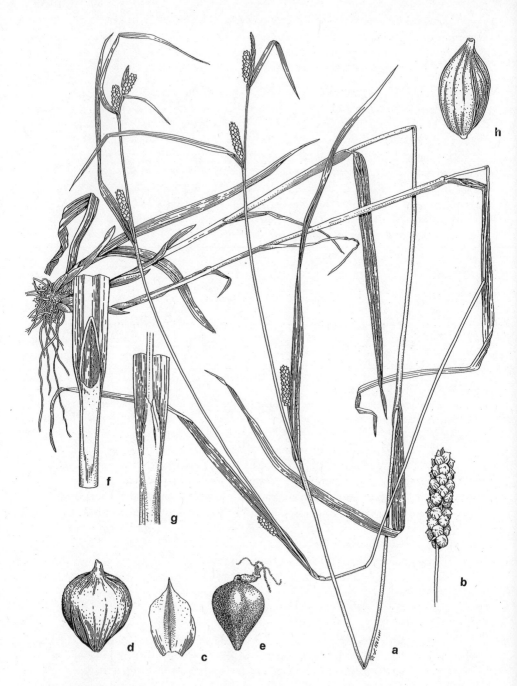

109. Carex granularis.
a. Habit.
b. Inflorescence.
c. Pistillate scale.

d. Perigynium.
e. Achene.
f. Sheath with ligule.

g. Sheath clasping culm.
var. **haleana**.
h. Achene.

109b. Carex granularis Muhl. var. **haleana** (Olney) Porter, Proc. Acad. Sci. Phila. 1887:74. 1888. Fig. 109h.
Carex haleana Olney, Car. Bor. Am. 6. 1871.
Carex granularis Muhl. var. *shriveri* Britt. in Britt. & Brown Ill. Fl. 1:322. 1896.
Carex shriveri (Britt.) Britt. Man. 208. 1901.

Pistillate spikes 3–5 mm thick; perigynia less inflated, ellipsoid, 2.3–3.0 mm long, 1.0–1.5 mm wide.

> *Common Name:* Hale's Sedge.
> *Habitat:* Woods, wet meadows.
> *Range:* Quebec to Saskatchewan, south to Kansas, Missouri, Tennessee, and Virginia.
> *Illinois Distribution:* Scattered throughout the state.

Variety *haleana* differs from var. *granularis* in its more slender pistillate spikes and its smaller and narrower perigynia.

The extreme specimens of var. *haleana* and var. *granularis* are quite different in appearance, but the two varieties may intergrade so that clear distinctions are not always possible. Some botanists do not even recognize var. *haleana* at the varietal level. This variety flowers during May and early June.

110. Carex crawei Dewey, Am. Journ. Sci. II. 2:246. 1846. Fig. 110.
Carex heterostachya Torr. Am. Journ. Sci. II. 2:248. 1846.

Plants perennial, with solitary culms arising from elongated rhizomes; culms up to 40 cm tall, obscurely triangular, smooth; leaves often in tufts, glaucous, sometimes folded, 1.5–3.5 mm wide, rather stiff, scabrous or smooth along the margins, shorter than the inflorescence; sheaths smooth, each cauline one with a pistillate spike; terminal spike staminate, up to 3 cm long, up to 3 mm thick, elevated on a scabrous peduncle up to 10 cm long; staminate scales obtuse, red-spotted; lateral spikes 2–4, pistillate, remote, up to 3 cm long, up to 6 mm thick, the uppermost often sessile, the others pedunculate, the lowest often near the base of the plant; pistillate scales acute to cuspidate, reddish brown with a green center, not reaching the beak of the perigynium; perigynia up to 45 per spike, ovoid, 3.0–3.5 mm long,

1.3–2.0 mm wide, glabrous, reddish punctate above the middle, strongly several-nerved, beakless or with a very short, hyaline beak; achenes trigonous, 1.7–2.0 mm long, yellow-brown, apiculate, substipitate; stigmas 3.

Common Name: Crawe's Sedge.
Habitat: Sandy flats, calcareous prairies, fens.
Range: Quebec to Alberta, south to Washington, Wyoming, Kansas, Alabama, and New Jersey.
Illinois Distribution: Occasional in the northern half of the state; also St. Clair County.

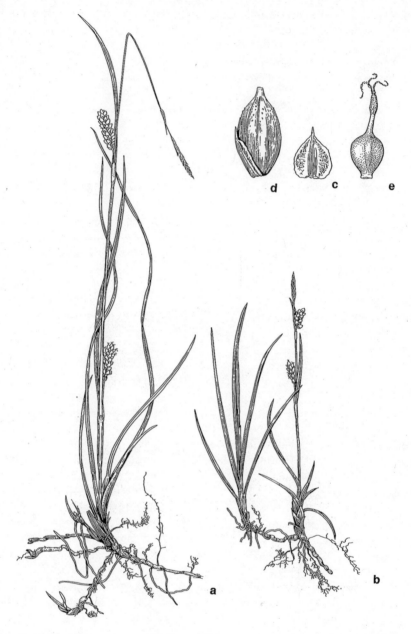

110. Carex crawei.
a, b. Habit.

c. Pistillate scale.

d. Perigynium with scale.
e. Achene.

This species is recognized by its long-stalked staminate spike; its red-dotted perigynia; and its remote pistillate spikes, one of which is usually near the base of the plant. John Torrey named this same species *C. heterostachya* in the same issue of the same journal that Dewey named *C. crawei*, but two pages later. Thus, Torrey's binomial is invalid by two pages! This species flowers from late April through May.

III. **Carex conoidea** Schkuhr ex Willd. Sp. Pl. 4:280. 1805. Fig. III.
Carex illinoensis Dewey, Am. Journ. Sci. II. 6:245. 1848.

Plants perennial, densely cespitose in tussocks, from short rhizomes; culms slender, triangular, to 50 cm tall, scabrous on the angles, brownish at the base; sterile shoots common; leaves up to 3 per culm, 3–5 mm wide, scabrous on the margins and the veins; sheaths red-dotted, at least where hyaline; terminal spike staminate, to 2 cm long, to 3 mm thick, on a scabrous peduncle up to 3 cm long, less commonly subsessile; staminate scales obtuse to acute, brownish with a green center; lateral spikes 1–3, pistillate, to 2.5 cm long, to 5 mm thick, sessile or on scabrous peduncles up to 2 cm long; bracts leaflike, often longer than the inflorescence; pistillate scales ovate, awned, brownish with a green center, shorter than to slightly longer than the perigynia; perigynia up to 25 per spike, oblongoid, 3–4 mm long, shiny, stramineous, finely impressed-nerved, glabrous, beakless or with a very short, entire beak; achenes trigonous, 2.0–2.2 mm long, yellow-brown, apiculate, stipitate; stigmas 3.

Common Name: Sedge.
Habitat: Wet meadows, wet prairies.
Range: Newfoundland to Ontario, south to Iowa, Illinois, Ohio, and Delaware.
Illinois Distribution: Occasional in the northern half of the state; also Massac County.

The distinguishing characteristics of this species are the long-stalked staminate spike; the upper bracts that are usually longer than the inflorescence; and the perigynia with numerous, impressed nerves.
Samuel B. Mead discovered this species in Hancock County sometime in the early 1800s, and it was named *C. illinoensis* by Dewey, thinking it to be a distinct species. However, *C. illinoensis* and *C. conoidea* are the same species. This species flowers from late April to mid-June.

112. **Carex amphibola** Steud. Syn. Pl. Cyp. 234. 1855. Fig. 112.
Carex grisea Wahl. var. *rigida* L. H. Bailey, Mem. Torrey Club 1:56. 1889.
Carex amphibola Steud. var. *rigida* (L.H.Bailey) Fern. Rhodora 44:315. 1942.

Plants perennial, cespitose, from short rhizomes; culms slender, triangular, to 80 cm tall, slightly scabrous on the angles, purplish at the base; sterile shoots common; leaves 3–7 mm wide, stiff, flat, scabrous along the margins; sheaths often red-dotted in the hyaline areas; terminal spike staminate, to 3 cm long, to 2.5 mm

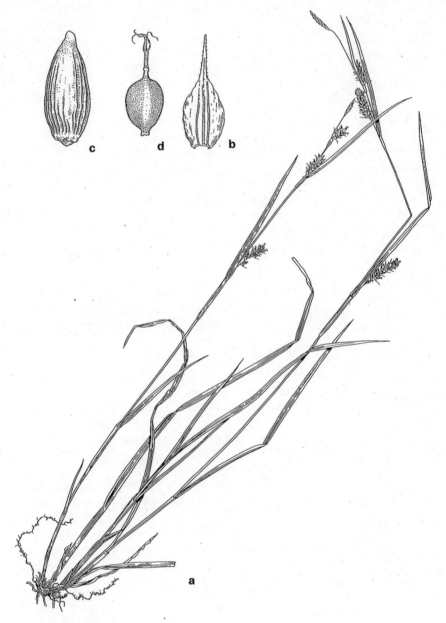

c   d   b

a

111. Carex conoidea.
a. Habit.

b. Pistillate scale.

c. Perigynium with scale.
d. Achene.

112. Carex amphibola.
a. Habit.

b. Pistillate scale.

c. Perigynium.
d. Achene.

thick, on a scabrous peduncle elevated above the pistillate spikes; staminate scales obtuse to acute, hyaline with a green center; lateral spikes 3–5, pistillate, to 2.5 cm long, 3–6 mm thick, the lowest often near the base of the plant; uppermost bracts leaflike; pistillate scales ovate, awned, hyaline and red-dotted with a green center, shorter than the perigynia; perigynia up to 12 per spike, oblongoid to obovoid, bluntly trigonous, scarcely inflated, 3.5–4.5 mm long, glabrous, with numerous impressed nerves, tapering to an entire tip; achenes trigonous, 2.4–2.5 mm long, yellow-brown, apiculate, stipitate; stigmas 3.

Common Name: Sedge.
Habitat: Forests, wet meadows.
Range: Massachusetts to Indiana and Missouri, south to Texas and Florida.
Illinois Distribution: Occasional throughout the state.

Carex amphibola is distinguished by its staminate spike elevated above the pistillate spikes, its awned pistillate scales, and its beakless perigynia that are 3.5–4.5 mm long. Some botanists consider our plants to be var. *rigida*, differing from the more eastern var. *amphibola* by its firmer, more scabrous leaves. *Carex amphibola* flowers from late April to mid-June.

113. Carex grisea Wahl. Kongl. Vet. Acad. Handl. II 24:154. 1803. Fig. 113.
Carex amphibola Steud. var. *turgida* Fern. Rhodora 44:311. 1942.

Plants perennial, cespitose, from short rhizomes; culms slender, triangular, to 50 cm tall, mostly smooth, brownish at the base; sterile shoots present; leaves 4–8 mm wide, soft, flat, scabrous on the veins; terminal spike staminate, sessile or nearly so, not elevated above the pistillate spike, to 3 cm long, to 3 mm thick; staminate scales acute to acuminate, hyaline with a green center; lateral spikes 3–5, pistillate, to 2.5 cm long, 4–7 mm thick, the lowest not near base of the plant; uppermost bracts leaflike; pistillate scales ovate, awned, hyaline and red-spotted with a green center, about as long as the perigynia; perigynia up to 15 per spike, oblong-cylindric, not trigonous, inflated, 4.5–5.5 mm long, glabrous, with numerous impressed nerves, tapering to an entire tip; achenes trigonous, 2.5–2.6 mm long, yellow-brown, apiculate, substipitate; stigmas 3.

Common Name: Sedge.
Habitat: Low woods, roadside ditches.
Range: New Brunswick to Ontario, south to Texas, Louisiana, and Georgia.
Illinois Distribution: Scattered throughout the state.

Carex grisea differs from the similar *C. amphibola* in its shorter stature; softer leaves; short-stalked staminate spike; and larger, inflated perigynia. Some botanists who do not recognize *C. grisea* as a distinct species call this plant *C. amphibola* var. *turgida. Carex grisea* flowers during May and June.

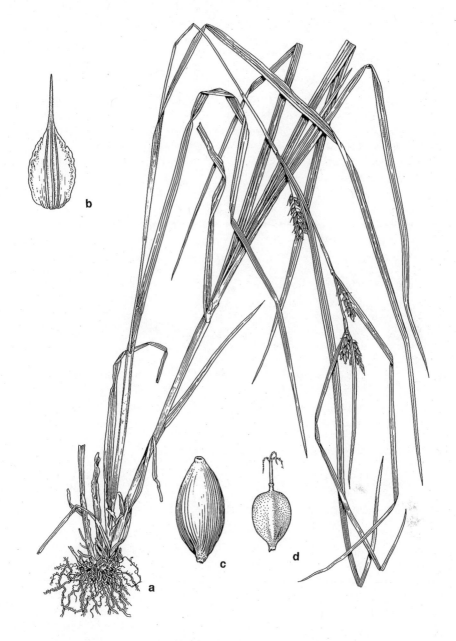

113. Carex grisea.
a. Habit.

b. Pistillate scale.

c. Perigynium.
d. Achene.

114. Carex flaccosperma Dewey, Am. Journ. Sci. II. 2:245. 1846. Fig. 114.

Plants perennial, cespitose, from short rhizomes; culms slender, triangular, to 60 cm tall, smooth, green or glaucous, brownish at the base; sterile shoots sometimes present, with overwintering leaves up to 1.5 cm wide; leaves up to 5 per culm, up to 1 cm wide, flat, soft, pale green or glaucous, scabrous on the margins and the veins; sheaths red-dotted; terminal spike staminate, to 2.5 cm long, to 3 mm thick, sessile or nearly so; staminate scales obtuse to acute, hyaline with a green center; lateral spikes 2–4, pistillate, to 3.5 cm long, to 8 mm thick, the lowest often near the base of the plant, short-stalked; bracts leaflike, the uppermost with expanded sheaths; pistillate scales ovate, awned, hyaline with a green center, about half as long as the perigynia; perigynia up to 60 per spike, crowded, oblong-cylindric, 4–6 mm long, 2.0–2.3 mm wide, yellow-brown, glabrous, not inflated, many-nerved, beakless; achenes trigonous, 2.4–2.5 mm long, yellow-brown, apiculate, substipitate; stigmas 3.

Common Name: Sedge.
Habitat: Wet woods, swamps.
Range: Virginia to Missouri, south to Texas and Florida.
Illinois Distribution: Mostly in the southern fourth of the state; also Fayette, Macon, McDonough, Menard, and St. Clair counties.

Carex flaccosperma is distinguished by its usually glaucous leaves, its oblong-cylindric perigynia that are 4–6 mm long, and its pistillate scales that are only half as long as the perigynia. Although this species is similar to *C. glaucodea,* it tends to occupy wetter habitats. *Carex flaccosperma* flowers from May to early June.

115. Carex glaucodea Tuckerm. ex Olney, Proc. Am. Acad. 7:395. 1868. Fig. 115.
Carex flaccosperma Dewey var. *glaucodea* (Tuckerm.) Kukenth. in Engler, Pflanzenr. 4 (20):518. 1909.

Plants perennial, cespitose, from short rhizomes; culms slender, triangular, to 60 cm tall, smooth, glaucous; sterile shoots sometimes present, with overwintering leaves up to 1 cm wide; leaves up to 7 per culm, up to 1 cm wide, flat, stiff, very glaucous, scabrous on the margins and the veins; sheaths red-spotted; terminal spike staminate, to 2.5 cm long, to 3 mm thick, sessile or nearly so; staminate scales obtuse to acute, hyaline with a green center; pistillate spikes 3–4, to 4 cm long, to 5 mm thick, the lowest often near the base of the plant and on slender peduncles; bracts leaflike; pistillate scales ovate, awned, hyaline with a green center, nearly as long as the perigynia; perigynia up to 45 per spike, oblong-cylindric, 3–5 mm long, 1.5–1.8 mm wide, yellow-brown, glabrous, not inflated, many-nerved, beakless; achenes trigonous, 2.0–2.3 mm long, brown, apiculate, stipitate; stigmas 3.

114. Carex flaccosperma.
a. Habit.

b. Pistillate scale.

c. Perigynium with scale.
d. Achene.

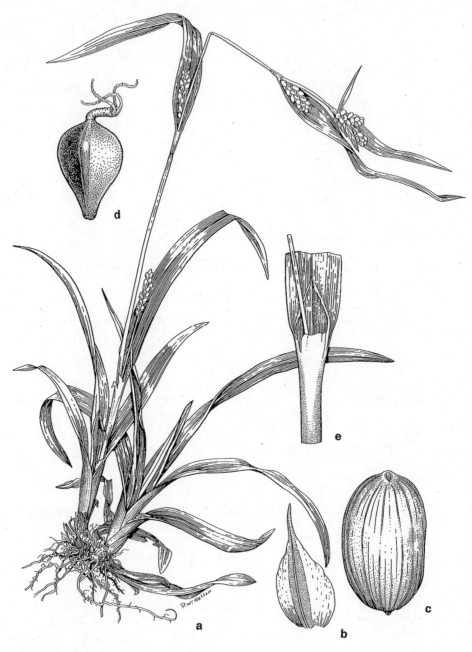

115. Carex glaucodea.
a. Habit.

b. Pistillate scale.
c. Perigynium.

d. Achene.
e. Sheath with ligule.

Common Name: Blue Sedge.
Habitat: Woods.
Range: Massachusetts to Ontario, south to Louisiana and Alabama.
Illinois Distribution: Confined to the southern half of the state.

Carex glaucodea is readily distinguished by its very glaucous leaves. It differs from *C. flaccosperma* in its smaller, narrower perigynia and in its pistillate scales that are nearly as long as the perigynia.

 Some botanists prefer to consider this plant as a variety of *C. flaccosperma*. This species flowers from May to early July.

116. Carex oligocarpa Schkuhr ex Willd. Sp. Pl. 4:279. 1805. Fig. 116.

Plants perennial, densely cespitose, from short rhizomes; culms slender, triangular, to 50 cm tall, stiff, scabrous beneath the inflorescence; sterile shoots present; leaves 1 or 2 per culm, 2–5 mm wide, dark green, scabrous along the margins and veins; lowest sheaths brown tinged with purple-red, papillate-scabrous, the uppermost green, with a well developed hyaline ventral band, none of them pubescent; uppermost bracts leaflike; terminal spike staminate, to 2.5 cm long, to 3 mm thick, sessile or short-pedunculate; staminate scales obtuse to acuminate; lateral spikes 2–4, pistillate, to 1.5 cm long, to 5 mm thick, loosely flowered, on short peduncles; pistillate scales lanceolate to ovate, awned, hyaline with a green center, shorter than to longer than the perigynia; perigynia up to 8 per spike, obovoid, trigonous, 3.5–4.0 mm long, many-nerved, glabrous, gray-green, tapering to a beak; achenes trigonous, 2.0–2.3 mm long, yellow-brown, apiculate, substipitate; stigmas 3.

Common Name: Sparse-fruited Sedge.
Habitat: Woods.
Range: Quebec to Ontario, south to Kansas, Illinois, Ohio, and Vermont.
Illinois Distribution: Scattered throughout the state.

The distinguishing features of this species are the few-flowered spikes, the awned pistillate scales, and the short-beaked perigynia (3.5–4.0 mm long). The similar *C. hitchcockiana* has pubescent sheaths and larger perigynia. *Carex oligocarpa* flowers during May and June.

117. Carex hitchcockiana Dewey, Am. Journ. Sci. 10:274. 1826. Fig. 117.

Plants perennial, densely cespitose, from short rhizomes; culms slender, triangular, to 50 cm tall, not stiff, scabrous beneath the inflorescence; sterile shoots present; leaves up to 4 per culm, 3–7 mm wide, light green, scabrous along the margins and veins; sheaths pubescent, the lowest brown, the uppermost green; uppermost bracts leaflike; terminal spike staminate, to 3 cm long, to 3 mm thick, usually on a short peduncle; staminate scales obtuse to acute; lateral spikes 3–4, pistillate, to 2.5 cm long, to 5 mm thick, loosely flowered, on short peduncles; pistil-

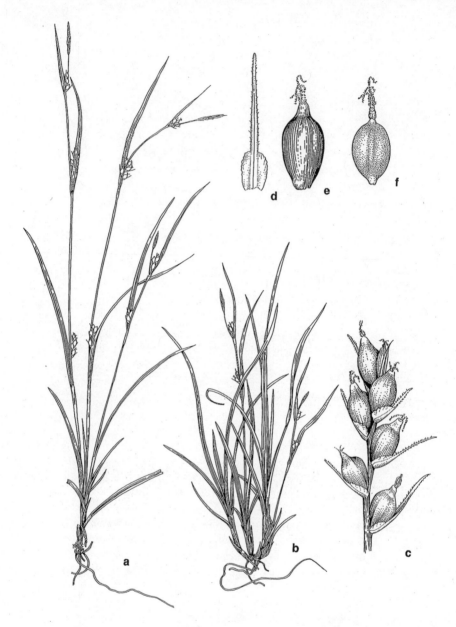

116. Carex oligocarpa.
a, b. Habit.
c. Pistillate spike.
d. Pistillate scale.
e. Perigynium.
f. Achene.

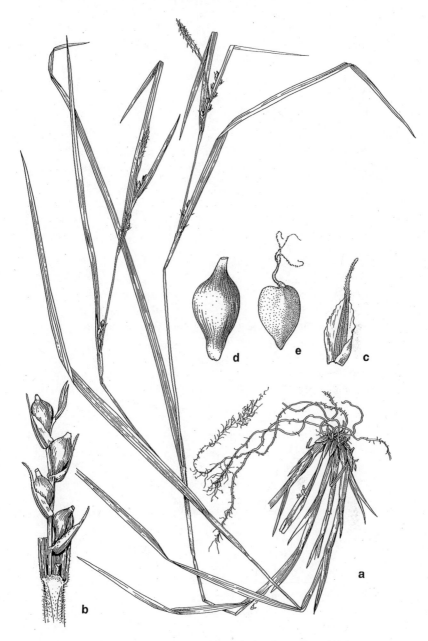

117. Carex hitchcockiana.
a. Habit.

b. Pistillate spike with sheath.
c. Pistillate scale.

d. Perigynia.
e. Achene.

late scales broadly ovate, awned, hyaline with a green center, about as long as or longer than the perigynia; perigynia up to 10 per spike, obovoid, trigonous, 4–5 mm long, many-nerved, glabrous, gray-green or yellow-green, tapering to a beak about 1 mm long; achenes trigonous, 2.7–3.0 mm long, yellow-brown, apiculate, stipitate; stigmas 3.

Common Name: Hitchcock's Sedge.
Habitat: Rich woods.
Range: Quebec to Ontario, south to Missouri, Tennessee, and Virginia.
Illinois Distribution: Scattered in the northern half of the state; also in Fayette County and at Little Grand Canyon in Jackson County.

This species is very similar to *C. oligocarpa*, differing in its pubescent sheaths and its larger perigynia and achenes. *Carex hitchcockiana* flowers during May and June.

118. Carex meadii Dewey, Am. Journ. Sci. 43:90. 1842. Fig. 118.
Carex panicea L. var. *meadii* (Dewey) Olney, Caric. Bor. Am. 2. 1871.
Carex tetanica Schkuhr var. *meadii* (Dewey) Bailey, Proc. Am. Acad. 22:118. 1886.
Carex tetanica Schkuhr var. *canbyi* Porter, Proc. Acad. Phila. 1887:76. 1887.

Plants perennial, with elongated, stout rhizomes; culms rather stout, triangular, up to 50 cm tall, scabrous, usually pale brown at the base; leaves up to 10 per culm, up to 7 mm wide, gray-green, stiff, scabrous along the margins; sheaths usually pale brown, the lowest usually blade-bearing; terminal spike staminate, up to 4 cm long, 3.5–4.5 mm thick, on a scabrous peduncle up to 4 cm long; staminate scales obtuse, purple-brown with a hyaline margin and a green center; lateral spikes 1–3, pistillate, up to 3.5 cm long, 5–7 mm thick, at least the lowest on scabrous peduncles; pistillate scales cuspidate to short-awned, purple-brown with a hyaline margin and a green center, shorter than to about as long as the perigynia; perigynia up to 30 per spike, obovoid, 3–5 mm long, trigonous, turgid, strongly nerved, glabrous, narrowed to a short, slightly curved, entire beak; achenes trigonous, 2.7–3.5 mm long, dark brown, apiculate; stigmas 3.

Common Name: Mead's Sedge.
Habitat: Prairies, barrens, fens, meadows.
Range: New Jersey to Saskatchewan, south to Texas and Georgia.
Illinois Distribution: Scattered in Illinois, but not common in the southern third of the state.

This species is characterized by its long-peduncled staminate spike, its stout rhizomes, its purple-brown scales, and its perigynia that taper to both ends. It differs from the similar *C. tetanica* in its wider leaves; thicker pistillate spikes; and larger, turgid perigynia. The type was collected by Dr. Samuel B. Mead from Augusta in Hancock County.

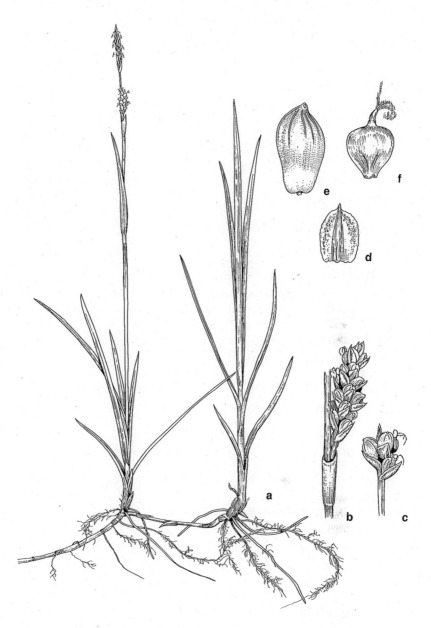

118. Carex meadii.
a. Habit.
b. Pistillate spike with sheath.
c. Spike.
d. Pistillate scale.
e. Perigynium.
f. Achene.

Some botanists consider this taxon to be a coarser variety of *C. tetanica*. Deam (1940) and Swink and Wilhelm (1994) note that intermediates between *C. meadii* and *C. tetanica* occur in moist calcareous prairies near Lake Michigan.

Early botanists at one time considered *C. meadii* to be a variety of the European *C. panicea*. *Carex meadii* flowers from mid-April to early June.

119. Carex tetanica Schkuhr, Riedgr. Nachtr. 68. 1806. Fig. 119.
Carex panicea L. var. *bebbii* Olney, Caric. Bor. Am. 2. 1871.
Carex panicea L. var. *tetanica* (Schkuhr) Olney, Caric. Bor. Am. 3. 1871.

Plants perennial, with elongated, slender rhizomes and numerous white stolons; culms slender, triangular, up to 60 cm tall, scabrous, usually purplish at the base; leaves up to 5 per culm, up to 5 mm wide, green, thin, scabrous along the margins; sheaths often reddish purple, the lowest usually blade-bearing; terminal spike staminate, up to 4 cm long, up to 3.5 mm thick, on a scabrous peduncle up to 4 cm long; staminate scales obtuse, red-purple with a hyaline margin and a green center; lateral spikes 1–3, pistillate, to 4 cm long, to 5 mm thick, at least some of them on scabrous peduncles; pistillate scales ovate, obtuse to acute, purple-brown with a hyaline margin and a green center, shorter than to about as long as the perigynia; perigynia up to 20 per spike, ovoid, 2.5–3.5 mm long, trigonous, not turgid, strongly nerved, glabrous, usually curved at the nearly beakless tip; achenes trigonous, 2.0–2.5 mm long, brown, apiculate; stigmas 3.

Common Name: Sedge.
Habitat: Wet prairies, fens, wet meadows.
Range: Massachusetts to Manitoba, south to South Dakota, Iowa, Illinois, Ohio, and Virginia.
Illinois Distribution: Occasional in the northern half of the state, rare elsewhere.

Carex tetanica differs from the similar *C. meadii* in its more slender rhizomes, narrower spikes, smaller perigynia and achenes, and lack of a true beak on the perigynium. From *C. woodii* it differs in its thicker, more crowded pistillate spikes and its broader, beakless perigynia.

Early Illinois botanists considered this species to be a variety of the European *C. panicea*. *Carex tetanica* flowers from late April to mid-June.

120. Carex woodii Dewey, Am. Journ. Sci. II 2:249. 1846. Fig. 120.
Carex tetanica Schkuhr var. *woodii* (Dewey) L. H. Bailey, Mem. Torrey Club 1:53. 1889.

Plants perennial, loosely cespitose, with slender rhizomes; culms slender, triangular, up to 50 cm tall, scabrous, usually purplish at base; leaves up to 4 per culm, up to 4 mm wide, light green, scabrous along the margins and veins, with translucent dots on the lower surface; sheaths reddish, the lowest not blade-bearing; terminal spike staminate, up to 3.5 cm long, up to 4 mm thick, on a short, scabrous

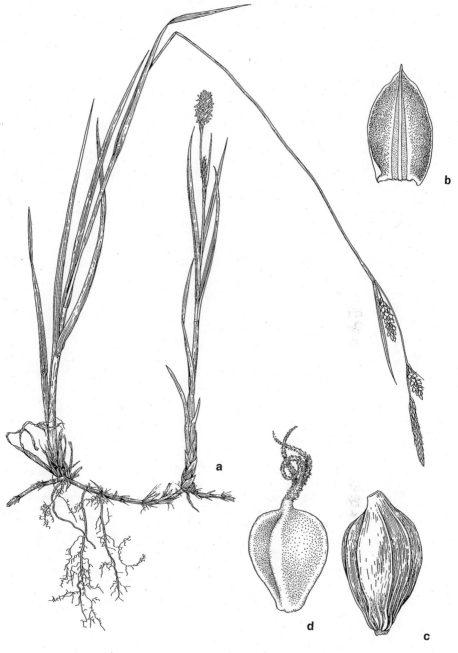

119. Carex tetanica.
a. Habit.

b. Pistillate scale.

c. Perigynium.
d. Achene.

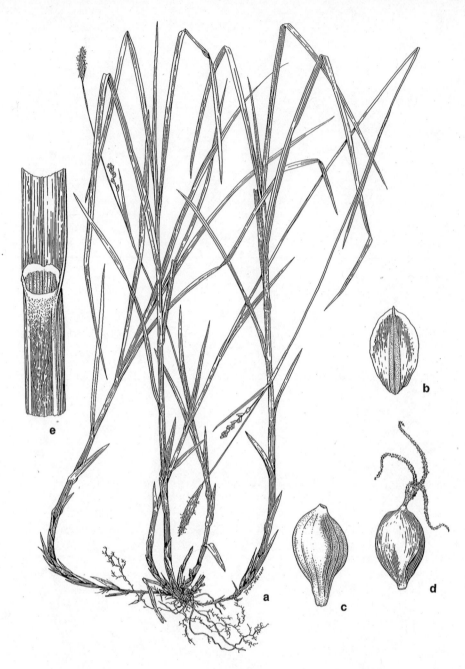

120. Carex woodii.
a. Habit.

b. Pistillate scale.

c. Perigynium.
d. Achene.

peduncle; staminate scales obtuse to acute, purple-brown with a hyaline margin and a green center; lateral spikes 2–3, pistillate, to 3.5 cm long, to 4 mm thick, loosely flowered, at least the lowest on a long, slender, scabrous peduncle; pistillate scales acute to cuspidate, purple-brown with a hyaline margin and green center, shorter than the perigynia; perigynia up to 15 per spike, broadly fusiform, 3–4 mm long, about 1.5 mm thick, trigonous, not turgid, faintly nerved, glabrous, tapering to a curved but beakless tip; achenes trigonous, 2.3–2.5 mm long, yellow-brown, apiculate; stigmas 3.

Common Name: Wood's Sedge.
Habitat: Mesic woods.
Range: Connecticut to Manitoba, south to Missouri, northern Illinois, Ohio, and West Virginia.
Illinois Distribution: Rare in the northeastern corner of Illinois, absent elsewhere except for Jo Daviess County.

This species differs from the similar appearing *C. meadii* and *C. tetanica* in its lack of blade-bearing lower sheaths and in its narrower perigynia.
    This species was originally thought to be a variety of *C. tetanica.*
The original Illinois collection was made by E. J. Hill in Kankakee County in 1870. *Carex woodii* flowers from early April to mid-May.

121. Carex plantaginea Lam. Encycl. 3:392. 1789. Fig. 121.

Plants perennial, cespitose, from short rhizomes; culms slender, wiry, up to 60 cm long, smooth, leafless but with tubular, concave, purple, bladeless sheaths; leaves basal, evergreen, up to 3 cm wide, scabrous along the margins, with several cross-veins; terminal spike staminate, purple, to 2 cm long, to 4 mm thick, on a long peduncle; staminate scales obtuse to acute, deep purple; lateral spikes 2–4, pistillate, remote, to 2.5 cm long, to 4 mm thick, at least the lowermost on a long, scabrous peduncle; pistillate scales acuminate to awned, purplish-tinged with a green center, shorter than to nearly as long as the perigynia; perigynia up to 12 per spike, oblong-ovoid, 3.5–5.0 mm long, trigonous, several-nerved, glabrous, with a straight or curved beak to 1 mm long, long-stipitate; achenes trigonous, with deeply concave sides, 2.5–2.7 mm long, brown, apiculate, stipitate; stigmas 3.

Common Name: Plantain-leaved Sedge.
Habitat: Rich woods.
Range: New Brunswick to Manitoba, south to Minnesota, Illinois, Pennsylvania, and Massachusetts; also in the mountains of North Carolina, Tennessee, Kentucky, and Alabama.
Illinois Distribution: Known only from Cook County in northeastern Illinois and Johnson County in extreme southern Illinois.

This is one of the most distinctive species of *Carex* in Illinois because of its leafless flowering culms and its basal leaves that are up to 3 cm wide.

121. Carex plantaginea.
a. Habit.

b. Pistillate scale.

c. Perigynium.
d. Achene.

The only Illinois collection for nearly a century was taken from woods in Cook County by H. C. Cowles in 1896. In 1993, it was discovered in a wooded ravine in Johnson County by Jody Shimp. This is one of the first species of *Carex* to flower, sometimes blooming during the latter part of March.

122. Carex careyana Torr. in Dewey, Am. Journ. Sci. 30:60. 1836. Fig. 122.

Plants perennial, cespitose, from short rhizomes; culms slender, erect or spreading, triangular, to 50 cm tall, smooth or somewhat scabrous, bright green, purple at the base, blade-bearing near the base; sterile shoots and basal leaves bright green with purple bases; leaves up to 3 cm wide, tapering to an acute tip, scabrous along the margins, purple at the base, some of them usually evergreen; sheaths purple; bracts leaflike but short; terminal spike staminate, to 2 cm long, to 4 mm thick, purplish, on a short peduncle; staminate scales obtuse to acute, purple with a green center; pistillate spikes 2–3, remote, to 2 cm long, to 5 mm thick, loosely few-flowered, the lowest on a slender peduncle up to 5 cm long; pistillate scales acute to awned, purple-tinged, shorter than the perigynia; perigynia up to 8 per spike, oblongoid, 5.0–6.5 mm long, trigonous, several-nerved, glabrous or rarely minutely hispid, with a straight beak up to 1.5 mm long, stipitate; achenes trigonous, with deep concave sides, 3.0–3.5 mm long, yellow-brown, apiculate, substipitate; stigmas 3.

Common Name: Carey's Sedge.
Habitat: Rich woods.
Range: Ontario to Michigan, south to Missouri and Virginia.
Illinois Distribution: Occasional in the southern fourth of the state; not common in the east-central counties; disjunct in Jo Daviess and Will counties.

This is the only broad-leaved species of *Carex* in Illinois with purple-based leaves and stems. *Carex careyana* flowers in May and June.

123. Carex platyphylla Carey, Am. Journ. Sci. II 4:23. 1847. Fig. 123.

Plants perennial, cespitose, from short rhizomes; culms slender, spreading, triangular, to 30 cm tall, somewhat scabrous, usually pale green, blade-bearing near the base; sterile shoots and basal leaves more or less glaucous, with brown bases; leaves up to 3 cm wide, tapering to an acute tip, scabrous along the margins, brown at the base; sheaths hyaline with minute red dots; bracts leaflike but short; terminal spike staminate, to 1.5 cm long, to 2.5 mm thick, pale, on a rather short peduncle; staminate scales obtuse to acute, reddish-brown with a green center and a hyaline margin; lateral spikes 2–4, pistillate, remote, to 2.5 cm long, to 5 mm thick, loosely few-flowered, nearly sessile or on scabrous peduncles; pistillate scales acuminate to awned, cinnamon with a green center and hyaline margins, shorter than the perigynia; perigynia up to 10 per spike, oblong-ovoid, 2.5–4.5 mm long, trigonous, several-nerved, glabrous, with a straight or curved beak to 1 mm long,

122. Carex careyana.
a. Habit.

b. Pistillate scale.

c. Perigynium.
d. Achene.

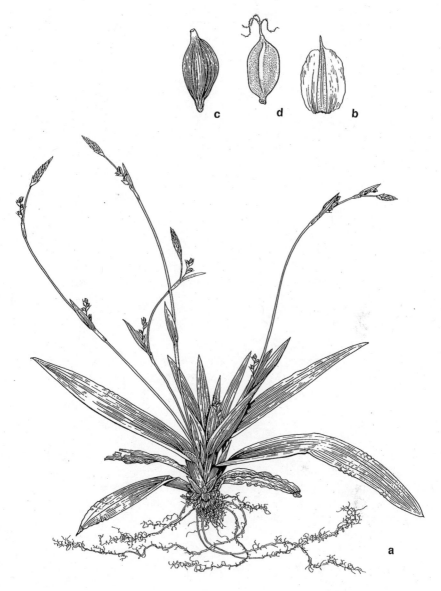

123. Carex platyphylla.
a. Habit.

b. Pistillate scale.

c. Perigynium.
d. Achene.

stipitate; achenes trigonous, with deep concave sides, 2.5–2.7 mm long, brown, apiculate, stipitate; stigmas 3.

Common Name: Broad-leaved Sedge.
Habitat: Rich woods.
Range: Quebec to Ontario, south to Illinois, Tennessee, and North Carolina.
Illinois Distribution: Known historically from McHenry County, where it was collected by George Vasey, and from St. Clair County, where it was collected by J. Macoun. Both collections were made during the nineteenth century. There is a collection in 1951 from Saline County by J. R. Swayne and W. Bailey.

This is one of four species of *Carex* that has at least some leaves up to 3 cm broad. The presence of some cauline leaves distinguishes it from *C. plantaginea.* It is very similar to *C. careyana,* but it differs in its glaucous basal leaves and the dull brown color at the base of the culms. *Carex albursina,* the fourth broad-leaved species, differs in its uppermost cauline leaf-blade that is many times longer than its sheath.

There is some doubt whether Vasey's specimen actually came from Illinois. This species flowers during May.

124. **Carex digitalis** Willd. Sp. Pl. 4:298. 1805. Fig. 124.
Carex digitalis var. *copulata* Bailey, Mem. Torrey Club 1:47. 1889.
Carex copulata (Bailey) Mack. N. Am. Fl. 18:251. 1935.
Carex laxiculmis Schwein. var. *copulata* (Bailey) Fern. Rhodora 8:183. 1906.

Plants perennial, cespitose, from short rhizomes; culms slender, erect to ascending, triangular, to 60 cm tall, somewhat scabrous, brownish at the base; leaves up to 5 mm wide, scabrous along the margins, dark green; at least the lowest sheaths pale or brownish, the ventral band broad and hyaline; terminal spike staminate, to 2 cm long, to 2 mm thick, pedunculate or rarely nearly sessile; staminate scales acute, usually yellow-brown; lateral spikes 2–5, pistillate, to 3 cm long, 3–4 mm thick, loosely flowered, the upper subsessile or short-stalked, the lower on slender peduncles up to 8 cm long, ascending; pistillate scales ovate, acute, brownish with a green center and hyaline margins, shorter than the perigynia; perigynia up to 12

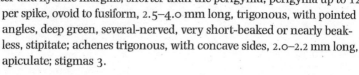

per spike, ovoid to fusiform, 2.5–4.0 mm long, trigonous, with pointed angles, deep green, several-nerved, very short-beaked or nearly beakless, stipitate; achenes trigonous, with concave sides, 2.0–2.2 mm long, apiculate; stigmas 3.

Common Name: Sedge.
Habitat: Dry woods.
Range: Maine and Ontario to Wisconsin, south to Missouri, Alabama, and Florida.
Illinois Distribution: Occasional in the southern sixth of Illinois; also Brown, Fayette, Fulton, and Vermilion counties.

124. Carex digitalis.
a. Habit.

b. Pistillate scale.

c. Perigynium.
d. Achene.

Of the eleven species of section Laxiflorae in Illinois, only *C. digitalis* and *C. laxiculmis* have perigynia that have pointed angles; all the others have perigynia with rounded angles. *Carex digitalis* differs from *C. laxiculmis* in its green rather than pale green or glaucous leaves, its narrower leaves, and its more slender pistillate spikes.

Specimens from Illinois with extremely scabrous leaves have been referred to as *Carex copulata*, but there seems to be little justification for maintaining this taxon. Fernald (1950) suspects *C. copulata* to be a hybrid between *C. digitalis* and *C. laxiculmis*. *Carex digitalis* flowers in April and May.

125. Carex laxiculmis Schwein. Ann. Lyc. N. Y. 1:70. 1824. Fig. 125.
Carex retrocurva Dewey in Wood, Class-book 423. 1845.

Plants perennial, cespitose, from short rhizomes; culms slender, spreading to ascending, usually slightly scabrous, cinnamon brown at the base; leaves up to 12 mm wide, scabrous along the margins, pale green or glaucous; sheaths pale to dark brown, red-dotted, the ventral band hyaline; terminal spike staminate, to 2 cm long, to 3 mm thick, pedunculate; staminate scales acute to acuminate, brownish with a green center and hyaline margins; lateral spikes 2–4, pistillate, to 2 cm long, 4–5 mm thick, loosely flowered, sometimes with a few staminate flowers at the base, all pedunculate, the lower on spreading to pendulous peduncles up to 10 cm long; pistillate scales ovate, acute to awned, brown with a green center and hyaline margins, shorter than the perigynia, the lowest on a culm empty; perigynia up to 12 per spike, ovoid to oblongoid, 3–4 mm long, trigonous with pointed angles, olive green, several-nerved, with a short, nearly straight beak up to 1 mm long, substipitate; achenes trigonous, with concave sides, yellow-brown, 2.2–2.5 mm long, apiculate, stipitate; stigmas 3.

Common Name: Sedge.
Habitat: Rich woods.
Range: Maine and Ontario to Wisconsin, south to Missouri, Tennessee, and North Carolina.
Illinois Distribution: Occasional but scattered throughout Illinois.

This species is similar to *C. digitalis*, but differs in its paler, often glaucous, broader leaves and its often pendulous lower pistillate spikes. The lowest pistillate scale on a culm is empty. *Carex laxiculmis* occupies a more mesic woods than *C. digitalis*. This species flowers in April and May.

126. Carex albursina Sheldon, Bull. Torrey Club 20:284. 1893. Fig. 126.
Carex laxiflora Lam. var. *latifolia* Boott, Ill. 38. 1858.

Plants perennial, cespitose, with short rhizomes; culms to 70 cm tall, triangular, with winged angles, scabrous on the angles; leaves up to 3.5 cm wide, pale green, usually somewhat scabrous along the margins, usually with sterile leafy

125. Carex laxiculmis.
a. Habit.

b. Pistillate scale.

c. Perigynium.
d. Achene.

126. Carex albursina.
a. Habit.

b. Pistillate scale.

c. Perigynium.
d. Achene.

tufts over-wintering; sheaths short, green; lower bracts leaflike; terminal spike staminate, to 1.7 cm long, to 1.5 mm thick, sessile or subsessile; staminate scales obtuse to acute, with a green center and hyaline margins; lateral spikes 2–4, pistillate, to 3.5 cm long, 3–4 mm thick, remotely flowered, sessile or on short, ascending peduncles; pistillate scales obovate, obtuse to acute, with a green center and hyaline margins, shorter than the perigynia; perigynia up to 15 per spike, obovoid, 3–4 mm long, trigonous, with rounded angles, yellow-green, several-nerved, with a short, curved beak, stipitate; achenes trigonous, with concave sides, yellow-brown, 2.3–2.5 mm long, apiculate, substipitate; stigmas 3.

Common Name: White Bear Sedge.
Habitat: Rich, wooded ravines; wooded slopes, often calcareous.
Range: Quebec to Minnesota, south to Arkansas and Virginia.
Illinois Distribution: Throughout the state.

This species, along with *Carex plantaginea, C. platyphylla,* and *C. careyana,* regularly has leaves 1.5–3.0 cm wide or wider, but *C. albursina* is the only one with the blade of the lowest bract many times longer than the sheath. *Carex albursina* was considered originally to be a broadleaved variety of *C. laxiflora.*

Early Illinois botanists such as Patterson (1876) and Brendel (1887) confused this species with *C. plantaginea.* This species flowers in April and May.

127. Carex laxiflora Lam. Encycl. 3:392. 1789. Fig. 127.

Plants perennial, cespitose, from short rhizomes; culms to 40 cm tall, triangular, smooth or slightly scabrous; sterile leafy tufts present, without prolonged culms; leaves up to 2 cm wide, pale green, smooth or slightly scabrous along the margins; sheaths of bracts smooth along the angles; lower leaf sheaths pale green to brownish green, with green cross-walls; terminal spike staminate, to 2 cm long, to 3 mm thick, pedunculate; staminate scales usually cuspidate, yellow-brown with a green center and hyaline margins; lateral spikes 3–4, pistillate, to 4.5 cm long, 3–4 mm thick, remotely flowered, the uppermost sessile or nearly so, the others on ascending peduncles; pistillate scales ovate, cuspidate to awned, with a green center and hyaline margins, usually shorter than the perigynia; perigynia up

to 15 per spike, obovoid to fusiform, 3.0–4.5 mm long, trigonous with rounded angles, usually pale green, several-nerved, with a short, usually straight beak, stipitate; achenes trigonous, with concave sides, yellow-brown, 1.8–2.0 mm long, apiculate, substipitate; stigmas 3.

Common Name: Sedge.
Habitat: Rich woods.
Range: Nova Scotia to Ontario and Wisconsin, south to Georgia.
Illinois Distribution: Known only from a few counties in the southern tip of the state and in northeastern Illinois; also Stark County.

127. Carex laxiflora.
a. Habit.

b. Pistillate scale.

c. Perigynium:
d. Achene.

This species is similar to *Carex albursina* but is more slender in all respects. In Illinois, the leaves rarely surpass 1.5 cm in width. *Carex laxiflora* is also similar to *C. striatula*, but *C. striatula* has thicker pistillate spikes, longer and narrower perigynia, and obtuse staminate scales.

Specimens in which the perigynia have curved beaks may be distinguished from *Carex blanda*, *C. gracilescens*, and *C. styloflexa* by their sterile leafy shoots that lack prolonged culms. *Carex laxiflora* flowers during April and May.

128. Carex striatula Michx. Fl. Bor. Am. 2:173. 1803. Fig. 128.

Plants perennial, cespitose, from rather long rhizomes; culms to 40 cm tall, triangular, scabrous; sterile leafy tufts present, without prolonged culms; leaves up to 1.5 cm wide, pale green, slightly scabrous along the margins; sheaths of bracts scabrous; lower leaf sheaths pale green to brownish green, usually with green cross-walls; terminal spike staminate, to 3.5 cm long, to 3 mm thick, pedunculate; staminate scales obtuse, yellow-brown with a green center and hyaline margins; lateral spikes 2–3, pistillate, to 5 cm long, 4–5 mm thick, remotely flowered, the uppermost sessile or nearly so, the others on ascending peduncles; pistillate scales ovate, acuminate to awned, with a green center and hyaline margins, usually shorter than the perigynia; perigynia up to 20 per spike, narrowly obovoid to fusiform, 4.0–5.5 mm long, trigonous with rounded angles, usually pale green, several-nerved, with a short, usually straight or sometimes curved beak, stipitate; achenes trigonous, with concave sides, brown, 2.2–2.5 mm long, apiculate, substipitate; stigmas 3.

Common Name: Sedge.
Habitat: Rich woods.
Range: Connecticut to Pennsylvania and southern Illinois, south to Texas and Florida.
Illinois Distribution: Known from a few locations in the southern three tiers of counties.

This species is similar to *C. laxiflora* but differs in its thicker pistillate spikes, longer and narrower perigynia, and obtuse rather than cuspidate staminate scales. Specimens in which the perigynia have curved beaks may be distinguished from *Carex blanda*, *C. gracilescens*, and *C. styloflexa* by their sterile leafy shoots that lack prolonged culms. *Carex striatula* flowers during April and May.

129. Carex blanda Dewey, Am. Journ. Sci. 10:45. 1826. Fig. 129.
Carex laxiflora Lam. var. *blanda* (Dewey) Boott, Illustr. Gen. Carex 37. 1858.

Plants perennial, cespitose, with short rhizomes; culms to 60 cm tall, triangular, scabrous, brownish at the base; sterile leafy shoots with blades up to 12 mm wide

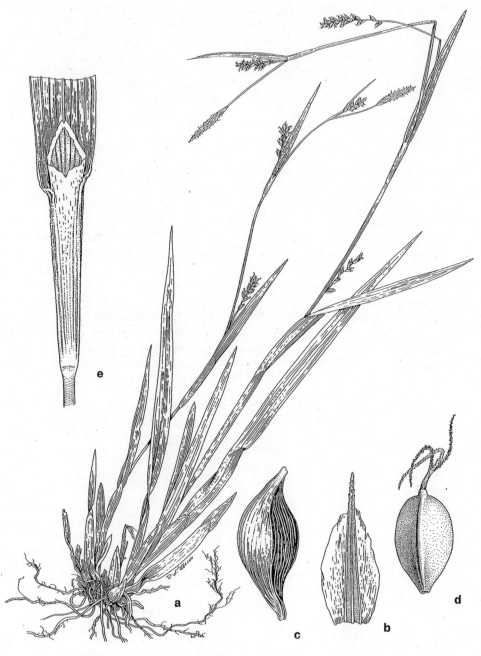

128. Carex striatula.
a. Habit.

b. Pistillate scale.
c. Perigynium.

d. Achene.
e. Sheath with ligule.

and with prolonged culms; cauline leaves up to 10 mm wide, pale green or occasionally somewhat glaucous, scabrous along the margins; sheath of bracts very scabrous; lower leaf sheaths pale to brownish green, loose, with green cross-walls; terminal spike staminate, to 2 cm long, to 3 mm thick, sessile or short-pedunculate; staminate scales obtuse to cuspidate, with a green center and hyaline margins; lateral spikes 2–5, pistillate, to 3 cm long, 3–5 mm thick, the uppermost nearly sessile, the others on longer peduncles, the perigynia crowded; pistillate scales ovate, usually awned, with a green center and hyaline margins, about as long as the perigynia; perigynia up to 25 per spike, ellipsoid to obovoid, 3.0–4.5 mm long, trigonous with rounded angles, usually pale green, several-nerved, with a short, distinctly curved beak, substipitate; achenes trigonous, with concave sides, yellow-brown, 2.2–2.5 mm long, apiculate, stipitate; stigmas 3.

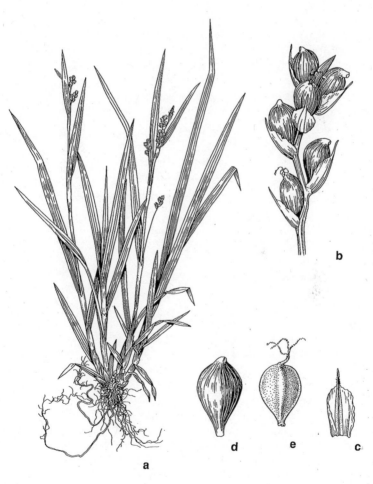

129. Carex blanda.
a. Habit.

b. Pistillate spike.
c. Pistillate scale.

d. Perigynium.
e. Achene.

Common Name: Sedge.
Habitat: Woods, meadows, mesic prairies.
Range: Quebec to North Dakota, south to Texas and Georgia.
Illinois Distribution: Common throughout the state.

Carex blanda and *C. gracilescens* always have perigynia with distinctly curved beaks. The base of the culms in *C. blanda* is brownish, while the base of the culms in *C. gracilescens* is purplish. *Carex blanda* flowers from April to June.

130. **Carex gracilescens** Steud. Syn. Cyp. 226. 1855. Fig. 130.

Plants perennial, cespitose, with short rhizomes; culms to 60 cm tall, triangular, scabrous, purplish at the base; sterile leafy shoots with blades to 8 mm wide and with prolonged culms; cauline leaves up to 6 mm wide, yellow-green, scabrous along the margins; sheaths of bracts scabrous or smooth; lower leaf sheaths purplish; terminal spike staminate, to 2.5 cm long, to 2.5 mm thick, usually pedunculate; staminate scales obtuse, reddish brown with a green center and hyaline margins; lateral spikes 2–4, pistillate, to 3 cm long, 3–4 mm thick, the uppermost sessile or nearly so, the lower on long, scabrous peduncles, the perigynia crowded; pistillate scales cuspidate or awned, obovate, usually reddish brown with a green center and hyaline margins, shorter than to longer than the perigynia; perigynia up to 20 per spike, ellipsoid to obovoid, 2.5–3.5 mm long, trigonous with rounded angles, yellow-green, several-nerved, with a distinctly curved beak up to 1 mm long, stipitate; achenes trigonous, with concave sides, yellow-brown, 2.0–2.5 mm long, apiculate, stipitate; stigmas 3.

Common Name: Sedge.
Habitat: Woods.
Range: Quebec to Ontario and Wisconsin, south to Texas, Louisiana, and North Carolina.
Illinois Distribution: Occasional throughout the state.

This species, with its distinctive curved beak of the perigynium, differs from the similar *C. blanda* in its purple-tinged leaf and stem bases. This species flowers from April to June.

131. **Carex styloflexa** Buckl. Am. Journ. Sci. 45:174. 1843. Fig. 131.
Carex laxiflora Lam. var. *styloflexa* (Buckl.) Boott, Ill. 37. 1858.

Plants perennial, cespitose, with long, slender rhizomes; culms to 80 cm tall, triangular, slightly scabrous, brownish at the base; sterile leafy shoots with blades to 7 mm wide and with prolonged culms; cauline leaves up to 3.5 mm wide, light green, slightly scabrous along the margins; sheaths yellow-brown, tight, slightly scabrous; terminal spike staminate, to 4 cm long, to 2.5 mm thick, on a scabrous

130. Carex gracilescens.
a. Habit.

b. Pistillate scale.

c. Perigynium.
d. Achene.

a

c

d

b

131. Carex styloflexa.
a. Habit.

b. Pistillate scale.

c. Perigynium.
d. Achene.

peduncle up to 8 cm long but often much shorter; staminate scales obtuse, brownish with a green center and hyaline margins; lateral spikes 1–4, pistillate, to 2 cm long, 4–5 mm thick, the uppermost on very short peduncles, the lower spikes on slender, pendulous, scabrous peduncles up to 8 cm long; pistillate scales acute to awned, brownish with a green center and hyaline margins, shorter than the perigynia; perigynia up to 15 per spike, fusiform, 3.5–4.5 mm long, trigonous with rounded angles, green to brown, several-nerved, with a straight or slightly curved beak up to 1 mm long, long-stipitate; achenes trigonous, with concave sides, yellow-brown, 2.0–2.5 mm long, apiculate, stipitate; stigmas 3.

Common Name: Sedge.
Habitat: Rich woods; low woods.
Range: Connecticut to Pennsylvania and southern Illinois, south to Texas and Florida.
Illinois Distribution: Apparently confined to the southern eighth of the state.

This species is distinguished by its fusiform perigynia with nearly straight or slightly curved prominent beaks and by the presence of prolonged culms with sterile leafy shoots. The staminate spike is usually on a long peduncle. Like several taxa in the section Laxiflorae, *Carex styloflexa* was considered to be a variety of *C. laxiflora* at one time. This species flowers during April and May.

132. **Carex cryptolepis** Mack. Torreya 14:156. 1914. Fig. 132.
Carex flava L. var. fertilis Peck, N. Y. State Mus. Rep. 48:197. 1896.

Plants perennial, cespitose, with short rhizomes; culms to 50 cm tall, triangular, usually smooth, pale brown at the base; sterile shoots usually conspicuous; leaves up to 4.5 mm wide, light green, slightly scabrous along the margins; sheaths convex at the summit; terminal spike staminate, or occasionally with a few perigynia at the top, sessile or on short peduncles; staminate scales acute, yellowish with a green center; pistillate spikes 2–5, globose or ovoid, crowded near tip of culm, with one other often remote toward middle of culm, 10–14 mm thick, each subtended by a leafy bract; pistillate scales ovate, acute to acuminate, pale green to stramine-

ous, the tip not reaching the base of the beak of the perigynium; perigynia up to 35 per spike, 3.2–4.5 mm long, narrowly ovoid, yellowish, spreading, or the lower ones reflexed, shiny, few-nerved, with a beak nearly as long as the body, the beak bidentate, greenish, smooth; achenes trigonous, with concave sides, nearly black, shiny, 1.3–1.5 mm long, short-apiculate; stigmas 3.

Common Name: Yellow Sedge.
Habitat: Fens.
Range: Newfoundland to Ontario and Minnesota, south to northeastern Illinois, northern Ohio, Pennsylvania, and New Jersey.

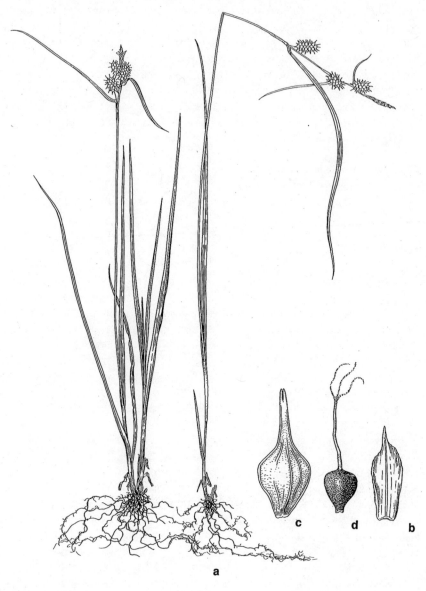

132. Carex cryptolepis.
a. Habit.

b. Pistillate scale.

c. Perigynium.
d. Achene.

Illinois Distribution: Confined to the extreme northeastern counties of Illinois.

This species is distinguished by its nearly globose, yellowish, pistillate spikes with several of the lower perigynia reflexed. The bidentate beak of the perigynium is about as long as the body. The very closely related *C. flava*, which has not been found in Illinois, is a taller plant with somewhat wider, yellow-green leaves; longer perigynia with serrulate beaks and with conspicuous pistillate scales; and smaller pistillate spikes. The achenes of *C. flava* are usually yellow-brown rather than nearly black.

Some botanists prefer to treat this plant as a variety of *C. flava*, calling it var. *fertilis*. Crins and Ball (1989) provide arguments for recognizing it as a distinct species. This species flowers during mid-May.

133. Carex viridula Michx. Fl. Bor. Am. 2:170. 1803. Fig. 133.
Carex flava L. var. *viridula* (Michx.) L. H. Bailey, Mem. Torrey Club 1:31. 1889.
Carex oederi Retz. var. *pumila* Fern. Rhodora 8:201. 1906.
Carex oederi Retz. f. *intermedia* Dudley, Bull. Cornell Univ. 2:117. 1886.
Carex viridula Michx. f. *intermedia* (Dudley) Herrm. in Deam, Fl. Indiana 256. 1940.

Plants perennial, cespitose, with short rhizomes, sometimes mat-forming; culms to 50 cm tall, wiry, triangular, smooth, pale brown at the base; sterile shoots sometimes present; leaves up to 3.5 mm wide, dull green, sometimes folded, rather stiff, slightly scabrous along the margins; sheaths green-nerved, with a broad, loose, hyaline ventral band; terminal spike staminate but often with a few perigynia at the tip, rarely nearly entirely pistillate; staminate scales obtuse, reddish brown with a green center and hyaline margins; pistillate spikes up to 6, broadly ovoid to short-cylindric, crowded near the summit of the culm, up to 1.5 cm long, 2–3 mm thick; pistillate scales obtuse, reddish brown with a green center and hyaline margins, about as long as the body of the perigynia; perigynia up to 30 per spike, 2–3 mm long, ovoid, green or yellow-green, horizontally spreading, or the lowest ones reflexed, few-nerved, with a minutely dentate beak about a third as long as the body; achenes trigonous, with concave sides, black, shiny, 1.2–1.3 mm long, substipitate; stigmas 3.

Common Name: Sedge.
Habitat: Calcareous pond shores, pannes, seeps, and fens; flat gravelly prairies.
Range: Newfoundland to Alaska, south to California, New Mexico, Minnesota, northern Illinois, Pennsylvania, and New Jersey.
Illinois Distribution: Confined to the extreme northeastern counties of Illinois.

This species is somewhat similar to *C. cryptolegis* in having some of the lowest perigynia in a spike reflexed. It differs in its more cylindrical pis-

133. Carex viridula.
a. Habit.
b. Pistillate scale.
c. Perigynium.
d. Achene.

tillate spikes and the beak of the perigynium that is only about a third as long as the body.

Occasional specimens with the terminal spike nearly entirely pistillate have been called f. *intermedia*. This species flowers from early May to late September.

134. Carex frankii Kunth, Enum. Pl. 2:498. 1837. Fig. 134.
Carex stenolepis Torr. Ann. Lyc. N. Y. 3:420. 1836, non Wahlenb. (1803).

Plants perennial, cespitose, with short, stout rhizomes; culms to 80 cm tall, triangular, usually slightly rough to the touch, usually purplish at the base; sterile shoots usually present; leaves up to 10 mm wide, deep green, firm, slightly scabrous along the margins and usually on the veins; sheaths tight, septate-nodulose, yellow-brown, hyaline at the summit, the lowermost often red-tinged, the ligule about as wide as long; terminal spike staminate or less commonly with pistillate flowers at tip, up to 3 cm long, up to 3 mm thick; staminate scales hyaline with a green center, awned; pistillate spikes 3–7, cylindric, up to 4 cm long, up to 1.2 cm thick, sessile or on short peduncles; bracts many times longer than the inflorescence; pistillate scales setaceous, slightly scabrous, green, much longer than the perigynia; perigynia up to 100 per spike, obconic, truncate across the top, 4–5 mm long, up to 2.5 mm broad, olive green, becoming brownish, many-nerved, abruptly contracted into a bidentate beak to 2.5 mm long; achenes trigonous, yellow-brown, about 1.5 mm long, substipitate; stigmas 3.

Common Name: Frank's Sedge.
Habitat: Moist woods, along streams, wet ditches.
Range: Western New York to Michigan, southwestward to Kansas, south to Texas and Georgia.
Illinois Distribution: Common to occasional throughout the state, except for the northernmost tier of counties.

This common species of low, moist areas, along with *C. squarrosa* and *C. typhina*, are grouped together into section Squarrosae. All three species have obconic or obovoid perigynia that are abruptly contracted to a beak. *Carex frankii* differs from the other two in its usually entirely staminate terminal spike. The bracts are often so long as to obscure the spikes.

During the nineteenth century, most Illinois botanists called this species *C. stenolepis*, but that binomial applies to a different species. *Carex frankii* flowers from late May to mid-September.

135. Carex squarrosa L. Sp. Pl. 973. 1753. Fig. 135.

Plants perennial, cespitose, from short rhizomes; culms to 90 cm tall, triangular, nearly smooth, usually dark brown at the base; sterile shoots usually present; leaves up to 6 mm wide, dark green, firm, slightly scabrous along the margins, longer than the culms; sheaths tight, pale brown, the lowermost fibrous and red-tinged, the ligule longer than wide; spike usually solitary, sometimes 2–4, up

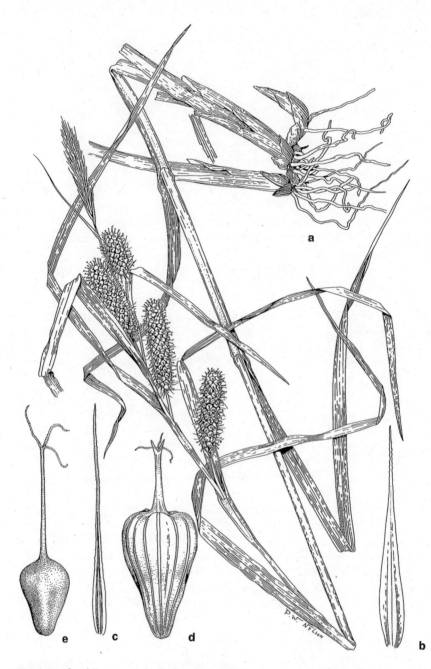

134. Carex frankii.
a. Habit.

b. Staminate scale.
c. Pistillate scale.

d. Perigynium.
e. Achene.

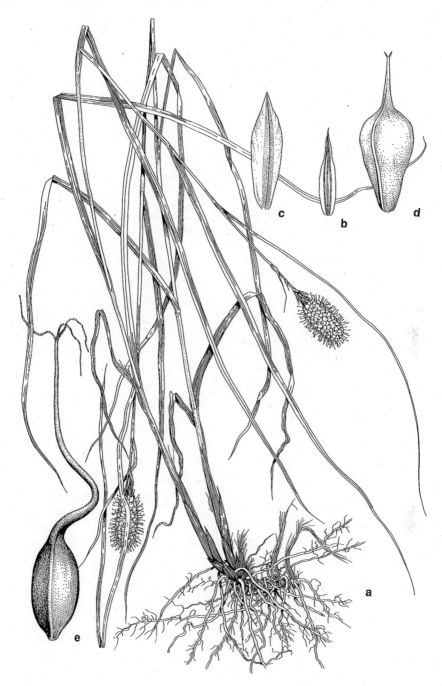

135. Carex squarrosa.
a. Habit.

b. Staminate scale.
c. Pistillate scale.

d. Perigynium.
e. Achene.

to 3 cm long, up to 2.2 cm thick, the terminal one subglobose to thick-cylindric, pistillate except for several staminate flowers at the base; staminate scales acute to awned, with a green center and hyaline margins; lateral spikes, if present, 2–4, entirely pistillate, subglobose to thick-cylindric; pistillate scales acute to acuminate to cuspidate, usually reaching the base of the beak of the perigynium; perigynia up to 100 or more per spike, obconic, rounded across the top, the lowermost reflexed, 4–5 mm long, up to 3 mm broad, pale green to pale brown, many-nerved, abruptly contracted into a bidentate beak to 3.5 mm long; achenes trigonous, 2.5–3.2 mm long, black, substipitate; stigmas 3.

Common Name: Squarrose Sedge.
Habitat: Low woods, swamps, along streams, wet meadows.
Range: Quebec to Ontario and Minnesota, south to eastern Nebraska, Arkansas, Tennessee, and North Carolina.
Illinois Distribution: Throughout the state, but apparently more common in the southern counties.

This species is similar in appearance to *C. typhina*, but the pistillate scales of *C. squarrosa* are acuminate to cuspidate, while those of *C. typhina* are obtuse. *Carex frankii*, which is somewhat similar, usually has a separated terminal staminate spike. I have seen two apparently anomalous specimens of *C. squarrosa* in Illinois that have a separate staminate spike. The flowers bloom from June through September.

136. Carex typhina Michx. Fl. Bor. Am. 2:169. 1803. Fig. 136.

Plants perennial, cespitose, from short rhizomes; culms to 1 m tall, triangular, nearly smooth, brown at the base; sterile shoots usually present; leaves up to 10 mm wide, dark green, firm, somewhat scabrous along the margins, longer than the culms; sheaths tight, pale brown, the lowermost red-tinged, the ligule longer than wide; spikes 2–5, up to 4 cm long, up to 1.7 cm thick, the terminal one thick-cylindric, gynecandrous; staminate scales obtuse, reddish brown with a green center; lateral spikes entirely pistillate, thick-cylindric; pistillate scales obtuse, or the uppermost acute, shorter than to just reaching the base of the beak of the perigynium; perigynia up to 100 or more per spike, obconic, rounded to nearly

truncate across the top, 4–5 mm long, up to 3 mm broad, none of them reflexed, pale green to pale brown, many-nerved, abruptly contracted into a bidentate beak to 3.5 mm long; achenes trigonous, 2.2–2.5 mm long, black, substipitate; stigmas 3.

Common Name: Sedge.
Habitat: Bottomland woods, swamps, wet meadows.
Range: Quebec to Wisconsin and Iowa, south to Louisiana and Georgia.
Illinois Distribution: Occasional in the southern half of the state, usually infrequent elsewhere.

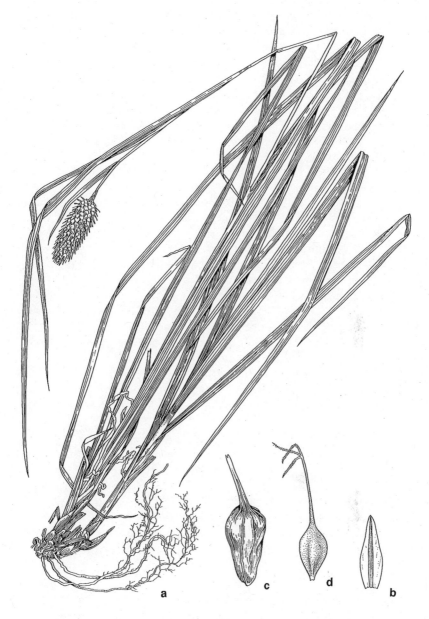

136. Carex typhina.
a. Habit.

b. Pistillate scale.

c. Perigynium.
d. Achene.

This species differs from the similar *C. squarrosa* in its several more slender spikes and its usually obtuse pistillate scales. It differs from *C. frankii* in its gynecandrous terminal spike. *Carex typhina* flowers from June through September.

137. Carex lacustris Willd. Sp. Pl. 4:306. 1805. Fig. 137.

Plants perennial, cespitose, with elongated rhizomes; culms to 1.3 m tall, stout, triangular, rough to the touch, at least beneath the inflorescence, usually purplish at the base; sterile shoots usually present; leaves up to 1.5 cm wide, glaucous at least below, firm, septate-nodulose, scabrous along the margins; lower sheaths red to purplish, bladeless, becoming fibrillose, the ligule much longer than wide; upper 2–5 spikes staminate, to 5 cm long, to 4 mm thick, sessile except for the uppermost; staminate scales obtuse to awned, reddish with usually hyaline margins; pistillate spikes 2–3, appressed, on short stalks, to 8 cm long, to 1.5 cm thick; pistillate scales lance-ovate, acute to acuminate, awned, reddish with hyaline margins and a green center, shorter than the perigynia; perigynia up to 150 per spike, lance-ovoid, 5–7 mm long, about 2 mm broad, ascending, coriaceous, glabrous, olive green, strongly nerved, tapering to a bidentate beak up to 1.5 mm long; achenes trigonous, 2.2–2.5 mm long, black; stigmas 3.

Common Name: Lake Sedge.
Habitat: Swampy woods, calcareous marshes, bogs, sometimes in standing water.
Range: Quebec and Nova Scotia to Manitoba, south to South Dakota, Iowa, Illinois, and Virginia; Idaho.
Illinois Distribution: Throughout the state but infrequent in the southern counties.

This stout species is very similar to *C. hyalinolepis*, but differs in its reddish or purple, bladeless, lower sheath and its strongly nerved perigynia.

The leaves of this species are glaucous, at least on the lower surface. There are 2–5 staminate spikes and 2–3 appressed pistillate spikes.

Several early Illinois botanists erroneously identified this species as *C. riparia*, a sedge that does not occur in Illinois. *Carex lacustris* flowers from mid-May to early September.

138. Carex hyalinolepis Steud. Syn. Cyp. 235. 1855. Fig. 138.
Carex lacustris Willd. var. *laxiflora* Dewey, Am. Journ. Sci. II, 35:60. 1863.
Carex riparia Curtis var. *impressa* S. H. Wright, Bull. Torrey Club 9:151. 1882.
Carex impressa (S. H. Wright) Mack. Bull. Torrey Club 37:236. 1910.

Plants perennial, cespitose, with elongated rhizomes; culms to 1.2 m tall, stout, triangular, rough to the touch, at least beneath the inflorescence, brown at the base; sterile shoots usually present; leaves up to 1.5 cm wide, glaucous, sometimes

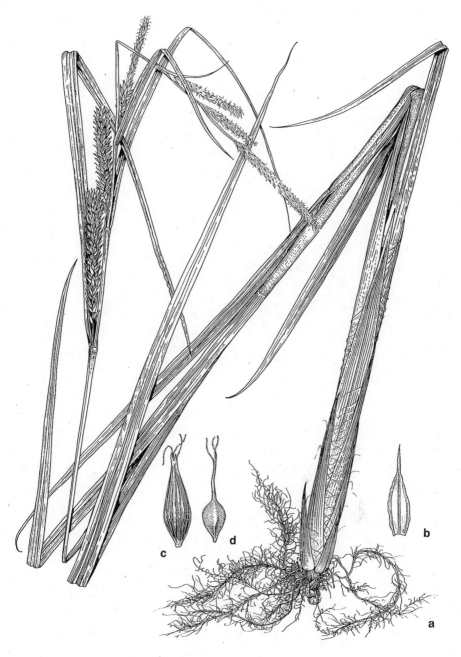

137. Carex lacustris.
a. Habit.

b. Pistillate scale.

c. Perigynium.
d. Achene.

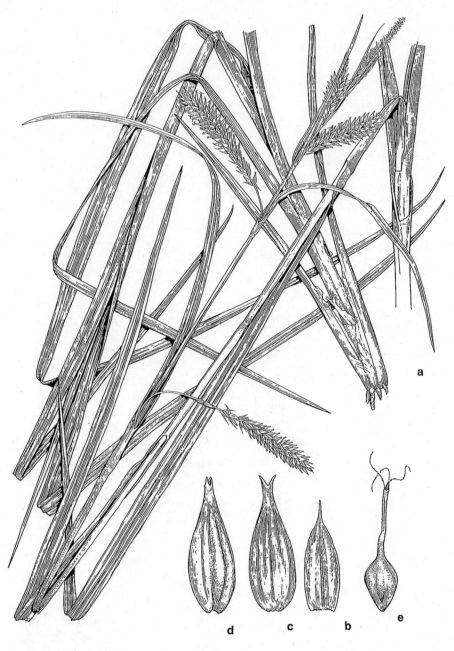

138. Carex hyalinolepis.
a. Habit.

b. Pistillate scale.

c. d. Perigynia.
e. Achene.

septate-nodulose, scabrous along the margins; lower sheaths white to pale brown, blade-bearing, rarely fibrillose, the ligule usually wider than long; upper 2–5 spikes staminate, up to 4 cm long, up to 5 mm thick, sessile except for the uppermost; staminate scales acute to short-awned, reddish brown with hyaline margins; pistillate spikes 2–4, appressed, up to 7.5 cm long, up to 1.5 cm thick; pistillate scales lance-ovate, acute to awned, reddish brown with hyaline margins, shorter than the perigynia; perigynia up to 150 per spike, ovoid to broadly ellipsoid, 6–8 mm long, 2.0–2.5 mm broad, ascending, coriaceous, glabrous, dull green, finely nerved, tapering to a bidentate beak up to 1 mm long; achenes trigonous, 2.0–2.5 mm long, black; stigmas 3.

Common Name: Sedge.
Habitat: Wet ditches, swamps.
Range: Ontario and New Jersey to central Illinois and Nebraska, south to Texas and Florida.
Illinois Distribution: Occasional to common in the southern half of the state, apparently absent elsewhere.

Although similar in appearance to *C. lacustris*, *C. hyalinolepis* differs in its blade-bearing, pale brown lower sheaths and its more finely nerved perigynia.

This species was called *C. impressa* for a number of years, but the binomial *C. hyalinolepis* has priority. *Carex hyalinolepis* flowers from mid-April to late July.

139. Carex atherodes Spreng. Syst. 3:827. 1826. Fig. 139.
Carex aristata R. Br. in Richards. Frankl. Journ. 751. 1823, non Honck. (1792).
Carex trichocarpa Muhl. var. *aristata* (R. Br.) Bailey, Bot. Gaz. 10:294. 1885.

Plants perennial, cespitose, from slender, long-creeping rhizomes; culms to 1.5 m tall, triangular, usually smooth to the touch, purplish at the base; sterile shoots usually present; leaves up to 1.2 cm wide, rather thin, dull green, septate-nodulose, scabrous above, sparsely hairy beneath; sheaths often pubescent, at least near the summit, the ventral band often brownish, the lower sheaths purple-tinged, becoming fibrillose, the ligule longer than wide; upper 2–5 spikes staminate, to 10 cm long, to 5 mm broad, sessile except for the terminal one; staminate scales acute, awned, ciliate, yellow-brown with hyaline margins; pistillate spikes 2–4, cylindric, appressed, sometimes with a few staminate flowers at the summit, up to 12 cm long, up to 15 mm broad; pistillate scales ovate, acute, awned, reddish brown with hyaline margins and a green center, shorter than the perigynia; perigynia up to 100 per spike, lance-ovoid, 7–9 mm long, about 2 mm wide, ascending, subcoriaceous, glabrous, pale brown or yellow-green, strongly nerved, tapering to a bidentate beak with teeth 1.2–3.0 mm long; achenes trigonous, yellow-brown, substipitate; stigmas 3.

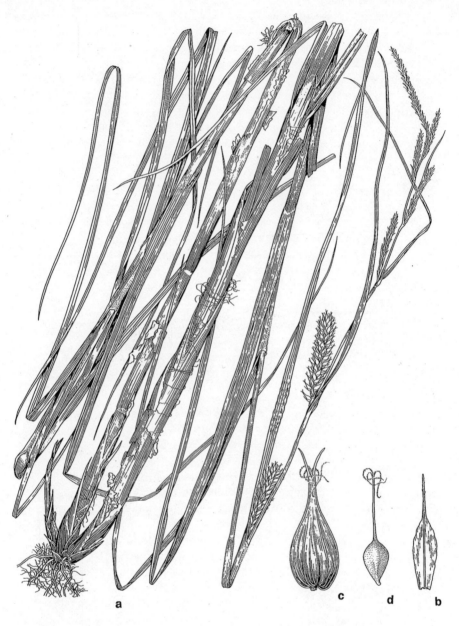

139. Carex atherodes.
a. Habit.

b. Pistillate scale.

c. Perigynium.
d. Achene.

Common Name: Sedge.
Habitat: Wet meadows, marshes.
Range: Ontario to Yukon, south to Oregon, Utah, Colorado, central Illinois, and New York; Maine.
Illinois Distribution: Occasional in the northern fourth of the state, less common in the central counties, apparently absent from the southern fourth of the state.

This species is readily distinguished by its several staminate spikes, its pubescent leaves and sheaths, and its strongly nerved perigynia with long-toothed beaks.

Carex atherodes often forms dense colonies. A few early Illinois botanists called this plant *C. trichocarpa* var. *aristata*. This species flowers from late May to mid-June.

140. **Carex heterostachya** Bunge, Enum. Pl. China Bor. 69. 1833. Fig. 140.
Carex X *fulleri* Ahles in Fell, Rhodora 58:318. 1956.

Plants perennial, cespitose, with long-creeping rhizomes; culms to 60 cm tall, triangular, scabrous at least beneath the inflorescence, brown at the base; sterile shoots usually present; leaves 1.5–3.2 mm wide, plicate, more or less glaucous, somewhat scabrous along the margins; sheaths tight, green to reddish brown, the inner band hyaline to pale brown, red-dotted, concave at the apex; upper 1–2 spikes staminate, up to 3 cm long, up to 3 mm thick; staminate scales acute to acuminate, reddish brown with hyaline margins and a green or brown center; pistillate spikes 1–4, ovate to short-cylindric, to 2 cm long, to 7.5 mm thick; pistillate scales ovate, acute to acuminate, awned, reddish brown with hyaline margins and a green center, shorter or longer than the perigynia; perigynia up to 30 per spike, ovoid, 3.0–4.5 mm long, 1.7–2.5 mm wide, coriaceous, glabrous, pale to dark brown, essentially nerveless, contracted to a bidentate beak 0.4–0.8 mm long, the teeth 0.2–0.3 mm long; achenes 2.0–2.8 mm long, trigonous, papillate, brown, with the lower part of the style persistent and bony; stigmas 3.

Common Name: Sedge.
Habitat: Gravel bluff (in Illinois).
Range: Adventive in Illinois; native to Asia.
Illinois Distribution: Known only from Winnebago County at north edge of Camp Grant.

This plant was first collected from a gravel bluff at Camp Grant, Winnebago County in 1949. After considerable deliberation, Ahles concluded that the plants represented an undescribed hybrid between *C. laeviconica* and *C. pensylvanica*, which he named *C.* X *fulleri*. Reznicek, however, has conclusively shown that the plants are actually *C. heterostachya*, an Asian species that apparently became adventive in Illinois at a military base shortly after World War II.

Carex heterostachya is related to *C. laeviconica* and other species of section

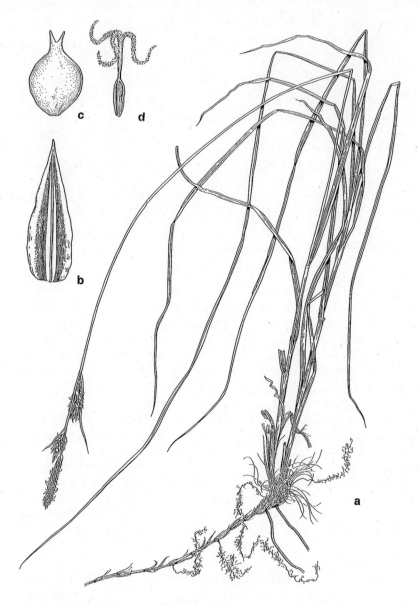

140. Carex heterostachya
a. Habit.

b. Pistillate scale.

c. Perigynium.
d. Achene.

Paludosae, but it has generally narrower leaves, reddish brown basal sheaths, and essentially nerveless perigynia. This species is known from no other place in North America. It flowers from late April to late July.

141. Carex laeviconica Dewey, Am. Journ. Sci. II, 24:47. 1857. Fig. 141.

Plants perennial, cespitose, with slender creeping rhizomes; culms up to 1.2 m tall, stout, triangular, usually somewhat scabrous beneath the inflorescence, purplish at the base; sterile shoots usually present; leaves up to 8 mm wide, dull green, glabrous, septate-nodulose, scabrous along the margins; sheaths tight, brown, the middle and upper ones cartilaginous, globose or slightly scabrous, the lower reddish, becoming fibrillose, the ligule as long as wide; upper 2–4 spikes staminate, up to 5 cm long, up to 4 mm thick, all but the uppermost sessile; staminate scales obtuse, usually awned, yellow-brown with hyaline margins; pistillate spikes 2–4, ascending, up to 7.5 cm long, up to 1.2 cm thick; bracts leaflike; pistillate scales ovate, acute and often awned, reddish brown with a hyaline margin, shorter than the perigynia; perigynia up to 50 per spike, ovoid, 5–7 mm long, ascending, coriaceous, yellow-green to stramineous, strongly nerved, tapering to a smooth or serrulate bidentate beak up to 1.6–2.0 mm long, the teeth 1–2 mm long; achenes trigonous, 2.2–2.5 mm long, brown; stigmas 3.

Common Name: Sedge.
Habitat: Wet prairies, marshes.
Range: Manitoba to Saskatchewan, south to Montana, Kansas, and Illinois.
Illinois Distribution: Occasional in the northern half of the state, rare elsewhere, and apparently absent from the southern fourth of Illinois.

This species is distinguished by the teeth of the perigynium being 1–2 mm long, the leaves being no more than 8 mm wide, and the ovoid perigynia being 5–7 mm long.

Illinois is at the easternmost limit of the range of this species. *Carex laeviconica* flowers during June and July.

142. Carex trichocarpa Muhl. ex Schkuhr, Nachtr. Riedgraeser. 47. 1806. Fig. 142.
Carex trichocarpa Muhl. var. *imberbis* Gray, Man. Bot., ed. 5, 597. 1867.

Plants perennial, cespitose, with slender rhizomes; culms to 1.2 m tall, triangular, scabrous, purplish at the base; sterile shoots usually present; leaves up to 8 mm wide, dull green, glabrous, septate-nodulose, scabrous along the margins; sheaths tight, green, septate, the ventral band reddish, the lowermost red and bladeless, the ligule as wide as long; upper 2–5 spikes staminate, up to 5 cm long, up to 4 mm thick, all sessile except sometimes the uppermost; staminate scales obtuse, awned, pale brown, with hyaline margins; pistillate spikes 2–4, elongate-cylindric, up to 9 cm long, up to 1.5 cm thick, the upper sessile, the lowest on a stiff, erect, scabrous peduncle; bracts leaflike; pistillate scales broadly ovate, acute to awned, red-

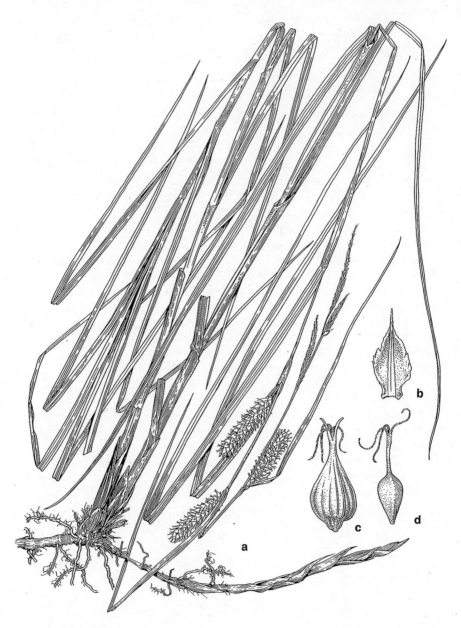

141. Carex laeviconica.
a. Habit.

b. Pistillate scale.

c. Perigynium.
d. Achene.

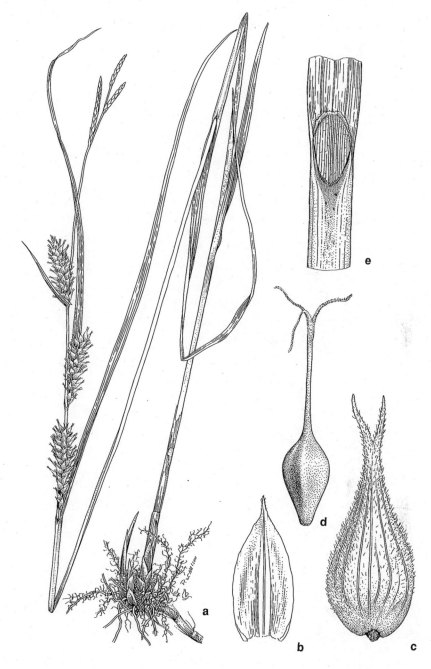

142. Carex trichocarpa.
a. Habit.
b. Pistillate scale.
c. Perigynium.
d. Achene.
e. Sheath with ligule.
f. Sterile shoot.

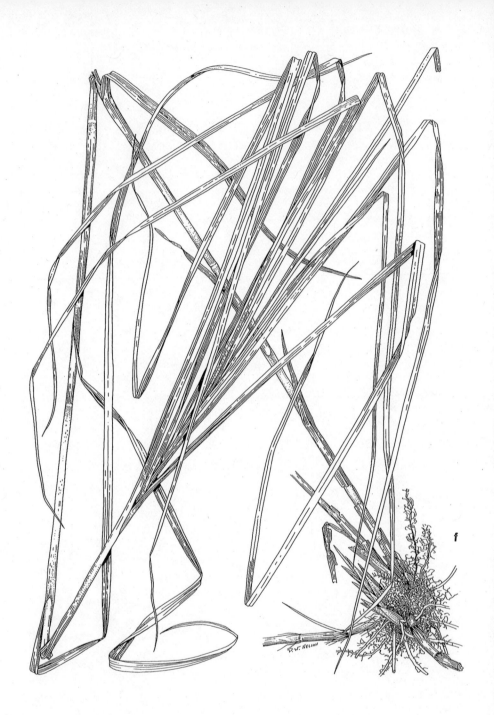

f

dish brown with hyaline margins, shorter than the perigynia; perigynia up to 40 per spike, ovoid, 5–10 mm long, 2.5–3.0 mm wide, ovoid, pubescent, strongly nerved, subcoriaceous, dull green or stramineous, tapering to a bidentate beak 3.0–3.5 mm long, the teeth 1.4–2.0 mm long; achenes trigonous, 2.2–2.5 mm long, yellowish, usually substipitate; stigmas 3.

Common Name: Sedge.
Habitat: Calcareous meadows, sloughs, seeps, marshes.
Range: Quebec to Ontario and Minnesota, south to northern Iowa, central Illinois, Ohio, Pennsylvania, and Delaware.
Illinois Distribution: Occasional to frequent in the upper half of the state; also Washington County.

Carex trichocarpa is distinguished by its several staminate spikes; its strongly nerved, pubescent perigynia; and the long teeth of the perigynial beaks. *Carex* X *subimpressa* is similar, even with its pubescent perigynia, but the nerves of that hybrid are obscure. This species flowers from June through August.

143. Carex X subimpressa Clokey, Rhodora 21:84. 1919. Fig. 143.
Carex impressa X *C. lanuginosa* Clokey, Torreya 16:199. 1916.

Plants perennial, cespitose, from long-creeping rhizomes; culms to 85 cm tall, triangular, scabrous beneath the inflorescence, with the remains of last year's leaves persistent; sterile shoots present; sheaths more or less tight, the inner band hyaline to pale brown, red-dotted, concave at the mouth; leaves up to 50 cm long, up to 10.5 mm wide, glabrous, plicate, scabrous along the margins; upper 2–4 spikes staminate, up to 6.8 cm long, up to 4.3 mm thick, sessile except for the uppermost; staminate scales stramineous to pale brown, awned; pistillate spikes 2–4, up to 8 cm long, up to 1.1 cm thick, the upper ones sessile, the lowest on a peduncle up to 9 cm long; lowest bract up to 4.5 cm long; pistillate scales stramineous to pale brown to purplish, acuminate, usually awned, usually shorter than the perigynia; perigynia up to 135 per spike, ovoid, 4.2–6.4 mm long, 1.8–2.7 mm wide, ascending to spreading, pubescent, brown, obscurely nerved, tapering to a bidentate beak 0.8–1.5 mm long, the teeth of the beak of the perigynium 1.2–2.0 mm long; achenes rarely formed, trigonous, obovoid, 1.5–2.2 mm long, pale brown; stigmas 3.

Common Name: Sedge.
Habitat: Marshes.
Range: Southeastern Michigan to eastern Kansas.
Illinois Distribution: Known only from Macon, Montgomery, and St. Clair counties.

The original collection of this plant was made by I. W. Clokey in 1915 from Macon County. It has subsequently been found in Montgomery

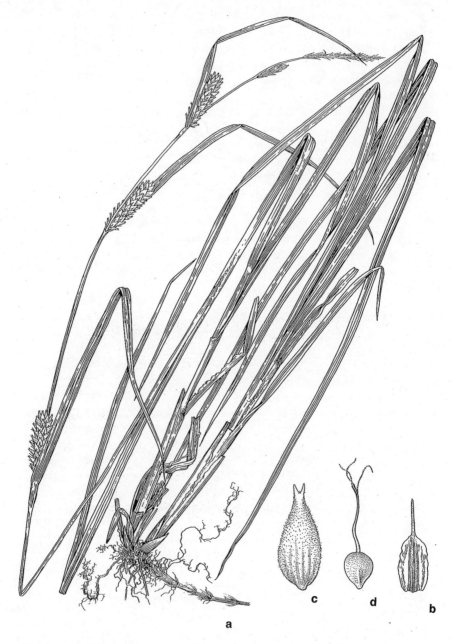

143. Carex X subimpressa.
a. Habit.

b. Pistillate scale.

c. Perigynium.
d. Achene.

and St. Clair counties. It is also known from a very few collections in Michigan, Indiana, Missouri, and Kansas. Clokey first described this plant as a hybrid between *C. impressa* (now *C. hyalinolepis*) and *C. lanuginosa* (now *C. pellita*), then later gave it the hybrid binomial *C. X subimpressa*. Most flowers do not produce achenes.

Carex X *subimpressa* is similar to *C. trichocarpa* but differs in its obscure nerves on the perigynia and its shorter teeth on the perigynial beak. This plant flowers during June.

144. Carex comosa Boott, Trans. Linn. Soc. 20:117. 1846. Fig. 144.
Carex pseudocyperus L. var. *americana* Hochst. ex Bailey, Mem. Torrey Club 1:54. 1869.

Plants perennial, densely cespitose, from short, stout rhizomes; culms to 1.5 cm tall, stout, triangular, usually scabrous beneath the inflorescence, pale brown at base with several of last year's leaves persistent; leaves up to 17 mm wide, septate-nodulose, glabrous, pale green, scabrous on the margins; sheaths pale yellow with hyaline margins, concave at the mouth, the ligule longer than wide; lower bracts leaflike; terminal spike 1, staminate, rarely partly pistillate, up to 7 cm long, up to 7 mm thick, sessile or nearly so; staminate scales awned, reddish brown with a darker center; lateral spikes 3–6, pistillate, up to 5.5 cm long, up to 1.7 cm thick, at least the lower ones pendulous or on flexuous but stout peduncles; pistillate scales narrowly lanceolate, serrulate-awned, reddish brown with a darker center, about half as long as the perigynia; perigynia up to 100 or more per spike, crowded, divaricately spreading to reflexed, lanceoloid, 5–7 mm long, up to 1.5 mm wide, not inflated, glabrous, strongly nerved, yellow-green, coriaceous, tapering to a bidentate beak 1.5–2.0 mm long, the teeth stiff, recurved-spreading, 1.2–2.0 mm long; achenes trigonous, 1.7–2.0 mm long, brown, continuous with the persistent, flexuous style; stigmas 3, pale brown.

Common Name: Porcupine Sedge.
Habitat: Swamps, boggy areas, wet ditches, pond margins.
Range: Nova Scotia and Quebec to Ontario, south to Minnesota, Louisiana, and Florida; Idaho to Washington, south to California.
Illinois Distribution: Frequent in the northern half of the state, becoming less common southward.

The common name refers to the prickly appearance of the pistillate spikes due to the recurved-spreading teeth of the perigynial beaks.

Carex comosa differs from the similar *C. hystericina* in having some of its perigynia reflexed in the spike and in having teeth of its perigynia that are 1.2–2.0 mm long. *Carex comosa* flowers from early May to late June.

145. Carex hystericina Muhl. ex Willd. Sp. Pl. 4:282. 1805. Fig. 145.

Plants perennial, densely cespitose, from short, stout rhizomes; culms to 1 m tall, stout, triangular, scabrous, at least beneath the inflorescence, purplish at the

144. Carex comosa.
a. Habit.

b. Pistillate scale.

c. Perigynium.
d. Achene.

145. Carex hystericina.
a. Habit.

b. Pistillate scale.

c. Perigynium.
d. Achene.

base, with last year's leaves persistent; sterile shoots present; leaves up to 10 mm wide, firm, scabrous on both surfaces, green, septate-nodulose, at least some of them overtopping the inflorescence; lower sheaths red, the lowest bladeless, the upper pale, greenish, septate-nodulose, the ligule as wide as long; terminal spike 1 or rarely 2, staminate, to 5 cm long, to 4 mm thick, pedunculate, usually subtended by a leaflike bract; staminate scales awned, reddish brown with a green center; lateral spikes 1–4, pistillate, elongate-cylindric, 2–5 cm long, 1.0–1.5 cm thick, with capillary, scabrous peduncles, the lowest spreading to pendulous; bracts leaflike; pistillate scales ovate, with long, serrulate awns, reddish brown with a green center, usually reaching at least to the base of the perigynium; perigynia up to 100 per spike, crowded, lance-ovoid to ovoid, 4–6 mm long, 1.5–2.0 mm thick, spreading but not reflexed, inflated, pale green to stramineous, strongly 15– to 20-nerved, abruptly tapering to a stout bidentate beak 1.8–2.2 mm long, the teeth of the beak 0.2–0.7 mm long; achenes 1.5–1.8 mm long, trigonous, brown, obovoid, continuous with the persistent, flexuous style; stigmas 3, reddish brown.

Common Name: Porcupine Sedge.
Habitat: Swamps, calcareous fens, wet ditches.
Range: Quebec to Alberta and Washington, south to California, Arizona, New Mexico, Texas, Tennessee, and New Jersey.
Illinois Distribution: Occasional to common in the northern two-thirds of the state, rare elsewhere.

Carex hystericina differs from the similar *C. comosa* in that none of the perigynia are reflexed. It differs from the somewhat similar *C. lurida* and *C. baileyi* in its narrower, lance-ovoid perigynia, its more numerous strong nerves on its perigynia, and its obovoid achenes. Most botanists in the past have spelled the epithet *hystricina*. *Carex hystericina* flowers from late May to early July.

146. Carex lurida Wahl. Sv. Vet. Akad. Nya. Handl. 24:153. 1803. Fig. 146.
Carex tentaculata Muhl. ex Willd. Sp. Pl. 4:266. 1805.

Plants perennial, densely cespitose, from short, stout rhizomes; culms to 1 m tall, stout, triangular, smooth or scabrous, purplish at the base, often with last year's leaves persistent; leaves up to 7 mm wide, glabrous, septate-nodulose, dull green, scabrous along the margins, some of them sometimes overtopping the culms; sheaths tight, pale or tan or the lower reddish, the ventral band weakly nerved, concave or truncate at the mouth, the ligule longer than wide; terminal spike 1, staminate, up to 7 cm long, up to 3 mm thick, sessile or short-pedunculate; staminate scales serrulate-awned, pale brown with hyaline margins; lateral spikes 1–4, pistillate, thick-cylindric to ovoid, 1.5–4.0 cm long, 1.5–2.0 cm thick, all but sometimes the lowest erect or ascending, on short, glabrous peduncles; bracts leaflike; pistillate scales linear, abruptly contracted to a serrulate awn, usually reaching the base of the beak of the perigynium; perigynia up to 100 or more per spike, crowded, 6–9 mm long, 3–4 mm thick, yellow-brown, membranous,

146. Carex lurida.
a. Habit.

b. Pistillate scale.
c. Perigynium.

d. Achene.
e. Sheath with ligule.

inflated, shiny, glabrous, strongly nerved, spreading to ascending but not reflexed, gradually tapering to a bidentate beak 3–4 mm long, the beak about as long as the body; achenes trigonous, 2.0–2.5 mm long, yellow-brown, substipitate, continuous with the persistent style; stigmas 3.

Common Name: Sedge.
Habitat: Swamps, wet woods, bogs, peaty fens, deep marshes.
Range: Nova Scotia to Ontario and Minnesota, south to Texas and Florida; Mexico.
Illinois Distribution: Common throughout the state.

This species differs from *C. comosa* in that none of its perigynia are reflexed. It differs from *C. hystericina* in its ovoid, thicker, inflated perigynia. It more closely resembles *C. baileyi* from which it differs in its more tapering perigynial beak and its thicker pistillate spikes.

A few early Illinois botanists erroneously called this plant *C. tentaculata. Carex lurida* flowers from May to July.

147. Carex baileyi Britt. Bull. Torrey Club 22:220. 1895. Fig. 147.

Plants perennial, densely cespitose, from short, stout rhizomes; culms to 75 cm tall, slender, triangular, smooth or scabrous, purplish at the base; leaves 2–4 mm wide, glabrous, septate-nodulose, green, scabrous along the margins, at least some of them overtopping the culms; sheaths tight, yellowish, the ventral band hyaline, concave or truncate at the mouth, the ligule as wide as long; terminal spike 1, staminate, up to 4 cm long, up to 3 mm thick, usually short-pedunculate; staminate scales usually serrulate-awned, stramineous; lateral spikes 1–2, pistillate, slender-cylindric, 1–4 cm long, 0.8–1.3 cm thick, erect or ascending on short, smooth peduncles; bracts leaflike; pistillate scales linear, serrulate-awned, yellow-brown with hyaline margins and a green center, nearly as long as the perigynia; perigynia up to 50 per spike, crowded, 5–7 mm long, 2.0–2.5 mm thick, yellow-brown, membranous, inflated, shiny, glabrous, strongly nerved, spreading to ascending but not reflexed, abruptly tapering to a bidentate beak 3–4 mm long, the beak as long as or longer than the body; achenes trigonous, 1.5–2.0 mm long, yellow-brown, substipitate, continuous with the persistent style; stigmas 3.

Common Name: Bailey's Sedge.
Habitat: Marshes.
Range: Quebec and New Hampshire to Michigan, south to southern Illinois, Tennessee, and Virginia.
Illinois Distribution: Known only from Jackson County.

This species is similar to *C. lurida*, but differs in its more slender pistillate spikes, its slightly narrower leaves, and its slightly smaller perigynia that abruptly taper into beaks that are as long as or longer than the bodies. *Carex baileyi* flowers from late May through August.

147. Carex baileyi.
a. Habit.

b. Pistillate scale.

c. Perigynium.
d. Achene.

148. Carex folliculata L. Sp. Pl. 978. 1753. Fig. 148.

Plants perennial, cespitose, from short, thick rhizomes; culms up to 1 m tall, slender, triangular, smooth, pale brown at the base, with last year's leaves persistent; leaves up to 18 mm wide, thin, lax, septate-nodulose, yellow-green; sheaths loose, pale, septate-nodulose, the ligule wider than long; terminal spike 1, staminate, to 2.5 cm long, to 3 mm thick, sessile or short-pedunculate; staminate scales acute, usually awned, yellow-brown with hyaline margins and a green center, the lowest bractlike; lateral spikes 2–5, pistillate, short-cylindric, up to 2.5 cm thick, on slender peduncles; bracts leaflike; pistillate scales acuminate, awned, yellow-brown with hyaline margins and a green center, three-fourths as long as or longer than the perigynia; perigynia up to 20 per spike, lanceoloid, 12–15 mm long, 2.5–3.5 mm thick, inflated, membranous, yellow-green, shiny, several-nerved, tapering to a bidentate, serrulate beak 3–5 mm long; achenes trigonous with deeply concave sides, 3.3–3.5 mm long, about 2 mm wide, yellow-brown, stipitate, continuous with the persistent style; stigmas 3.

Common Name: Sedge.
Habitat: Swamps.
Range: Newfoundland to Wisconsin, south to northern Illinois, Tennessee, and North Carolina.
Illinois Distribution: Apparently found once in Cook County but no longer known from Illinois.

This species is readily distinguished by its large, lanceoloid perigynia arranged in spikes that are about as wide as they are long.

There are no specimens of this plant that have been preserved from Illinois. Its inclusion in this work is based on a report by Pepoon that prior to 1926 it occurred at Bowmanville in Cook County. This station in north Chicago has since been destroyed. Since *C. folliculata* is such an easily recognized species, I have no doubt that Pepoon knew it when he saw it. *Carex folliculata* flowers from mid-May to mid-June.

149. Carex grayi Carey, Am. Journ. Sci. II. 4:22. 1847. Fig. 149.
Carex intumescens Rudge var. *globularis* Gray, Ann. Lyc. N. Y. 3:236. 1835.
Carex grayi Carey var. *hispidula* Bailey, Mem. Torrey Club 1:54. 1889.
Carex grayi Carey var. *rariflora* Farwell, Rep. Mich. Acad. 22:181. 1921.

Plants perennial, cespitose, growing in large clumps, with fibrous roots; culms up to 1 m tall, stout, much exceeded by upper leaves and bracts, triangular, scabrous beneath the spikes, light brown or green, purplish at base; leaves up to 30 cm long, 4.5–8.5 mm wide, tapering gradually to an attenuated tip, septate-nodulose, flat, dull green, the margins scabrous; sheaths concave or short-prolonged at the mouth, white-hyaline ventrally, the ligule usually wider than long; staminate spike single, linear, 1–7 cm long, up to 4 mm broad, borne on a scabrous peduncle; pistillate spikes 1–2, globose or subglobose, 1.5–4.5 cm long, 2.3–4.3 cm thick, erect, ses-

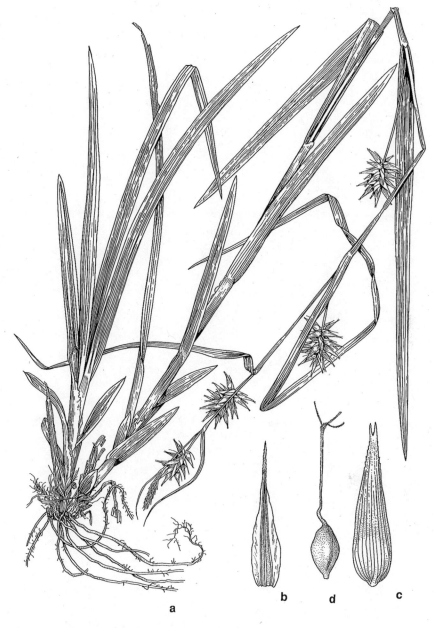

148. Carex folliculata.
a. Habit.

b. Pistillate scale.

c. Perigynium.
d. Achene.

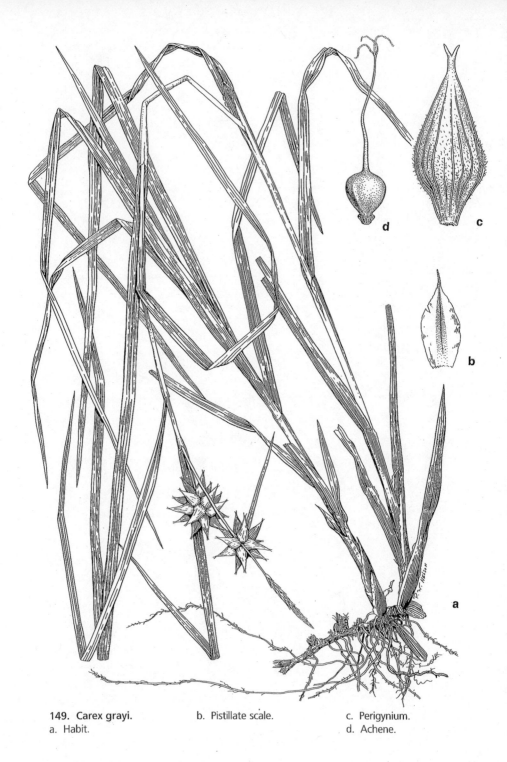

149. Carex grayi.
a. Habit.

b. Pistillate scale.

c. Perigynium.
d. Achene.

sile or on peduncles about as long as the spikes, densely flowered; bracts leaflike, much exceeding the culm; staminate scales oblong-obovate to lanceolate, obtuse to awned, light yellowish red with hyaline margins and a green center; pistillate scales lance-ovate to nearly orbicular, obtuse, often awned, shorter than the perigynia; perigynia 6–40 per spike, radiating at all angles from a central point; perigynia narrowly ovoid, 11.0–18.5 mm long, 3.0–6.5 mm broad, strongly inflated, glabrous to hispidulous, dull green, cuneate, sessile, with 15–20 strong nerves, tapering gradually to a beak a fourth or less the total length of the perigynium, the beak bidentate, with teeth 1–2 mm long, hispid within; achenes obovoid, ovoid, or subglobose, longer than wide to as long as wide, 3.0–4.5 mm long and 2.25–3.75 mm broad, the sides usually convex, rarely concave or flat, with rounded angles, golden brown at maturity, continuous with persistent lower half of the straight to slightly curved style; stigmas 3, slender, short, blackish.

Common Name: Gray's Sedge.
Habitat: Wet woods, in floodplains, along streams, wooded swamps, freshwater marshes, meadows.
Range: Quebec and Vermont to Michigan, Wisconsin, and Iowa, south to Arkansas, Mississippi, Alabama, and North Carolina.
Illinois Distribution: Throughout the state.

Within section Lupulinae, *C. grayi* and *C. intumescens* are very closely related. They both have achenes with rounded angles and flat to convex sides, perigynia with beaks that are at most only a third the length of the entire perigynia, and pistillate spikes that are about as wide as they are long. *Carex grayi* may be distinguished from *C. intumescens* by the cuneate base of the perigynium and the perigynia radiating at all angles from the center of the pistillate spike.

Variation exists in the surface of the perigynium, with some of the perigynia glabrous and others hispidulous. This difference does not seem to merit recognition as a variety. The flowers bloom from mid-May through September.

150. **Carex intumescens** Rudge, Trans. Linn. Soc. 8:97. 1804. Fig. 150.
Carex folliculata L. var. *major* Pursh, Fl. Am. Sept. 1:42. 1814.
Carex intumescens Rudge var. *fernaldii* Bailey, Bull. Torrey Club 20:418. 1893.

Plants perennial, cespitose, growing in large clumps, from fibrous roots, with spreading rhizomes; culms up to 80 cm tall, slender, much exceeded by the upper leaves and bracts, triangular, scabrous beneath the spikes, light brown or green, becoming purplish at the base; leaves up to 30 cm long, 3–8 mm wide, tapering gradually to an attenuated tip, septate-nodulose, flat, dull green, the margins scabrous; sheaths short-prolonged at the mouth, white-hyaline ventrally, the ligule usually wider than long; staminate spike 1, linear, 1–7 cm long, 1–4 mm broad, borne on a scabrous peduncle; bract longer than the staminate spike; pistillate spikes 1–4, globose or subglobose, 1–3 cm long, 0.5–3.3 cm broad, erect, sessile or on peduncles about as long as the spikes, loosely flowered; bracts leaflike, exceeding the

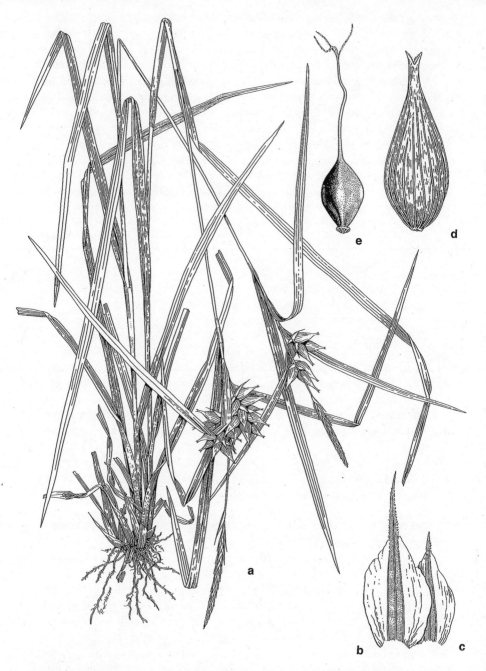

150. Carex intumescens.
a. Habit.

b, c. Pistillate scales.

d. Perpgynium.
e. Achene.

inflorescence, sheathing, the sheaths with short extensions prolonged at the mouth beyond base of blade; staminate scales loose, lanceolate, obtuse to cuspidate, light yellowish red with hyaline margins and a green center; pistillate scales ovate to ovate-lanceolate, awned or awnless, hyaline with a several-nerved green center, shorter than the perigynia; perigynia 1–20 per spike, diverging from the axis of the pistillate spike at a combination of angles rarely including retrorse; perigynia ovoid to ovoid-lanceoloid, 9.0–17.5 mm long, 2.5–6.5 mm broad, strongly to barely inflated, glabrous, shining, green, membranous, rounded at base, sessile, with 15–20 strong nerves, tapering gradually to a poorly defined beak up to a fourth the length of the total perigynia, the beak bidentate, with teeth 1 mm long, hispid within; achenes obovoid to ovoid, longer than wide to as long as wide, 3–6 mm long, 2.0–3.5 mm broad, the sides usually flat but sometimes concave or convex, with rounded angles, golden brown at maturity, continuous with the persistent lower half of the straight to fully coiled style; stigmas 3, slender, short, blackish.

Common Name: Sedge.
Habitat: Wet woods, swamps, marshes, bogs.
Range: Newfoundland to Ontario and Minnesota, south to southeastern Oklahoma, eastern Texas, and central Florida.
Illinois Distribution: Scattered in Illinois, but rare.

Carex intumescens is closely related to *C. grayi,* but the perigynia of *C. intumescens* are rounded at the base rather than cuneate. In addition, rays of the pistillate spikes radiate in all directions except retrorsely, while the rays of *C. grayi* also radiate retrorsely.

Specimens with barely inflated, narrowed perigynia have been called var. *fernaldii* Bailey. They do not seem worthy of recognition. The flowers bloom from mid-May to late September.

151. Carex louisianica Bailey, Bull. Torrey Club 20:428. 1893. Fig. 151.
Carex halei Carey in Chapm. Fl. S. U. S. 543. 1860, non Dewey (1846).

Plants perennial, from fibrous roots, with long connected rhizomes; culms up to 70 cm tall, slender, triangular, smooth throughout, exceeded by the upper leaves and bracts, light brown or green, becoming purplish red at the base; leaves up to 30 cm long, 2.5–6.0 mm wide, tapering gradually to an attenuated tip, septate-nodulose, flat, dull green, with the margins rough and slightly revolute; sheaths concave or short-prolonged at the mouth, hyaline ventrally, the ligule usually longer than wide; staminate spikes 1–2, linear, 1.5–6.4 cm long, 2–3 mm broad, on peduncles 3–10 cm long; bracts short; pistillate spikes 1–4, oblongoid to subovoid, 1.1–3.9 cm long, 1.3–3.0 cm broad, erect, sessile or on peduncles up to 12.5 cm long, loosely flowered; bracts leaflike, exceeding the inflorescence, sheathing, the sheaths short-prolonged and convex at the mouth; staminate scales stramineous, with hyaline margins and a several-nerved green center; perigynia 2–46 per spike, diverging from the axis of the pistillate spike at an angle of 30–90 degrees; pistillate scales ovate to lanceolate, acute to acuminate, stramineous, with a sev-

151. Carex louisianica.
a. Habit.

b. Pistillate scale.

c. Perigynium.
d. Achene.

eral-nerved green center and hyaline margins, much shorter and narrower than the perigynia; perigynia subuloid to ovoid, 9–14 mm long, 3.0–5.5 mm broad, strongly inflated, glabrous, dull green, sessile, rounded at the base, with 15–20 nerves gradually contracted into a beak half the total length of the perigynium, the beak bidentate, with the teeth 0.5–1.5 mm long, glabrous or nearly so within; achenes rhomboid to obovoid, longer than wide to as wide as long, 2.0–3.5 mm long, 1.25–2.50 mm broad, the sides concave to flat, with obtuse to rounded angles, tapering to the base, contracted into and continuous with the flexuous or completely coiled, persistent style; stigmas 3, slender, blackish, short.

Common Name: Louisiana Sedge.
Habitat: Wet woods, wooded swamps, floodplains, meadows.
Range: Southern Virginia to southern Indiana, southern Illinois, southeastern Missouri, and southeastern Oklahoma, south to eastern Texas, northwestern Florida, and Georgia; also District of Columbia.
Illinois Distribution: Confined to extreme southern Illinois; also Wabash County.

Carex louisianica appears to be intermediate between *C. grayi* and *C. intumescens* on the one hand and *C. lupulina*, *C. lupuliformis*, and *C. gigantea* on the other. It differs from *C. grayi* and *C. intumescens* in being smooth on the inner faces of the teeth of the bidentate beak of the perigynium. It differs from *C. gigantea* in its shorter beak of the perigynium, from *C. lupuliformis* in its rounded or obtuse angles of the achene, and from *C. lupulina* in its more loosely flowered pistillate spike and its narrower leaves. Of all the species in section Lupulinae, *C. louisianica* tends to be the smallest in stature and usually grows singly or with only a few culms together. This species flowers from late May to mid-October.

152. Carex lupulina Willd. Sp. Pl. 4:266. 1805. Fig. 152.
Carex lupulina Willd. var. *pedunculata* Gray in Beck, Bot. U.S. 438. 1833.
Carex canadensis Dewey, Am. Journ. Sci. II. 41:229. 1866.
Carex lupulina Muhl. var. *androgyna* Wood, Bot. & Fl. 376. 1870.
Carex lupulina Muhl. var. *longipedunculata* Sart. ex Dudley, Bull. Cornell Univ. 2:119. 1886.
Carex gigantea Rudge var. *lupulina* (Muhl.) Farwell, Ann. Rep. Comm. Parks Detroit II:50. 1900.
Carex lupulina Muhl. var. *albomarginata* Sherff, Bull. Torrey Club 38:482. 1911.

Plants perennial, cespitose, from fibrous roots, with long connected rhizomes; culms up to 1.26 m tall, stout, much exceeded by the upper leaves and bracts, triangular, smooth throughout, light brown or green, becoming purplish near the base; leaves up to 60 cm long, 4.5–8.5 mm wide, tapering gradually to an attenuated tip, septate-nodulose, flat, dull green, with the margins rough; sheaths short-prolonged at the mouth, white-hyaline ventrally, the ligule usually longer than wide; staminate spike 1, rarely 2–3, linear, 2–9 cm long, 2.0–4.5 mm broad, nearly sessile to

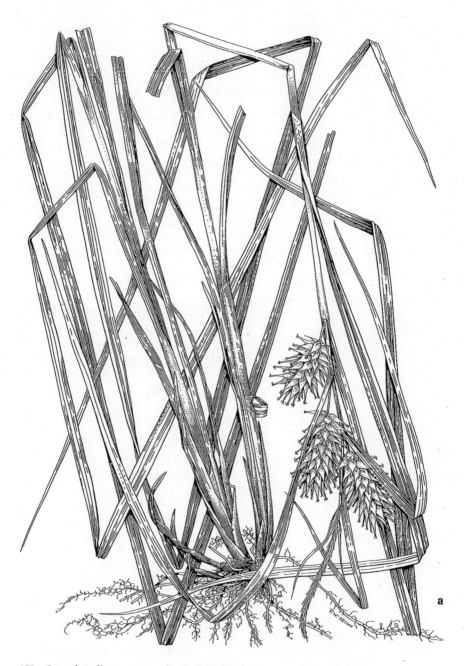

a

152. Carex lupulina.
a. Habit.

b. Pistillate scale.

c. Perigynium.
d. Achene.

long peduncled; bracts somewhat shorter than to much exceeding the spike; pistillate spikes 1–6, oblongoid to oblongoid-cylindric, 1.9–7.0 cm long, 1.2–3.0 cm broad, erect, sessile or with peduncles up to 17.2 cm long, densely flowered; bracts leaflike, exceeding the inflorescence, sheathing; staminate scales linear to lanceolate, acuminate to aristate, stramineous, with hyaline margins and a strongly nerved green center; perigynia diverging from the axis of the pistillate spike at an angle of 30–60 degrees; pistillate scales lanceolate, rough-awned or acuminate, straw-colored with a several-nerved green center and hyaline margins, much narrower than and usually shorter than the perigynia; perigynia subuloid, 10–17 mm long, 2.5–5.0 mm broad, rounded at the base, strongly inflated, glabrous, green or brownish yellow at maturity, sessile or short-stipitate, with 18–22 nerves, tapering to a beak half or slightly less than half the total length of the perigynia, the beak bidentate, with the teeth up to 2 mm long, smooth within; achenes rhomboid to trulloid, longer than wide to as long as wide, 2.0–4.5 mm long, 1.25–3.00 mm broad, the sides usually concave, occasionally flat, with obtuse to acute angles and a slight suggestion of a knob, tapering to the base, tapering into and continuous with the persistent abruptly bent or fully coiled style; stigmas 3, short, slender, blackish.

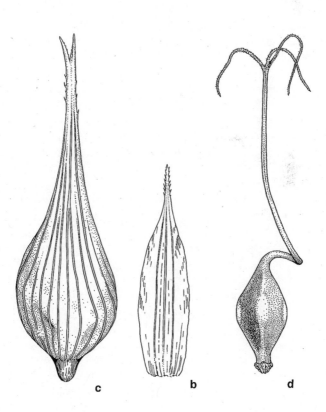

c   b   d

Common Name: Thick Sedge.
Habitat: Wet woods, wooded swamps, meadows, wet prairies, bogs, roadside ditches.
Range: Nova Scotia to Ontario and Minnesota, south to eastern Kansas, eastern Oklahoma, southeastern Texas, and Florida.
Illinois Distribution: Throughout the state.

Carex lupulina is distinguished by its large pistillate spikes that are longer than they are broad; its large, beaked, crowded perigynia; and its achenes that usually have blunt angles and slightly concave to flat sides.

There is considerable variation in the length of the peduncle of the pistillate spikes. Although some of these differences have been given nomenclatural standing, there is too much overlap to recognize these variations.

Some pistillate spikes were found with staminate flowers at the apex. The flowers bloom from June through October.

153. Carex lupuliformis Sartw. ex Dewey, Am. Journ. Sci. II. 9:29. 1850. Fig. 153.
Carex lupulina Muhl. var. *polystachia* Schw. & Torr. Ann. Lyc. N. Y. 1:337. 1825.
Carex lurida Wahlenb. var. *polystachya* (Schw. & Torr.) L. H. Bailey, Proc. Am. Acad. 22:63. 1886.
Carex eggertii L. H. Bailey, Bot. Gaz. 21:6. 1896.

Plants perennial, cespitose or growing singly, from fibrous roots, with long, slender, connected rhizomes; culms up to 1.06 m tall, stout, much exceeded by the upper leaves and bracts, triangular, smooth throughout, light brown or green, becoming purplish at base; leaves up to 60 cm long, 5.0–11.5 mm wide, tapering gradually to an attenuated tip, septate-nodulose, flat, dull green, with rough margins; sheaths short prolonged at mouth, white-hyaline ventrally; staminate spikes 1–3, linear, 2.0–10.5 cm long, 2–6 mm broad, on short or long peduncles, the scales linear to lanceolate to ovate, strongly awned to acuminate, stramineous with a several-nerved green center with hyaline margins; bracts shorter to much exceeding the spikes; pistillate spikes 2–7, oblong-cylindric, 1.5–9.6 cm long, 1.5–3.5 cm broad, erect, densely flowered, sessile or on peduncles up to 9 cm long; bracts leaflike, exceeding the inflorescence, with sheaths strongly prolonged at the mouth; perigynia 10–87, diverging from the axis of the pistillate spike at an angle of 30–90 degrees; pistillate scales lanceolate, strongly awned to acute, stramineous, with a several-nerved green center and hyaline margins, much narrower than and usually shorter than the perigynia; perigynia subuloid, 10–19 mm long, 3.0–5.5 mm broad, strongly inflated, glabrous, dull green or brownish yellow at maturity, sessile, rounded at the base, with 18–30 nerves, tapering to a beak half or slightly less than half the length of the total perigynia, the beak bidentate, with teeth up to 2 mm long, smooth within; achenes rhomboid to obtrulloid, wider than long to longer than wide, 2–4 mm long, 2.0–3.5 mm broad, the sides usually concave, occasionally flat to slightly convex, with the angles acute, upturned, or with well defined knobs, tapering to the base, tapering into and continuous with the persistent abruptly bent or fully coiled style; stigmas 3, short, slender, blackish.

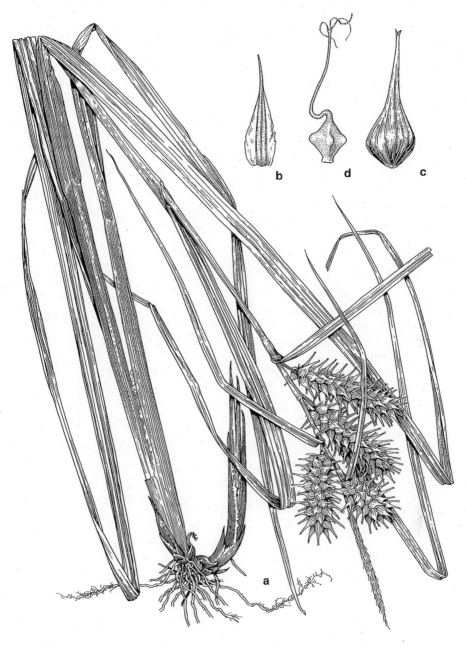

153. Carex lupuliformis.
a. Habit.

b. Pistillate scale.

c. Perigynium.
d. Achene.

Common Name: Sedge.

Habitat: Wet woods, wooded swamps, marshes, meadows, roadside ditches.

Range: Vermont to Ontario and Wisconsin, south to Missouri, Illinois, Indiana, Ohio, and Delaware; also uncommon in Virginia, North Carolina, Florida, Alabama, Arkansas, and Louisiana.

Illinois Distribution: Scattered throughout the state.

There is apparent difficulty in distinguishing *Carex lupuliformis* from *C. lupulina.* The most reliable difference is in the achenes that have the angles knobbed or acute and upturned and with usually concave sides. Plants from the southern United States, differing by achenes which appear to be 6-sided, have sometimes been called *C. eggertii,* but Hoffman (1974) does not believe this variation deserves nomenclatural recognition. The flowers bloom from late June through October.

154. **Carex gigantea** Rudge, Trans. Linn. Soc. 7:99. 1804. Fig. 154.
Carex lacustris Willd. var. *gigantea* (Rudge) Pursh, Fl. Am. Sept. 1:45. 1814.
Carex grandis L. H. Bailey, Mem. Torrey Club 1:13. 1889.
Carex gigantea Rudge var. *grandis* (L. H. Bailey) Farwell, Rhodora 23:87. 1921.
Carex lupulina Muhl. var. *gigantea* (Rudge) Britt. Mem. Torrey Club 5:84. 1894.

Plants perennial, cespitose or growing singly, from fibrous roots, with long, connected rhizomes; culms up to 1.06 m tall, stout, much exceeded by the upper leaves and bracts, triangular, smooth throughout, light brown or green, becoming purplish near the base; leaves up to 60 cm long, 5–14 mm wide, tapering gradually to an attenuated tip, septate-nodulose, flat, dull green, with margins rough; sheaths little or not at all prolonged at the mouth, yellowish tinged to white-hyaline ventrally; staminate spikes 1–7, linear, 1.1–9.0 cm long, 1.5–4.0 mm broad, borne on short or long peduncles, with scales linear to lanceolate to obovate, strongly awned to acuminate, stramineous with a several-nerved green center and hyaline margins; bracts half as long to exceeding the spikes; pistillate spikes 2–6, oblongoid-cylindric, 1–8 cm long, 1.7–3.0 cm broad, erect, densely flowered, sessile or on peduncles up to 15 cm long; bracts leaflike, exceeding the inflorescence, with sheaths short-prolonged at the mouth; perigynia 9–92, diverging from the axis of the pistillate spikes at angles of 60–90 degrees; pistillate scales lanceolate, acuminate, or the lower awned, stramineous with a several-nerved green center and hyaline margins, much narrower than and usually shorter than the perigynia; perigynia subuloid, 8–16 mm long, 2–5 mm broad, round-truncate at the base, barely inflated, glabrous, deep green or brownish yellow at maturity, with 16–20 nerves, abruptly contracted or tapering into a well defined beak usually more than half the length of the total perigynia, the beak bidentate, with teeth up to 1.5 mm long, smooth within; achenes depressed, obovoid to ovoid, wider than long to as wide as long, 1.5–2.5 mm long, 2.0–3.5 mm broad, the sides deeply concave, the angles obtuse and flattened into broad winglike projections, abruptly contracted into a stipitate base, abruptly contracted into and continuous with the persis-

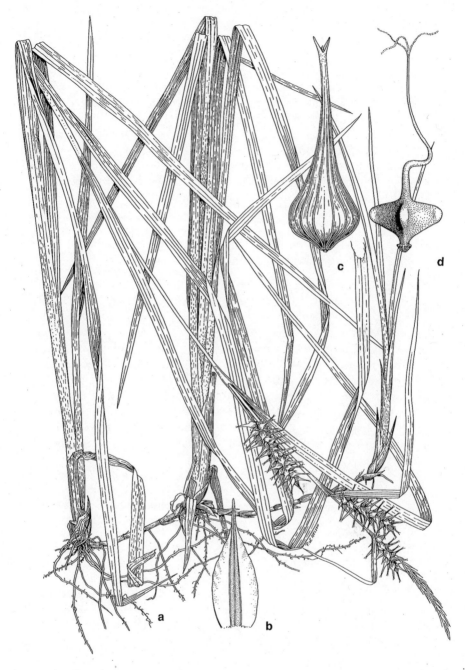

154. Carex gigantea.
a. Habit.

b. Pistillate scale.

c. Perigynium.

d. Achene.

tent, slender, barely flexuous to completely coiled style; stigmas 3, short, slender, blackish.

Common Name: Sedge.
Habitat: Wet woods, swampy woods, meadows.
Range: Southeastern Virginia to southern Illinois to southeastern Missouri, south to southeastern Texas and Florida.
Illinois Distribution: Known only from Jackson, Johnson, Pulaski, and Union counties.

The distinctive feature of this species is the achenes that have the angles flattened into broad wings and the sides deeply concave. The flowers bloom from late May through September.

155. Carex retrorsa Schw. Ann. Lyc. N. Y. 1:71. 1824. Fig. 155.

Plants perennial, cespitose, from short rhizomes; culms to 1 m tall, stout, triangular, smooth or usually slightly scabrous beneath the inflorescence; leaves up to 10 mm wide, lax, septate-nodulose, dark green, longer than the culms; sheaths loose, septate-nodulose, pale green to brownish, the lowermost often red-tinged, the ligule as long as or longer than wide; upper 1–4 spikes staminate, occasionally with a few perigynia near the base, up to 6 cm long, up to 4 mm thick, sessile to short-pedunculate; staminate scales obtuse to acute, yellow-brown with a green center; lower 3–6 spikes pistillate, thick-cylindric, crowded except for the lowermost, to 8 cm long, to 2 cm thick; lowest bracts much longer than the inflorescence; pistillate scales lanceolate, acuminate, yellow-brown or purplish with a green center, much shorter than the perigynia; many per spike, crowded, spreading to reflexed, 7–10 mm long, 2.5–3.5 mm thick, yellow-green to golden brown, the body suborbicular, inflated, membranous, shiny, strongly nerved, tapering to a smooth, bidentate, green beak 2.0–3.5 mm long; achenes trigonous, 2.2–2.5 mm long, 1.2–1.5 mm wide, brown, stipitate, continuous with the persistent, twisted or bent style; stigmas 3.

Common Name: Sedge.
Habitat: Stream terraces, alluvial woods.
Range: Quebec to British Columbia, south to Washington, Colorado, Illinois, Pennsylvania, and New Jersey.
Illinois Distribution: Confined mostly to the northern half of Illinois; also Coles, Richland, and Union counties.

This coarse sedge is recognized by its crowded pistillate spikes that bear some perigynia that are reflexed. The lower bracts are much longer than the inflorescence. Although this is primarily a northern species, there is one record from Union County in the extreme southern end of the state. Carex retrorsa flowers from late May to early July.

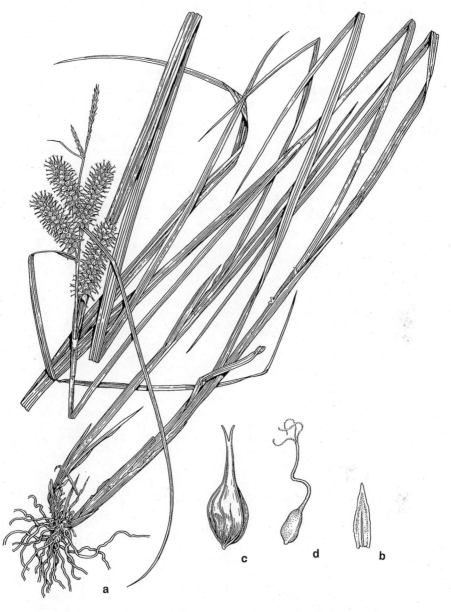

155. Carex retrorsa.
a. Habit.

b. Pistillate scale.

c. Perigynium.
d. Achene.

156. Carex utriculata Boott in Hook. F. Bor. Am. 2:221. 1839. Fig. 156.
Carex rostrata Stokes var. *utriculata* (Boott) Bailey, Proc. Am. Acad. 22:67. 1886.

Plants perennial, cespitose, from long, stout rhizomes; culms up to 1.2 m tall, triangular, stout, usually not scabrous, pale brown or sometimes purplish at the base, with last year's leaves persistent; leaves 6–12 mm wide, firm, septate-nodulose, smooth or scabrous along the margins, yellow-green, usually as long as or longer than the culms; upper sheaths septate-nodulose except for the hyaline ventral band, the lower sheaths spongy, pale brown to red-tinged, the ligule wider than long; staminate spikes 2–4, to 6 cm long, to 3 mm thick, pedunculate; staminate scales acute, yellow-brown with hyaline margins; pistillate spikes 2–5, thick-cylindric to ovoid, to 15 cm long, to 1.5 cm thick, sessile or the lowest one pedunculate; bracts leaflike, longer than the inflorescence; pistillate scales linear-lanceolate to narrowly ovate, acuminate, sometimes awned, purplish brown with hyaline margins, as long as or longer than the perigynia; perigynia up to 150 per spike, ovoid, 4–7 mm long, pale yellow to golden brown, inflated, few-nerved, shiny, contracted to a smooth, bidentate beak 1–2 mm long; achenes trigonous, 2.0–2.2 mm long, yellow-brown, substipitate, continuous with the twisted or bent style; stigmas 3.

Common Name: Beaked Sedge.
Habitat: Marshes, bogs, sometimes in standing water.
Range: Greenland and Labrador to British Columbia, south to California, New Mexico, South Dakota, northern Illinois, eastern Tennessee, and West Virginia.
Illinois Distribution: Confined to northeastern Illinois; also St. Clair County.

Many botanists consider our plants to be a variety of *C. rostrata*, a more northern species that has narrower leaves and shorter, awnless, pistillate scales.

Carex utriculata differs from *C. retrorsa* in the lack of reflexed perigynia, and from *C. tuckermanii*, *C. oligosperma*, and *C. vesicaria* in its spongy-based culms and its smooth or only slightly scabrous culms. It also differs from *C. tuckermanii* and *C. vesicaria* in its long, deep rhizomes. This species flowers during May and early June.

157. Carex tuckermanii Boott ex Dewey, Am. Journ. Sci. 49:48. 1845. Fig. 157.

Plants perennial, loosely cespitose, with short, stout rhizomes; culms to 1 m tall, triangular, slender, usually scabrous beneath the inflorescence, purplish red at the base; sterile shoots present; leaves 3–5 mm wide, septate-nodulose, dark green, scabrous along the margins; sheaths septate-nodulose, the lower reddish, the ventral band nerveless, the ligule as wide as long; staminate spikes 2–3, to 5 cm long, to 2.5 mm thick, stramineous, separated, sessile or short-pedunculate; staminate scales obtuse, yellow-brown to purple-brown, with hyaline margins and a green center; pistillate spikes 2–3, thick-cylindric, up to 6 cm long, up to 1.8 cm thick, sessile or on slender peduncles; bracts leaflike; pistillate scales broadly lanceolate,

156. Carex utriculata.
a. Habit.

b. Pistillate scale.

c. Perigynium.
d. Achene.

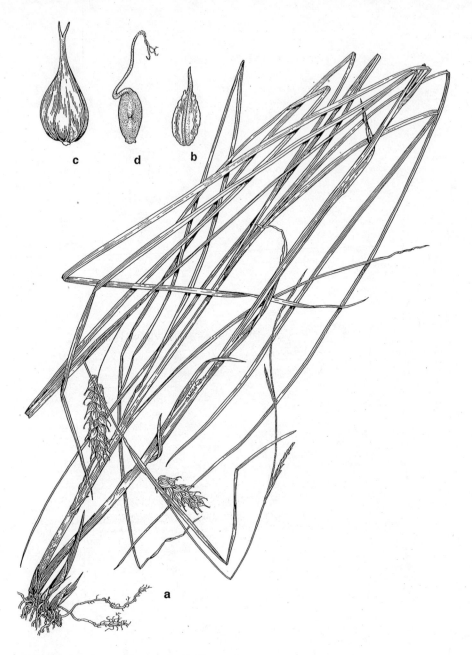

157. Carex tuckermanii.
a. Habit.

b. Pistillate scale.

c. Perigynium.
d. Achene.

acuminate, often aristate, yellow-brown to purple-brown, with hyaline margins and a green center, just reaching the base of the beak of the perigynium; perigynia up to 30 per spike, 7–10 mm long, 4.5–6.5 mm wide, the body suborbicular to broadly ovate, membranous, inflated, golden brown, shiny, strongly few-nerved, tapering to a smooth, bidentate beak 2.5–3.0 mm long, the teeth 1–2 mm long; achenes trigonous, 3–4 mm long, 2.0–2.5 mm wide, yellowish, deeply grooved on one side, substipitate, continuous with the persistent, twisted or bent style; stigmas 3.

Common Name: Tuckerman's Sedge.
Habitat: Upland depressions in wet savannas.
Range: Quebec to Ontario and Minnesota, south to Iowa, northern Illinois, Ohio, Pennsylvania, and New Jersey.
Illinois Distribution: Confined to the northeastern corner of Illinois; also Hancock County.

This species is readily distinguished by its inflated, bladderlike perigynia and its achenes that are deeply grooved on one side. *Carex tuckermanii* flowers from June to early April.

158. Carex oligosperma Michx. Fl. Bor. Am. 2:174. 1803. Fig. 158.

Plants perennial, loosely cespitose, with long, slender, deep rhizomes; culms to 1 m tall, somewhat triangular, firm, filiform, scabrous at least beneath the inflorescence, purple-red at the base, with last year's leaves persistent; leaves filiform, involute, to 40 cm long, to 3 mm wide, firm, septate-nodulose, scabrous along the margins; sheaths tight, green to pink-tinged, the ligule wider than long; terminal spike 1, staminate, to 4 cm long, to 1.5 mm thick, on a scabrous peduncle; staminate scales obtuse, yellow-brown with hyaline margins; lateral spikes 1–3, pistillate, short-cylindric to ovoid, to 2 cm long, to 1 cm thick, sessile; lowest bract leaflike; pistillate scales broadly ovate, acute to cuspidate, brown with hyaline margins and a green center, shorter than the perigynia; perigynia up to 15 per spike, ovoid, 4–7 mm long, 2.5–3.0 mm wide, inflated, subcoriaceous, shiny, yellow-green, strongly nerved, tapering to a smooth, bidentate beak 1–2 mm long; achenes trigonous, 2.8–3.0 mm long, about 2 mm wide, brown, substipitate, continuous with the style; stigmas 3.

Common Name: Few-seeded Sedge.
Habitat: Bogs.
Range: Labrador to Mackenzie, south to Minnesota, northern Illinois, northern Ohio, Pennsylvania, and Massachusetts.
Illinois Distribution: Confined to the northeastern corner of Illinois.

This species is readily identified by its involute-filiform leaves; its inflated perigynia (up to fifteen); its slender staminate spike; and its few, widely separated pistillate spikes. This species is confined to sphagnum bogs. *Carex oligosperma* flowers from late May to mid-June.

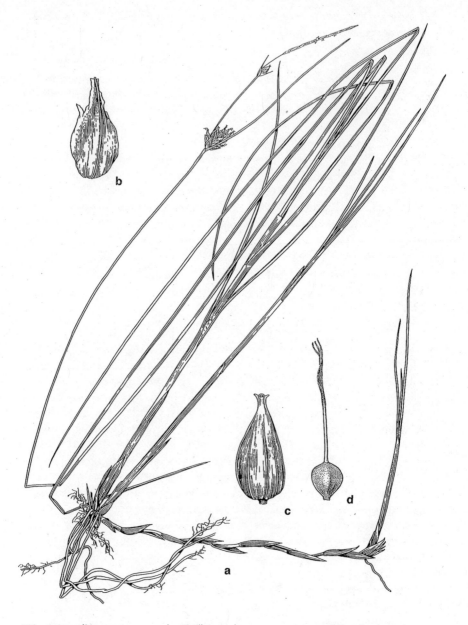

158. Carex oligosperma.
a. Habit.

b. Pistillate scale.

c. Perigynium.
d. Achene.

159. Carex vesicaria L. var. **monile** (Tuckerm.) Fern. Rhodora 3:53. 1901. Fig. 159.
Carex monile Tuckerm. Enum. Caric. 20. 1843.

Plants perennial, densely cespitose, with short, stout rhizomes; culms to 1 m
tall, triangular, slender, usually scabrous beneath the inflorescence, purplish at
the base; sterile shoots present; leaves up to 40 cm long, 4–7 mm wide, flat, septate-
nodulose, green, scabrous along the margins; sheaths yellow-brown, septate-nodulose, the ligule longer than wide; staminate spikes 2–4, to 4 cm long, to 4 mm
thick, well above the pistillate spikes; staminate scales acicular, yellow-brown;
pistillate spikes 1–4, to 7 cm long, to 1.5 cm thick, sessile or short-pedunculate;
bracts leaflike; pistillate scales lanceolate to ovate-lanceolate, acuminate, sometimes awned, yellow-brown to red-brown with hyaline margins, reaching the
base of the beak of the perigynium; perigynia up to 100 per spike, globose-ovoid,
5.5–8.0 mm long, 3.5–4.0 mm broad, half to two-thirds as thick as long, inflated,
membranous, yellow-green or golden brown, shiny, strongly nerved, rather
abruptly tapered to a smooth, bidentate beak 1.8–2.0 mm long; achenes trigonous,
2.2–2.5 mm long, 1.7–2.0 mm wide, yellowish, substipitate, continuous with the
flexuous or bent style; stigmas 3.

Common Name: Sedge.
Habitat: Upland swamps and depressions, wet meadows.
Range: Quebec to Ontario, south to Missouri, Illinois, Ohio, Pennsylvania, and Delaware.
Illinois Distribution: Occasional in the northern half of the state,
extending southward to Lawrence, St. Clair, Wabash, and Washington
counties.

This taxon is recognized by its dense clumps, its several staminate
spikes, and its few pistillate spikes with inflated perigynia.

There is some justification for recognizing this plant as a distinct species from
typical *C. vesicaria*. The typical variety has narrower, longer perigynia that taper
gradually to the beaks, and it has slightly wider leaves. *Carex vesicaria* var. *monile*
flowers from late May to early August.

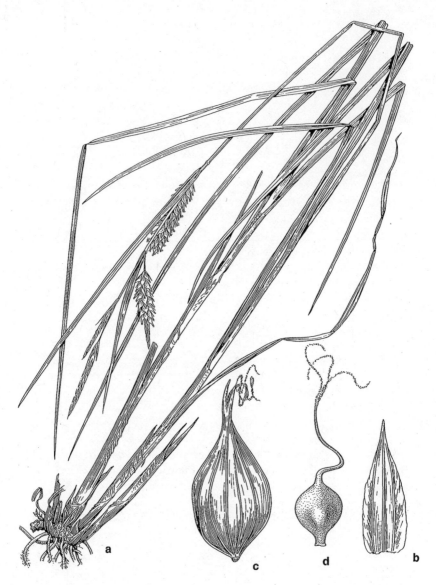

159. Carex vesicaria var.
monile.

a. Habit.
b. Pistillate scale.

c. Perigynium.
d. Achene.

Appendix: Additions and Changes to the First Edition

After studying and teaching about *Carex* thoroughly in the field for the past twenty years particularly, I have reached several conclusions that do not necessarily agree with the concepts of some botanists today. However, seeing these plants on a regular basis in the field substantiates my views on certain complexes of species. These observations are reflected in this book.

Page 7. The following species should be added to the paragraphs on that page:

Under "Hairs Present on Leaves, Sheaths, and/or Culms," add: *C. complanata*.

Under "Perigynia Pubescent," add: *C. complanata, C. houghtoniana, C. peckii,* and *C. rugosperma*.

Under "Some or All of the Leaves More than 10 mm Wide," add: *C. gynandra* and *C. kraliana*.

Under "Culms Purple, Red-Brown, or Purple-Brown at the Base," add: *C. arctata, C. aureolensis, C. copulata, C. corrugata, C. debilis, C. flexuosa, C. gynandra, C. houghtoniana, C. peckii, C. planispicata, C. rugosperma, C. timida,* and *C. triangularis; remove C. jamesii*.

Under "Some or All of the Pistillate Scales Longer than the Perigynia," add: *C. annectens, C. aureolensis, C. fissa, C. gynandra, C. haleana, C. setacea, C. timida, C. triangularis. and C. vulpinoidea*.

Pages 11–27. Substitute the following keys:

Page 11. Since 34 additional species are reported for Illinois, a new key to the species is provided to replace the keys on pages 11–27:

Key to the Species of Carex in Illinois

1. Hairs present on leaves, sheaths, and/or culms (this lead does not include plant parts that are merely scabrous or perigynia that may be papillose or granular). .... Group I
1. Leaves, sheaths, and/or culms all glabrous.
  2. Perigynia pubescent. ....Group II
  2. Perigynia glabrous.
    3. Stigmas 3; achenes trigonous.
      4. Spike 1, androgynous (with staminate flowers at tip, pistillate below). .... 62. *C. leptalea*
      4. Spikes more than 1 (sometimes only 1 in *C. squarrosa*), some of them entirely staminate or gynecandrous (with pistillate flowers at tip, staminate flowers below), or androgynous in rare specimens of *C. sprengelii* and *C. retrorsa.*
        5. Lowest scale of pistillate flower leaflike, 15–50 mm long (do not confuse this with bract that subtends the spike); pistillate spike with 2–4 perigynia. ....Group III
        5. Lowest scale of pistillate flower less than 20 mm long, not leaflike; pistillate spike usually with more than 4 perigynia.
          6. Terminal spike entirely staminate.
            7. Perigynia 10 mm long or longer. ....Group IV
            7. Perigynia up to 10 mm long, usually much shorter.
              8. Staminate spikes more than 1 per culm. ....Group V
              8. Staminate spike 1 per culm.
                9. Leaves threadlike, 0.5–1.0 mm wide; perigynia 1.5–2.0 mm long. .... 78. *C. eburnea*
                9. Leaves filiform or wider, some or all of them more than 1 mm wide; perigynia usually more than 2 mm long.
                  10. Flowering culms hidden among the leaves; plants to 15 cm tall. .... 75. *C. tonsa*
                  10. Flowering culms not hidden among the leaves; plants usually more than 15 cm tall.
                    11. Perigynia prominently 2-toothed at the tip, or prolonged into a conspicuous bidentate beak. .... Group VI
                    11. Perigynia ending abruptly at the tip, either without teeth or with merely a small notch.
                      12. Most or all of the leaves, particularly on sterile shoots, 15 mm wide or wider. .... Group VII
                      12. Most of the leaves less than 15 mm wide.
                        13. At least the lowest pistillate spikes on flexuous or pendulous peduncles. ....Group VIII
                        13. All pistillate spikes ascending to erect.
                          14. Beak or tip of perigynium bent or curved to one side. ....Group IX
                          14. Beak or tip of perigynium not bent nor curved to one side. .... Group X
           6. Terminal spike androgynous or gynecandrous. ....Group XI
    3. Stigmas 2; achenes lenticular.
      15. Some or all the spikes pedunculate; staminate flowers usually on separate spikes.... ....Group XII
      15. All spikes sessile, essentially alike, with staminate flowers either at the tip or at the base of each spike (exceptional specimens of *C. sterilis, C. bromoides,* and *C. praegracilis* may have a few all-staminate or all-pistillate flowers).

16. Culms solitary, or forming stoloniferous or rhizomatous colonies. .... Group XIII
16. Culms cespitose, not arising singly from extensive stolons and/or rhizomes.
    17. Spikes androgynous.. Group XIV
    17. Spikes gynecandrous.
        18. Perigynia plano-convex, with rounded margins or with narrowly rimmed
            margins... Group XV
        18. Perigynia more or less flat and scalelike, with winged margins... Group XVI

Group I
Hairs present on leaves, sheaths, and/or culms.

1. Perigynia pubescent.
  2. Terminal spike(s) entirely staminate.
    3. Staminate spike 1, sessile or nearly so; perigynia 3.5–5.0 mm long, with teeth of the beak minute.. 79. *C. hirtifolia*
    3. Staminate spikes 1–3, on long peduncles; perigynia 4.5–9.0 mm long, with teeth of the beak 1–2 mm long.
        4. Leaves 2–4 mm wide; perigynia up to 35 per spike, 5–9 mm long; achenes 2.7–3.0 mm long. .... 95. *C. hirta*
        4. Leaves 3–8 mm wide; perigynia up to 70 per spike, 4.5–6.5 mm long; achenes 1.6–2.0 mm long... A18. *C. houghtoniana*
  2. Terminal spike gynecandrous.
    5. Spikes linear-cylindric; perigynia ellipsoid-ovoid to lance-ellipsoid, prominently nerved.
        6. Perigynia ellipsoid-ovoid, 2.0–2.5 mm long, more or less compressed..........................
        ....100. *C. virescens*
        6. Perigynia lance-ellipsoid, 3.5–5.0 mm long, angled. .... 104. *C. oxylepis*
    5. Spikes short-cylindric to subglobose; perigynia broadly ovoid, finely nerved.....................
    ....101. *C. swanii*
1. Perigynia glabrous.
  7. At least the terminal spike staminate.
    8. Staminate spikes 2–5; remainder of spikes usually androgynous; pistillate scales shorter than the perigynia; perigynia numerous per spike, lance-ovoid, 7–9 mm long, with a beak 1.5–3.0 mm long.. 139. *C. atherodes*
    8. Staminate spike 1; lateral spikes entirely pistillate; pistillate scales as long as or longer than the perigynia; perigynia up to 40 per spike, ellipsoid-triangular to obovoid, up to 5 mm long, beakless or with a beak about 1 mm long.
        9. Leaves and sheaths pubescent; leaves up to 3 mm wide; perigynia up to 40 per spike, ellipsoid-triangular, up to 3 mm long, beakless....................................... 96. *C. pallescens*
        9. Only the sheaths pubescent; leaves 3–7 mm wide; perigynia up to 10 per spike, obovoid, 4–5 mm long, with a beak about 1 mm long. .... 117. *C. hitchcockiana*
  7. Terminal spike gynecandrous.
    10. Spikes narrowly cylindric, not clavate at the base, the lowest on long, flexuous or pendulous peduncles; perigynia ellipsoid to narrowly ovoid to oblong-ovoid, more or less trigonous, short-beaked.
        11. Leaves up to 1.8 mm wide; terminal spike up to 2 mm thick; pistillate scales obtuse to acute; perigynia up to 20 per spike, ellipsoid... 105. *C. formosa*
        11. Leaves 3–8 mm wide; terminal spikes more than 2 mm thick; pistillate scales awned; perigynia up to 40 per spike, narrowly ovoid to oblong-ovoid.
            12. Pistillate spikes 5–7 mm thick; pistillate scales as long as or longer than the perigynia; perigynia oblong-ovoid, loose around the achene. .... 106. *C. davisii*
            12. Pistillate spikes 3–5 mm thick; pistillate scales shorter than the perigynia; perigynia narrowly ovoid, closely enveloping the achene. .... 104. *C. oxylepis*

10. Spikes broadly cylindric, clavate at the base, on short, ascending peduncles; perigynia obovoid, rounded or bluntly angled on all faces, or with a flat, inner face, pointed at the tip.
    13. Perigynia trigonous in cross-section.
        14. Leaves densely pubescent; pistillate scales acute to cuspidate. ....97. *C. hirsutella*
        14. Leaves sparsely pubescent; pistillate scales acute to short-awned. .... A19. *C. complanata*
    13. Perigynia circular in cross-section.
        15. Pistillate scales awned, longer than the perigynia; leaves densely pubescent; achenes 2.0–2.5 mm long. ....99. *C. bushii*
        15. Pistillate scales acute to acuminate, shorter than the perigynia; leaves sparsely pubescent; achenes 1.5–2.2 mm long. .... 98. *C. caroliniana*

Group II
Leaves, stems, and sheaths glabrous; perigynia pubescent.

1. Terminal spike androgynous. ....76. *C. pedunculata*
1. Terminal spike gynecandrous, or entirely staminate.
    2. Terminal spike gynecandrous. .... 98. *C. caroliniana*
    2. Terminal spike entirely staminate.
        3. Staminate spikes 2 or more per culm.
            4. Lower spikes sometimes androgynous; pistillate scales lanceolate to broadly lanceolate, slightly shorter than to longer than the perigynia; perigynia 2–5 mm long, the teeth of the beak up to 1 mm long.
                5. Culms smooth; leaves up to 2 mm wide, convolute, without a midvein; pistillate scales slightly shorter than the perigynia; perigynia obscurely nerved, 3–5 mm long, the teeth of the perigynia up to 0.6 mm long. ....93. *C. lasiocarpa*
                5. Culms usually somewhat scabrous; leaves 2–5 mm wide, flat, with a conspicuous midvein; pistillate scales as long as or slightly longer than the perigynia; perigynia obviously nerved, 2.0–3.5 mm long, the teeth of the perigynia about 1 mm long.. ....94. *C. pellita*
            4. Lower spikes almost always pistillate; pistillate scales broadly ovate, shorter than the perigynia; perigynia 4.2–10.0 mm long, the teeth of the beak 1.2–2.0 mm long.
                6. Perigynia strongly nerved, up to 40 per pistillate spike, 5–10 mm long. ....142. *C. trichocarpa*
                6. Perigynia obscurely nerved, up to 135 per pistillate spike, 4.2–6.4 mm long. ....143. *C. X subimpresa*
        3. Staminate spike 1 per culm.
            7. Perigynia 12 mm long or longer, crowded into a globose spike. .... 149. *C. grayi*
            7. Perigynia up to 4.5 mm long, not forming a globose spike.
                8. Plants more than 60 cm tall; terminal staminate spike up to 4 cm long; perigynia 40–75 per spike. ....94. *C. pellita*
                8. Plants up to 60 cm tall, usually much shorter; terminal staminate spike up to 2 cm long; perigynia up to 25 per spike.
                    9. Perigynia spongy at base; culms capillary to slender and flexuous; leaves pale green or glaucous.
                        10. Culms and peduncles capillary; perigynia 1–8 per spike, 3.8–4.5 mm long. ... . 76. *C. pedunculata*
                        10. Culms and peduncles slender and flexuous, but not capillary; perigynia 10–25 per spike, 2.0–3.5 mm long. .... 77. *C. richardsonii*

9. Perigynia firm at base; culms neither slender nor flexuous but not capillary; leaves green to dark green.
  11. Some or all the fertile culms hidden among the leaf bases.
    12. Some fertile culms sometimes elevated above the leaves; plants with stolons; body of perigynium ellipsoid, yellow-green; beak about one-third as long as the body of the perigynium..........................72. *C. nigromarginata*
    12. All spikes hidden at base of plant; plants without stolons; body of perigynium orbicular, not yellow-green, the beak about one-half as long as the body of the perigynium.
      13. Perigynia very sparsely pubescent.....................................75. *C. tonsa*
      13. Perigynia uniformly pubescent.
        14. Beak of perigynium 2.5–4.0 mm long, about as long as the body of the perigynium; leaves bright green........... A16. *C. rugosperma*
        14. Beak of perigynium 0.8–1.7 mm long, one-half to three-fourths as long as the body of the perigynium; leaves green.
          15. Beak of perigynium about three-fourths as long as the body of the perigynium; perigynia 2.2–3.3 mm long; pistillate scales acuminate. ....73. *C. umbellata*
          15. Beak of perigynium one-half to two-thirds as long as the body of the perigynium; perigynia 3.2–4.5 mm long; pistillate scales acute.. 74. *C. abdita*
  11. All fertile culms protruding above the leaves, not hidden among the leaf bases.
    16. Perigynia ovoid to obovoid.
      17. Beak at least two-thirds the length of the body of the perigynium; perigynia yellow-green...66. *C. lucorum*
      17. Beak one-fourth to one-third as long as the body of the perigynium; perigynia pale green or green.
        18. Perigynia more or less terete in cross-section; leaves usually stiff. .... 67. *C. heliophila*
        18. Perigynia trigonous in cross-section; leaves usually soft.
          19. Plants with slender stolons; leaves 1.5–3.6 mm wide............. .... 65. *C. pensylvanica*
          19. Plants without slender stolons; leaves 1.8–5.0 mm wide. ....68. *C. communis*
    16. Perigynia ellipsoid.
      20. Plants with long stolons; beak one-third to one-half as long as the body of the perigynium. ....71. *C. physorhyncha*
      20. Plants without stolons; beak up to one-third as long as the body of the perigynium.
        21. Culms and leaves weak, often reclining; leaves up to 1.5 mm wide, bright green... 69. *C. emmonsii*
        21. Culms and leaves more or less erect or ascending, not weak; leaves more than 1.5 mm wide, green or pale green.
          22. Pistillate scales shorter than the perigynia, exposing the perigynia; achenes 2.0–2.5 mm long. ....A15. *C. peckii*
          22. Pistillate scales about as long as the perigynia, not exposing the perigynia; achenes 1.3–2.0 mm long.
            23. Leaves 1.8–5.0 mm wide; staminate scales acute to short-acuminate; perigynia 1.2–2.6 mm wide. ...68. *C. communis*
            23. Leaves 1.0–2.5 mm wide; staminate scales obtuse to acute; perigynia 0.8–1.1 mm wide. .... 70. *C. albicans*

Group III

Leaves, sheaths, and/or culms all glabrous; perigynia glabrous; stigmas 3; achenes trigonous; perigynia less than 10 mm long; lowest scale of pistillate flower leaflike, 15–50 mm long; pistillate spike with 2–4 perigynia.

1. Culms green at base; staminate part of spike often more than 6 mm long; staminate scales 1–2 mm long.
    2. Perigynia orbicular, 2.3–2.5 mm wide; achenes 2.5 mm wide; staminate scales truncate.. .... 63. *C. jamesii*
    2. Perigynia mostly ellipsoid, about 1.5 mm wide; achenes 1.5–2.0 mm wide; staminate scales acute.. 64. *C. willdenowii*
1. Culms red-purple at base; staminate part of spike 3.3–6.0 mm long; staminate scales 2.0–3.2 mm long..A14. *C. timida*

Group IV

Leaves, sheaths, and/or culms all glabrous; perigynia glabrous; stigmas 3; achenes trigonous; terminal spike 1, entirely staminate; perigynia 10 mm long or longer.

1. Perigynia not inflated, lanceoloid, up to 3.5 mm wide at its widest point................................ .... 148. *C. folliculata*
1. Perigynia strongly inflated, subuloid to lanceoloid, usually 3.5 mm wide or wider at its widest point.
    2. Pistillate spikes globose or subglobose.
        3. Perigynia radiating all directions in a spike; perigynia not shiny, cuneate at base. .... 149. *C. grayi*
        3. Perigynia radiating in all directions except retrorse; perigynia shiny, rounded at base... ....150. *C. intumescens*
    2. Pistillate spikes thick-cylindric, longer than broad.
        4. All perigynia horizontally spreading; beak of perigynium 2–3 times longer than the body; achenes truncate at summit, broader than long. .... 154. *C. gigantea*
        4. Perigynia ascending; beak of perigynium somewhat shorter than or slightly longer than the body; achenes narrowed at summit, longer than broad or about as long as broad.
            5. Pistillate spikes more than twice as long as broad; achenes about as long as broad, with conspicuously knobby angles and concave sides...................153. *C. lupuliformis*
            5. Pistillate spikes up to twice as long as broad; achenes longer than broad, without conspicuous knobby angles and concave sides.
                6. Perigynia crowded, 13–20 mm long; pistillate scales long-acuminate to awned; leaves 5 mm wide or wider; culms several in a tuft. ....152. *C. lupulina*
                6. Perigynia loosely arranged, 10–14 mm long; pistillate scales obtuse to acute; leaves up to 6 mm wide; culms few or solitary, never in a tuft............... 151. *C. louisianica*

Group V

Leaves, sheaths, and/or culms all glabrous; perigynia glabrous; stigmas 3; achenes trigonous; terminal spike entirely staminate, usually more than one per culm; perigynia up to 10 mm long, usually much shorter.

1. Style disarticulating from mature achene; perigynia abruptly contracted to a beak as long as the body; lowest spikes often on pendulous peduncles. ....108. *C. sprengelii*
1. Style continuous and persistent on mature achene; perigynia tapering to the beak, the beak usually somewhat shorter than the body of the perigynium; lowest spikes erect or ascending.

2. Plants growing in large colonies from extensive rhizomes; perigynia thick and firm, not inflated.
    3. Culms red-purple at base; leaves neither glaucous nor blue-green; nerves of perigynia conspicuous and slightly elevated.
        4. Teeth of the beak of the perigynium up to 1 mm long; most of the leaves 8 mm wide or wider.. 137. *C. lacustris*
        4. Teeth of the beak of the perigynium 1.0–2.2 mm long; leaves mostly 2–6 mm wide. . ...141. *C. laeviconica*
    3. Culms brown at base; leaves glaucous or blue-green; nerves of perigynia finely impressed, or absent.
        5. Leaves 8–15 mm wide, flat; perigynia up to 150 per spike, 6–8 mm long..................... ....138. *C. hyalinolepis*
        5. Leaves 1.5–3.2 mm wide, plicate; perigynia up to 30 per spike, 3.0–4.5 mm long........ .... 140. *C. heterostachya*
2. Plants growing in dense clumps, usually without extensive rhizomes; perigynia thin and papery, inflated.
    6. Perigynia horizontally spreading or even reflexed at maturity; sheaths loose. .... 155. *C. retrorsa*
    6. Perigynia spreading to ascending at maturity; sheaths tight.
        7. Culms spongy at the base; some or all the leaves 7–12 mm wide. ..... 156. *C. utriculata*
        7. Culms firm at the base; leaves 2–7 mm wide.
            8. Achenes deeply grooved on one side; perigynia 7–10 mm long.............................. .... 157. *C. tuckermanii*
            8. Achenes not deeply grooved on one side; perigynia 5–8 mm long....................... .... 159. *C. vesicaria*

Group VI

Leaves, sheaths, and/or culms all glabrous; perigynia glabrous; stigmas 3; achenes trigonous; terminal spike 1, entirely staminate; perigynia up to 10 mm long, usually much shorter; perigynia prominently 2-toothed at the tip, or prolonged into a conspicuous, bidentate beak.

1. Pistillate spikes 2–4 mm thick.
    2. Pistillate spikes linear-cylindric, up to 8 cm long; lowest spikes on pendulous peduncles; perigynia 3.0–8.5 mm long, all ascending.
        3. Perigynia up to 45 per spike; pistillate scales usually with awns up to 1.5 mm long; perigynia 3–5 mm long, with several faint nerves...................................... A20. *C. arctata*
        3. Perigynia up to 25 per spike; pistillate scales obtuse to acute to cuspidate; perigynia 5.0–8.5 mm long, with 2 strong ribs and several faint nerves.
            4. Perigynia strongly overlapping, fusiform, gradually tapering to an elongated beak nearly one-half as long as the perigynium. ....107. *C. flexuosa*
            4. Perigynia barely overlapping, lance-ellipsoid, abruptly narrowed to an elongated beak about one-third as long as the perigynium. ....A21. *C. debilis*
2. Pistillate spikes short-cylindric, up to 3 cm long; lowest spikes ascending; perigynia 2–5 mm long, spreading to reflexed to ascending.
    5. Perigynia up to 30 per spike, spreading to reflexed, 2–3 mm long............ 133. *C. viridula*
    5. Perigynia up to 15 per spike, ascending, 3.5–5.0 mm long..................... A26. *C. vaginata*
1. Pistillate spikes at least 1 cm thick (sometimes slightly more slender in *C. sprengelii*).
    6. Some of the perigynia reflexed.
        7. Perigynia up to 35 per spike, 3.2–6.3 mm long, yellowish to bright yellow, the beak at least half as long as the body of the perigynium.

8. Perigynia yellowish, 3.2–4.5 mm long, the beak glabrous, usually straight, about as long as the body of the perigynium; leaves up to 4.5 mm wide.......................... .... 132. *C. cryptolepis*
8. Perigynia bright yellow, 4.0–6.3 mm long, the beak scabrous, curved, over half as long as the body of the perigynium; leaves up to 6 mm wide................. A32. *C. flava*
7. Perigynia usually 50–100 per spike, 5–10 mm long, yellow-green, the beak up to half as long as the body of the perigynium.
  9. Lowest spikes on pendulous peduncles; perigynia 3.5–7.0 mm long, the beak 1–2 mm long; pistillate scales serrulate-awned.
    10. Teeth of beak of perigynium nearly straight, 0.7–1.2 mm long. .... A34. *C. pseudocyperus*
    10. Teeth of beak of perigynium curved outward, 1.2–2.0 mm long. ....144. *C. comosa*
  9. Lowest spikes ascending; perigynia 7–10 mm long, the beak 2.5–3.5 mm long; pistillate scales acuminate. .... 155. *C. retrorsa*
6. All perigynia spreading to ascending, never reflexed.
  11. Culms filiform; leaves 1–3 mm wide; perigynia up to 15 per spike. ....158. *C. oligosperma*
  11. Culms broader than filiform; leaves 2.5–12.0 mm wide (if narrower, then plicate in *C. heterostachya*); perigynia 30–100 per spike.
    12. Culms purple-red at base; perigynia 30–100 per spike.
      13. Pistillate scales much longer than the perigynia; spikes subtended by bracts many times their length; perigynia obconic, broadest above the middle.
        14. Rhizomes short, not forming colonies; staminate scales 0.5–1.0 mm wide; beak of perigynium up to 2.5 mm long. .... 134. *C. frankii*
        14. Rhizomes long, forming colonies; staminate scales 1.0–1.5 mm wide; beak of perigynium 1.2–2.0 mm long.................................A33. *C. aureolensis*
      13. Pistillate scales not longer than the perigynia; spikes not subtended by exceptionally long bracts; perigynia ovoid, broadest near the base.
        15. Lowest spikes on slender, often pendulous peduncles; perigynia 4–6 mm long, 1.5–2.0 mm wide, the beak 1.8–2.2 mm long..... 145. *C. hystericina*
        15. Lowest spikes ascending; perigynia 5–9 mm long, 2–4 mm wide, the beak 3–4 mm long.
          16. Leaves up to 7 mm wide; perigynia 3–4 mm wide, the beak about equaling the body..146. *C. lurida*
          16. Leaves 2–4 mm wide; perigynia 2.0–2.5 mm wide, the beak longer than the body. .... 147. *C. baileyi*
    12. Culms brown at base; perigynia 30–50 per spike.
      17. Perigynia 5.0–6.5 mm long, the beak half as long to equaling the body of the perigynium; leaves 2.5–7.0 mm wide, usually flat.
        18. Beak of the perigynium as long as the body; leaves 2.5–4.0 mm wide; perigynium strongly nerved. ....108. *C. sprengelii*
        18. Beak of the perigynium about half as long as the body; leaves up to 7 mm wide; perigynia finely nerved. ....A22. *C. cherokeensis*
      17. Perigynia 3.0–4.5 mm long, the beak much shorter than the body of the perigynium; leaves 1.5–3.2 mm wide, often plicate. .... 140. *C. heterostachya*

Group VII

Leaves, sheaths, and/or culms all glabrous; perigynia glabrous; stigmas 3; achenes trigonous; terminal spike 1 per culm, entirely staminate; perigynia up to 10 mm long, usually much shorter; perigynia ending abruptly at the tip, either without teeth or with merely a small notch; most or all the leaves, particularly on sterile shoots, more than 15 mm wide.

1. Culms leafless, bearing only long-tubular purplish sheaths; staminate spike purple.............
....121. *C. plantaginea*
1. Culms leafy; staminate spike brown to brown-purple.
  2. Blade of uppermost leaf at most only three times longer than its sheath.
    3. Basal leaves green; culms purple at base; perigynia 5.0–6.5 mm long............................
.... 122. *C. careyana*
    3. Basal leaves glaucous; culms brown at base; perigynia 2.5–4.5 mm long. ....
.... 123. *C. platyphylla*
  2. Blade of uppermost leaf many times longer than its sheath.
    4. Uppermost bract much longer than the inflorescence; culms somewhat winged.
      5. Leaves pale green; pistillate scales obtuse to acute; perigynia obovoid.........................
.... 126. *C. albursina*
      5. Leaves dark green; pistillate scales acute to short-awned; perigynia ovoid to oblongoid..A30. *C. kraliana*
    4. Uppermost bract about as long as or barely longer than the inflorescence; culms wingless... 127. *C. laxiflora*

Group VIII
Leaves, sheaths, and/or culms all glabrous; perigynia glabrous; stigmas 3; achenes trigonous; terminal spike 1, entirely staminate; perigynia up to 10 mm long, usually much shorter; perigynia ending abruptly at the tip, either without teeth or with merely a small notch; most or all the leaves less than 15 mm wide; at least the lowest pistillate spikes on flexuous or pendulous peduncles.

1. Perigynia up to 15 per spike.
  2. Perigynia fusiform.
    3. Perigynia 3.5–4.5 mm long, curved at tip, with rounded angles............131. *C. styloflexa*
    3. Perigynia 2.5–3.3 mm long, straight at tip, with pointed angles.............. 124. *C. digitalis*
  2. Perigynia ellipsoid to ovoid to oblongoid.
    4. Culms purple-red at base; leaves yellow-green................................... 130. *C. gracilescens*
    4. Culms brown at base; leaves pale green, bright green, or glaucous.
      5. Staminate spike sessile or nearly so; pistillate scales about as long as the perigynia; perigynia strongly curved at tip; angles of perigynia rounded............... 129. *C. blanda*
      5. Staminate spike pedunculate; pistillate scales shorter than the perigynia; perigynia straight at tip; angles of perigynia pointed.
        6. Some or all the leaves up to 12 mm wide; pistillate scales sometimes short-awned.
          7. Leaves bright green, harshly scabrous; leaves up to 8 mm wide. ... A27. *C. copulata*
          7. Leaves green, slightly scabrous; leaves up to 12 mm wide........ 125. *C. laxiculmis*
        6. Leaves 2.0–5.3 mm wide; pistillate scales all acute.
          8. Perigynia 2.8–3.5 mm long, with 7–10 nerves; leaves 2.0–2.9 mm wide; peduncle of staminate spike 8–16 mm long, surpassing uppermost bract. ....
.... A28. *C. macropoda*
          8. Perigynia 2.5–3.3 mm long, with 11–15 nerves; leaves 2.5–5.3 mm wide; peduncle of staminate spike 1–7 cm long, surpassed by uppermost bract... 124. *C. digitalis*
1. Perigynia 15–50 per spike.
      5. Leaves up to 2.5 mm wide; plants stoloniferous; perigynia glaucous-green. ... 90. *C. limosa*
      5. Leaves 3–6 mm wide; plants not stoloniferous; perigynia green. .... 102. *C. prasina*

Group IX

Leaves, sheaths, and/or culms glabrous; perigynia glabrous; stigmas 3; achenes trigonous; terminal spike 1, staminate; perigynia up to 10 mm long, usually much shorter; perigynia ending abruptly at the tip, either without teeth or with a small notch; most or all the leaves less than 15 mm wide; all pistillate spikes ascending to erect; beak or tip of perigynium bent or curved to one side.

1. Perigynia crowded in pistillate spikes, at least always overlapping.
  2. Perigynia rounded at the base, olive-green to brown; leaves more or less glaucous.
    3. Perigynia ellipsoid, not inflated, 2.3–3.0 mm long, 1.0–1.5 mm wide; cauline leaves up to 16 mm wide...A23. *C. haleana*
    3. Perigynia obovoid, inflated, 2.5–4.0 mm long, 1.5–2.5 mm wide; cauline leaves up to 12 mm wide .... 109. *C. granularis*
  2. Perigynia tapering to the base, yellow-green to green to pale green; leaves pale green, green, or gray-green, but not glaucous.
    4. Perigynia with 2–3 conspicuous nerves ....A31. *C. leptonervia*
    4. Perigynia with 4 or more faint or somewhat conspicuous nerves, or nerveless.
      5. Perigynia distichously arranged .... A24. *C. planispicata*
      5. Perigynia spirally arranged.
        6. Leaves gray-green, up to 7 mm wide; pistillate spikes 5–7 mm thick; pistillate scales cuspidate to short-awned; perigynia 3–5 mm long, turgid .... 118. *C. meadii*
        6. Leaves green, up to 5 mm wide; pistillate spikes about 5 mm thick; pistillate scales obtuse to acute; perigynia 2.5–3.5 mm long, not turgid .... 119. *C. tetanica*
1. Perigynia loosely or remotely arranged in pistillate spikes, scarcely overlapping.
  7. Leaves up to 4 mm wide; plants stoloniferous; pistillate scales with purple-brown margins...120. C. woodii
  7. Leaves 5–12 mm wide; plants tufted; pistillate scales without purple-brown margins.
    8. Pistillate spikes 4–5 mm thick; perigynia 4–5 mm long........................... 128. *C. striatula*
    8. Pistillate spikes 3–4 mm thick; perigynia 3.0–4.5 mm long .... 127. *C. laxiflora*

Group X

Leaves, sheaths, and/or culms glabrous; perigynia glabrous; stigmas 3; achenes trigonous; terminal spike 1, entirely staminate; perigynia up to 10 mm long, usually much shorter; perigynia ending abruptly at the tip, either without teeth or with merely a small notch; most or all of the leaves less than 15 mm wide; all pistillate spikes ascending to erect, not on pendulous peduncles; beak or tip of perigynium not bent or curved to one side.

1. Pistillate scales dark red-purple; perigynia more biconvex than trigonous............................
  ....89. *C. buxbaumii*
1. Pistillate scales not dark red-purple; perigynia trigonous.
  2. Staminate spike strongly elevated above the pistillate spikes on peduncles 2 cm long or longer.
    3. Pistillate spikes loosely or remotely flowered.
      4. Perigynia with rounded angles; leaves 6–20 mm wide, pale green.
        5. Achenes 1.8–2.0 mm long; staminate scales cuspidate; pistillate spikes 3–4 mm thick; perigynia up to 15 per spike.. 127. *C. laxiflora*
        5. Achenes 2.2–2.5 mm long; staminate scales obtuse; pistillate spikes 4–5 mm thick; perigynia up to 20 per spike ....127. *C. striatula*
      4. Perigynia with pointed angles; leaves up to 8 mm wide, green or dark green.

6. Peduncle of staminate spike up to 7 cm long, surpassed by the uppermost bract; perigynia usually fusiform, with 11–15 nerves; leaves 2.5–3.5 mm wide ....
.... 124. *C. digitalis*

6. Peduncle of staminate spike 8–16 cm long, surpassing the uppermost bract; perigynia never fusiform, with 6–10 nerves; leaves 2.0–2.9 mm wide ....A28. *C. macropoda*

3. Pistillate spikes with crowded perigynia.

7. Perigynia distichously arranged.. A24. *C. planispicata*

7. Perigynia spirally arranged.

8. Leaves 1.5–3.5 mm wide; perigynia up to 45 per spike, strongly nerved ....
.... 110. *C. crawei*

8. Leaves 3–8 mm wide; perigynia up to 27 per spike, finely nerved.

9. Staminate spike 0.5–1.5 mm thick; pistillate scales acute........ A29. *C. abscondita*

9. Staminate spike 2.5–3.0 mm thick; pistillate scales awned.

10. Culms brown at base; leaves up to 5 mm wide; perigynia 3–4 mm long; achenes 2.0–2.2 mm long.. 111. *C. conoidea*

10. Culms purple-red at base; many or all the leaves 5–8 mm wide; perigynia 4.0–7.5 mm long; achenes 2.5–3.3 mm long.

11. Sheaths loose; pistillate spikes 3–6 mm thick; perigynia up to 12 per spike, oblongoid to obovoid, dull, the angles pointed...............112. *C. amphibola*

11. Sheaths tight; pistillate spikes up to 10 mm thick; perigynia up to 27 per spike, ellipsoid, shiny, the angles rounded......................A25. *C. corrugata*

2. Staminate spike not strongly elevated above the pistillate spikes, sessile or on peduncles less than 2 cm long.

12. Perigynia up to 10(–12) per spike; leaves green or glaucous, 2–8 mm wide.

13. Pistillate spikes loosely flowered, up to 1.5 cm long ....116. *C. oligocarpa*

13. Pistillate spikes with crowded perigynia, (1.0–) 1.5–2.5 cm long.

14. Leaves more or less glaucous or strongly glaucous.

15. Pistillate scales up to one-half as long as the perigynia; perigynia 2.0–2.3 mm wide; plants more or less glaucous; plants usually in rather wet habitats....
.... 114. *C. flaccosperma*

15. Pistillate scales nearly as long as the perigynia; perigynia 1.5–1.8 mm wide; plants strongly glaucous; plants usually in rather dry habitats ....
.... 115. *C. glaucodea*

14. Leaves green.

16. Perigynia distichously arranged .... A24. *C. planispicata*

16. Perigynia spirally arranged.

17. Perigynia 3.5–4.7 mm long; one or more pistillate spikes usually near base of plant.

18. Sheaths loose; pistillate spikes 3–6 mm thick; perigynia up to 12 per spike, oblongoid to obovoid, dull, the angles pointed.......................
....112. *C. amphibola*

18. Sheaths tight; pistillate spikes up to 10 mm thick; perigynia up to 27 per spike, ellipsoid, shiny, the angles rounded.......A25. *C. corrugata*

17. Perigynia 4.5–5.5 mm long; no pistillate spikes near base of plant. ....
.... 113. *C. grisea*

12. Perigynia 10–60 per spike; leaves more or less glaucous (green in *C. grisea*), up to 10 mm wide.

19. Pistillate scales one-half as long as the perigynia or nearly as long; leaves more or less glaucous or strongly glaucous.

20. Perigynia 2.3–4.0 mm long, usually strongly nerved; sheaths tight; leaves more or less glaucous.

21. Perigynia ellipsoid, not inflated, 2.3–3.0 mm long, 1.0–1.5 mm wide; cauline leaves up to 16 mm wide..A23. *C. haleana*
21. Perigynia obovoid, inflated, 2.5–4.0 mm long, 1.5–2.5 mm wide; cauline leaves up to 12 mm wide.. 109. *C. granularis*
20. Perigynia 3.5–6.0 mm long, finely impressed-nerved; sheaths loose; plants strongly glaucous.
22. Pistillate scales up to one-half as long as the perigynia; perigynia 2.0–2.3 mm wide; plants in rather wet habitats........................114. *C. flaccosperma*
22. Pistillate scales nearly as long as the perigynia; perigynia 1.5–1.8 mm wide; plants in rather dry habitats...115. *C. glaucodea*
19. Pistillate scales as long as or longer than the perigynia; leaves green ....
.... 113. *C. grisea*

Group XI

Leaves, sheaths, and/or culms glabrous; perigynia glabrous; stigmas 3; achenes trigonous; terminal spike androgynous or gynecandrous.

1. Terminal spike androgynous.
    2. Pistillate spikes up to 3.5 cm long, up to 1 cm thick, the lowest on pendulous peduncles; perigynia loosely arranged, 5.0–6.5 mm long, 1.8–2.0 mm wide, 2-ribbed but otherwise nerveless, the beak as long as the body; leaves 2.5–4.0 mm wide, shorter than the culms ....108. *C. sprengelii*
    2. Pistillate spikes up to 8 cm long, up to 2 cm thick, none of them on pendulous peduncles; perigynia crowded, 7–10 mm long, 2.5–3.5 mm wide, strongly nerved, the beak prominent but usually not one-half as long as the body; leaves mostly 4–10 mm wide, longer than the culms .... 155. *C. retrorsa*
1. Terminal spike gynecandrous.
    3. Perigynia obconic; pistillate scales setaceous, much longer than the perigynia; spikes located about midway on plant, much surpassed by leaves and leaflike bracts.
        4. Rhizomes short, not forming colonies; staminate scales 0.5–1.0 mm wide; perigynia up to 100 per spike; beak of perigynium up to 2.5 mm long........................... 134. *C. frankii*
        4. Rhizomes long, forming colonies; staminate scales 1.0–1.5 mm wide; perigynia up to 80 per spike; beak of perigynium 1.2–2.0 mm long.................................A33. *C. aureolensis*
    3. Perigynia narrowly lanceoloid to narrowly ellipsoid to ovoid to obovoid (sometimes obconic in *C. typhina*); pistillate scales lanceolate to ovate, shorter than to about as long as the perigynia; most spikes in upper part of plant or even surpassing the leaves and bracts.
        5. Pistillate spikes (or terminal spike if only one spike present) 10–22 mm thick.
            6. Perigynia horizontally spreading, or the lowest one reflexed; pistillate scales acute to acuminate to cuspidate.. 135. *C. squarrosa*
            6. Perigynia ascending, none of them reflexed; pistillate scales usually obtuse, less commonly acute .... 136. *C. typhina*
        5. Pistillate spikes up to 10 mm thick.
            7. At least the lowest spikes pendulous or spreading on long, slender peduncles.
                8. Perigynia narrowly lanceoloid or fusiform, 4.0–8.5 mm long.
                    9. Perigynia strongly overlapping, fusiform, 5.5–8.5 mm long, gradually tapering to an elongated beak nearly half as long as the perigynium; pistillate scales obtuse to acute, white or pale...A21. *C. debilis*
                    9. Perigynia barely overlapping, narrowly lanceoloid, 4–7 mm long, abruptly narrowed to an elongated beak about one-third as long as the perigynium; pistillate scales acute to cuspidate, with reddish streaks...........................107. *C. flexuosa*
                8. Perigynia narrowly ellipsoid to ovoid, 2–4 mm long.

10. Perigynia narrowly ellipsoid, beakless; leaves up to 8 mm wide ....
....103. *C. gracillima*
10. Perigynia ovoid, with a beak up to 2 mm long; leaves 3–6 mm wide ....
....102. *C. prasina*
7. All spikes sessile, or on short, ascending peduncles.
11. Culms purple-red at base; pistillate scales dark red-purple; perigynia more or less
biconvex, glaucous-green; leaves glaucous below ....89. *C. buxbaumii*
11. Culms usually brownish at base; pistillate scales brown, red-brown, or greenish;
perigynia trigonous, yellow, olive-green, or brown; leaves not glaucous.
12. Perigynia squarrose, none of them reflexed; leaves 4–9 mm wide; perigynia
nerveless.
13. Pistillate spikes 4–6 mm thick; beak of perigynium absent or up to 0.5 mm
long...91. *C. shortiana*
13. Pistillate spikes 7–8 mm thick; beak of perigynium 1.0–1.5 mm long ....
....92. *C. X deamii*
12. Perigynia not squarrose, the lowest ones usually reflexed; leaves up to 4.5 mm
wide; perigynia with a few nerves.
14. Pistillate spikes 6–10 mm thick; pistillate scales acute to acuminate, shorter
than the perigynia; perigynia 3.2–4.5 mm long, narrowly ovoid, yellowish,
the beak about as long as the body ....132. *C. cryptolepis*
14. Pistillate spikes 2–3 mm thick; pistillate scales obtuse, about as long as the
perigynia; perigynia 2–3 mm long, ovoid, green or yellow-green, the beak
about one-third as long as the body.................................. 133. *C. viridula*

Group XII
Leaves, sheaths, and/or culms all glabrous; perigynia glabrous; stigmas 2;
achenes lenticular; none or all the spikes pedunculate; staminate flowers often
on separate spikes.

1. Terminal spike gynecandrous.
2. Culms to 50 cm tall, slender; leaves 1.0–3.4 mm wide; pistillate scales obtuse to acute,
shorter than to about as long as the perigynia; perigynia up to 30 per spike, yellow to orange
to white-papillate; pistillate spikes to 2 cm long.
3. Perigynia yellow to orange..81. *C. aurea*
3. Perigynia white-papillate ....80. *C. garberi*
2. Culms to 1.5 m tall, stout; leaves 4–14 mm wide; pistillate scales long-awned, much longer
than the perigynia; perigynia more than 30 per spike, green to pale brown to stramineous;
pistillate spikes up to 12 cm long.
4. Lower sheaths smooth; perigynia 2.0–3.5 mm long, inflated; achenes often crimped
along one margin ....82. *C. crinita*
4. All sheaths scabrous; perigynia 3–4 mm long, scarcely inflated; achenes not crimped
along one margin .... A17. *C. gynandra*
1. Terminal spike entirely staminate.
5. Culms to 50 cm tall, slender; leaves 1–3 mm wide; perigynia yellow to orange.... 81. *C. aurea*
5. Culms to 1.5 m tall, usually stout; leaves 3–14 mm wide; perigynia brown to olive-green
to deep green.
6. Pistillate scales long-awned, longer than the perigynia.
7. Lower sheaths smooth; perigynia 2.0–3.5 mm long, inflated; achenes often with a crimp
along one margin..82. *C. crinita*
7. All sheaths scabrous; perigynia 3–4 mm long, scarcely inflated; achenes not crimped
along one margin .... A17. *C. gynandra*

6. Pistillate scales obtuse to acuminate, shorter than to barely longer than the perigynia.

    8. Lowest leaf sheath with well-developed blades; perigynia broadest above the middle . ... 83. *C. aquatilis*

    8. Lowest leaf sheath bladeless; perigynia broadest at or below the middle.

        9. Perigynia with a distinct beak 0.3–1.0 mm long; pistillate scales purple-black to purple-brown.

            10. Beak of perigynium twisted, 0.3–1.0 mm long; pistillate scales purple-black, shorter than the perigynia; leaves 3–5 mm wide .... 88. *C. torta*

            10. Beak of perigynium straight, 0.3–0.6 mm long; pistillate scales purple-brown, about as long as the perigynia; leaves up to 12 mm wide .... 85. *C. nebrascensis*

        9. Perigynia beakless, or the beak up to 0.3 mm long; pistillate scales red-brown, not purplish.

            11. Ligule longer than the width of the blade, V-shaped; perigynia nerveless or faintly nerved.

                12. Pistillate scales shorter than the perigynia; pistillate spikes to 10 cm long; perigynia more or less flat, 1.7–3.4 mm long; lowest sheath deep red or purple .... 84. *C. stricta*

                12. Pistillate scales longer than the perigynia; pistillate spikes to 5 cm long; perigynia biconvex, 1.5–2.8 mm long; lowest sheath red-brown .... 87. *C. haydenii*

            11. Ligule very short, not forming a V; perigynia distinctly nerved..................... .... 86. *C. emoryi*

Group XIII

Leaves, sheaths, and/or culms all glabrous; perigynia glabrous; stigmas 2; achenes lenticular; all spikes sessile, essentially alike, with staminate flowers either at the tip or at the base of each spike; culms solitary, or forming stoloniferous or rhizomatous colonies.

1. Culms arising from axils of last year's dead leaves along prostrate or reclining stems; culms usually smooth; tip of perigynium smooth, not serrulate .... 3. *C. chordorrhiza*
1. Culms arising directly from the rhizomes.

    2. Inflorescence headlike, with all spikes crowded; leaves filiform, up to 1.5 mm wide........... .... 1. *C. duriuscula*

    2. Inflorescence elongated, often the lowest spikes not contiguous with the ones above; leaves 1–3 mm wide, not filiform.

        3. Perigynia biconvex; sheaths tight; inflorescence usually 5–10 cm long; rhizomes short.

            4. Sheaths copper-colored at the summit; perigynia 2.5–4.0 mm long, 1.2–1.3 mm wide, stramineous to brown...25. *C. prairea*

            4. Sheaths not copper-colored at the summit; perigynia 2–3 mm long, about 1 mm wide, dark brown to olive-black .... 24. *C. diandra*

        3. Perigynia plano-convex; sheaths loose and open (except in *C. socialis*); inflorescence to 6 cm long; rhizomes elongated.

            5. Inflorescence 5–12 mm wide; perigynia narrowly winged, conspicuously nerved ventrally (except in a variety of *C. siccata*); pistillate scales usually red-brown.

                6. Rhizomes slender, cordlike, brown; inflorescence 1–4 cm long; perigynia 4.5–6.0 mm long, the beak 2–3 mm long.. 5. *C. siccata*

                6. Rhizomes thick, black; inflorescence 2.5–6.5 mm long; perigynia 2.0–4.6 mm long, the beak about 0.75 mm long...4. *C. sartwellii*

            5. Inflorescence 3–5 mm wide; perigynia wingless or nearly so, usually nerveless or obscurely nerved ventrally; pistillate scales pale brown to green and hyaline.

7. Rhizomes stout; pistillate scales pale brown, longer than the perigynia; perigynia 8–12 per spike, ovate-lanceoloid, not spongy at base .... 2. *C. praegracilis*

7. Rhizomes slender; pistillate scales green and hyaline, shorter than the perigynia; perigynia 1–9 per spike, narrowly lanceoloid, spongy at base .... 10. *C. socialis*

Group XIV

Leaves, sheaths, and/or culms all glabrous; perigynia glabrous; stigmas 2; achenes lenticular; all spikes sessile, essentially alike, with staminate flowers at the tip of each spike; culms cespitose, not arising singly from extensive stolons and/or rhizomes.

1. Perigynia 1–3 per spike; culms capillary.. 31. *C. disperma*

1. Perigynia 4–many per spike (sometimes 1–3 perigynia per spike in *C. socialis*); culms not capillary.

  2. Culms conspicuously wing-angled; sheaths loose and open.

    3. Culms firm, not easily compressed .... A8. *C. oklahomensis*

    3. Culms soft, spongy, easily compressed.

      4. Beak of perigynium one-half as long as to nearly as long as the body; sheaths with red-brown dots; spikes 8–12 per inflorescence.

        5. Sheaths septate-nodulose; pistillate scales green and hyaline; beak of perigynium as long as the body .... 27. *C. conjuncta*

        5. Sheaths usually not septate-nodulose; pistillate scales brownish; beak of perigynium about one-half as long as the body .... 26. *C. alopecoidea*

      4. Beak of perigynium one-and-one-half to three times longer than the body; sheaths without red-brown dots; spikes 10–25 per inflorescence.

        6. Pistillate scales shorter than the perigynia; perigynia yellow to brown; inflorescence up to 20 cm long, up to 6 cm thick; spikes 15–25 per inflorescence.

          7. Perigynia 3.5–6.0 mm long, strongly nerved on the inner face, the base tapering to the beak ....28. *C. stipata*

          7. Perigynia 6–8 mm long, nerveless or faintly nerved on the inner face, the base abruptly enlarged below the beak..30. *C. crus-corvi*

        6. Pistillate scales about as long as the perigynia; perigynia greenish; inflorescence up to 6 cm long, up to 1.5 cm thick; spikes 10–15 per inflorescence ....
.... 29. *C. laevivaginata*

  2. Culms not conspicuously wing-angled; sheaths loose or tight.

    8. Inflorescence consisting of 10 or more spikes, usually compound at the lower nodes.

      9. Sheaths not septate-nodulose; leaves 1.0–3.1 mm wide; perigynia biconvex, 1.0–1.3 mm wide; pistillate scales acute to cuspidate.

        10. Sheaths copper-colored at summit; perigynia 2.5–4.0 mm long, 1.2–1.3 mm wide, stramineous to brown ....25. *C. prairea*

        10. Sheaths not copper-colored at summit; perigynia 2–3 mm long, about 1 mm wide, dark brown to olive-black... 24. *C. diandra*

      9. Sheaths usually septate-nodulose; leaves 2–8 mm wide; perigynia plano-convex (biconvex in *C. decomposita*), 1.2–2.0 mm wide; pistillate scales awned or at least mucronate.

        11. Perigynia plano-convex, green to stramineous to yellow-brown to golden brown, not spongy at base; inflorescence up to 10 cm long, up to 1.5 cm thick.

          12. Spikelets and perigynia greenish to stramineous at maturity.

            13. Leaves longer than the culms.

              14. Spikelets not subtended by setaceous bristles much longer than the perigynia, not appearing bristly; spikes usually up to 40 per inflorescence; perigynia lance-ovoid to ovoid, 1.2–1.8 mm wide....... 21. *C. vulpinoidea*

14. Spikelets subtended by setaceous bristles much longer than the perigynia, appearing bristly; spikes usually more than 40 per inflorescence; perigynia lanceoloid, 1.2–1.4 mm wide........................... A4. *C. setacea*

13. Leaves shorter than the culms.

   15. Spikelets not subtended by setaceous bristles much longer than the perigynia, not appearing bristly; perigynia more or less orbicular, 3–4 mm long, 2.0–2.5 mm wide; achenes 1.5–2.0 mm long................A5. *C. fissa*

   15. Spikelets subtended by setaceous bristles much longer than the perigynia, appearing bristly; perigynia lanceoloid, 2.2–2.8 mm long, 1.2–1.4 mm wide; achenes 1.2–1.6 mm long .... A4. *C. setacea*

12. Spikelets and perigynia yellow-brown or golden brown at maturity.

   16. Spikes and perigynia golden brown; perigynia 2.2–2.6 mm long, 1.4–1.8 mm wide, obscurely nerved, the beak 0.3–0.5 mm long ....
....22. *C. brachyglossa*

   16. Spikes and perigynia yellow-brown; perigynia 2.5–5.0 mm long, 1.7–3.0 mm wide, with 2–5 nerves on the dorsal face, the beak 0.5–1.2 mm long.

      17. Pistillate scales awn-tipped; perigynia 2.5–3.0 mm wide, the beak 0.8–1.2 mm long .... A7. *C. triangularis*

      17. Pistillate scales long-awned; perigynia 1.7–2.4 mm wide, the beak 0.5–0.7 mm long.. A6. *C. annectens*

11. Perigynia biconvex, olive-black, spongy at base; inflorescence up to 18 cm long, up to 4 cm thick ....23. *C. decomposita*

8. Inflorescence consisting of up to 10 spikes (occasionally up to 15 in *C. gravida*, *C. lunelliana*, and *C. sparganioides*), not compound at the lower nodes.

18. Sheaths loose, with green mottling or green septations; leaves 4–10 mm wide (sometimes only 3 mm wide in *C. aggregata*).

   19. Pistillate scales as long as or longer than the perigynia, awn-tipped ....
.... A3. *C. gravida*

   19. Pistillate scales one-half as long as the perigynia to nearly as long, acute (rarely awned in *C. aggregata*).

      20. Inflorescence 5–15 cm long, the lowest spikes remote from the others; leaves up to 10 mm wide; perigynia 15–50 per spike...................20. *C. sparganioides*

      20. Inflorescence up to 5 cm long, the spikes all contiguous; leaves 3–8 mm wide; perigynia 5–20 per spike.

         21. Leaves 3–5 mm wide; beak of the perigynium 1.0–1.5 mm long; summit of sheaths concave ....18. *C. aggregata*

         21. Leaves 4–8 mm wide; beak of the perigynium 0.5–1.3 mm long; summit of sheaths truncate.

            22. Leaves dark green; bracts subulate or absent; pistillate scales acute; perigynia narrowly ovoid, 3.3–4.5 mm long, 1.5–2.5 mm wide, obscurely nerved dorsally .... 19. *C. cephaloidea*

            22. Leaves light green; bracts setaceous; pistillate scales acuminate to cuspidate; perigynia broadly ovoid to orbicular, 3.7–5.0 mm long, 3–5 mm wide, prominently nerved dorsally .... 17. *C. lunelliana*

18. Sheaths tight, without green mottling or green septations; leaves up to 4 mm wide (sometimes up to 5 mm wide in *C. cephalophora*).

23. All spikes crowded into a head; perigynia usually not spongy-thickened at the base.

   24. Leaves 5–10 per culm; perigynia 2.7–4.5 mm long, 2–3 mm wide; ligule as wide as or wider than long; plants to 1 m tall.

      25. Pistillate scales about equaling the perigynia; inflorescence up to twice as long as thick; ligule wider than long .... 12. *C. mesochorea*

25. Pistillate scales shorter than the perigynia; inflorescence 2–4 times longer than thick; ligule as wide as long.

  26. Perigynia with conspicuous nerves on one or both faces, 3.0–4.5 mm long.

    27. Perigynia 3.0–3.5 mm long, 2.0–2.5 mm wide; achenes 2.0–2.2 mm long .... A1. *C. muehlenbergii*

    27. Perigynia 3.5–4.5 mm long, 2.5–3.0 mm wide; achenes 2.2–2.5 mm long ....:................A2. *C. austrina*

  26. Perigynia nerveless on both faces, 2.7–3.1 mm long .... 14. *C. plana*

24. Leaves 3–5 per cùlm; perigynia 2.5–3.2 mm long, 1.2–2.0 mm wide; ligule longer than wide; plants up to 75 cm tall.

  28. Leaves 2–5 mm wide; perigynia broadest above the base, 1.5–2.0 mm wide ....11. *C. cephalophora*

  28. Leaves up to 3 mm wide; perigynia broadest at the base, 1.2–2.0 mm wide ....13. *C. leavenworthii*

23. Spikes in an elongated, often interrupted, inflorescence; perigynia spongy-thickened at the base (except in *C. muehlenbergii*, *C. australis*, and *C. plana*).

  29. Perigynia 2–3 mm wide; inflorescence up to 1 cm thick, often capitate above.

    30. Perigynia not spongy-thickened at the base; leaves 5–10 per culm; beak of the perigynium up to 1 mm long.

      31. Perigynia with conspicuous nerves on one or both faces, 3.0–4.5 mm long.

        32. Perigynia 3.0–3.5 mm long, 2.0–2.5 mm wide; achenes 2.0–2.2 mm long .... A1. *C. muehlenbergii*

        32. Perigynia 3.5–4.5 mm long, 2.5–3.0 mm wide; achenes 2.2–2.5 mm long ....A2. *C. austrina*

      31. Perigynia nerveless on both faces, 2.7–3.1 mm long .... 14. *C. plana*

    30. Perigynia spongy-thickened at the base; leaves 3–6 per culm; beak of the perigynium 1.0–1.8 mm long.

      33. Bracts leaflike, up to 25 cm long; leaves 1.0–2.5 mm wide; perigynia 3.5–4.0 mm long, pale green; beak of the perigynium 1.0–1.2 mm long ....:............................16. *C. arkansana*

      33. Bracts setaceous, up to 2 cm long; leaves 2–4 mm wide; perigynia 4.0–4.5 mm long, green-brown to black; beak of the perigynium 1.5–1.8 mm long ....15. *C. spicata*

  29. Perigynia 0.7–2.0 mm wide; inflorescence up to 5 mm thick, never capitate at the tip.

    34. Beak of the perigynium smooth; bracts absent or only poorly developed; some perigynia reflexed.

      35. Perigynia broadly ovoid, 1.3–2.0 mm wide, greenish brown at the base; leaves up to 2.5 mm wide....................................6. *C. retroflexa*

      35. Perigynia narrowly lanceoloid, 0.7–1.2 mm wide, pale green throughout; leaves up to 1.5 mm wide.. 7. *C. texensis*

    34. Beak of the perigynium serrulate; bracts usually present, setaceous, up to 15 cm long; perigynia spreading to ascending.

      36. Stolons present, the plants colonial; perigynia narrowly lanceoloid ....10. *C. socialis*

      36. Stolons absent, the plants not colonial; perigynia ovoid-lanceoloid.

        37. Stigmas twisted; leaves up to 3 mm wide; beak of the perigynium 0.8–1.5 mm long... 8. *C. rosea*

        37. Stigmas more or less straight; leaves up to 1.9 mm wide; beak of the perigynium 0.5–1.0 mm long .... 9. *C. radiata*

Group XV

Leaves, sheaths, and/or culms all glabrous; perigynia glabrous; stigmas 2; achenes lenticular; all spikes sessile, essentially alike, with staminate flowers at the base of each spike; culms cespitose, not arising singly from extensive stolons and/or rhizomes; perigynia plano-convex, with rounded margins or with narrowly rimmed margins.

1. Perigynia with rounded margins, ascending.
   2. Plants weak and often reclining; spikes 1–3 (–4) per culm; perigynia 1–5 per spike, oblongoid .... 32. *C. trisperma*
   2. Plants firm or, if weak, scarcely reclining; spikes usually 4 or more per culm; perigynia 5 or more per spike (rarely only 4 in *C. bromoides*), lanceoloid to ellipsoid to narrowly ovoid.
      3. All spikes overlapping; perigynia 3.5–4.5 mm long; beak of the perigynium 1.25–1.50 mm long.
         4. Perigynia lanceoloid, nerved on both faces; most leaves up to 2 mm wide.................. ....35. *C. bromoides*
         4. Perigynia ovoid-lanceoloid, nerveless or faintly nerved only on the convex face; most leaves more than 2 mm wide... 36. *C. deweyana*
      3. Some or all of the spikes separated from each other; perigynia 1.8–3.0 mm long; beak of the perigynium up to 1.2 mm long.
         5. Leaves up to 2 mm wide, dark green; perigynia 5–10 per spike; ventral band of sheath red-brown dotted.. 34. *C. brunnescens*
         5. Leaves 2–4 mm wide, glaucous to pale green; perigynia 10–30 per spike; ventral band of sheath not red-brown dotted... 33. *C. canescens*
1. Perigynia with narrowly rimmed margins, spreading to reflexed.
   6. Terminal spike usually unisexual, either entirely staminate or entirely pistillate; perigynia castaneous to nearly black ....38. *C. sterilis*
   6. Terminal spike gynecandrous; perigynia green to dark brown.
      7. Ventral face of the perigynium nerveless or nearly so; staminate part of the terminal spike proportionally longer than in the next two species ....37. *C. interior*
      7. Ventral face of the perigynium usually with up to 10–12 nerves; staminate part of the terminal spike proportionally shorter than in *C. interior.*
         8. Perigynia broadly ovoid to suborbicular, 2.0–3.5 mm long, 1.3–2.8 mm wide, the beak 0.5–1.2 mm long..40. *C. atlantica*
         8. Perigynia lanceoloid to ovoid, 2.5–4.5 mm long, 1–2 mm wide, the beak 1–2 mm long... 39. *C. echinata*

Group XVI

Leaves, sheaths, and/or culms all glabrous; perigynia glabrous; stigmas 2; achenes lenticular; all spikes sessile, essentially alike, with staminate flowers at the base of each spike; culms cespitose, not arising singly from extensive stolons and/or rhizomes; perigynia more or less flat and scalelike, with winged margins.

1. Perigynia 7–10 mm long, the beak 4.0–4.8 mm long; spike 12–27 mm long ....41. *C. muskingumensis*
1. Perigynia up to 7 mm long (to 7.5 mm in *C. bicknellii*), usually much shorter, the beak rarely as much as 4 mm long; spikes up to 12 (rarely to 15) mm long.
   2. Tip of pistillate scales exceeding, or at least equaling, the tip of the perigynium.
      3. Tip of pistillate scales exceeding the tip of the perigynium; perigynia 4.5–6.5 mm long, 1.5–2.0 mm wide, not broadly winged ....61. *C. praticola*

3. Tip of pistillate scales as long as the tip of the perigynium but not exceeding it; perigynia 4–5 mm long, 2.5–4.0 mm wide, broadly winged .... 60. *C. alata*

2. Tip of pistillate scales not reaching the tip of the perigynium.

4. Perigynia up to 2 mm wide.

5. Perigynia 0.8–1.0 mm wide, the wing very narrow to almost absent.......................... ....44. *C. crawfordii*

5. Perigynia 1.2 mm wide or wider, obviously winged.

6. Perigynia widest at the middle.

7. Perigynia evenly winged all the way to the base.

8. Perigynia appressed or ascending.

9. Perigynia lanceolate.

10. Perigynia usually nerved only on the dorsal face; sterile tufts of leaves usually not present; perigynia 3.8–5.5 mm long.............. 42. *C. scoparia*

10. Perigynia nerved on both faces; sterile tufts of leaves present; perigynia 3.0–3.6 mm long.. 48. *C. bebbii*

9. Perigynia ovate to obovate to ovate-lanceolate to broadly elliptic.

11. Perigynia faintly nerved; wing of perigynium not reaching base of the beak of the perigynium; perigynia 1.2–1.5 mm wide ....A9. *C. sangamonensis*

11. Perigynia strongly nerved; wing of perigynium reaching base of the beak of the perigynium; perigynia 1.5–2.6 mm wide.

12. Perigynia ovate; pistillate scales usually awn-tipped......................... ....46. *C. straminea*

12. Perigynia obovate; pistillate scales acute to acuminate, not awn-tipped...53. *C. longii*

8. Perigynia spreading.

13. Wing of perigynium not reaching tip of perigynium; perigynia obovate; spikes usually tapering to tip .... 52. *C. albolutescens*

13. Wing of perigynium extending to tip of perigynium; perigynia ovate; spikes usually subglobose, rounded at tip .... 49. *C. normalis*

7. Wing of perigynium narrowed above the base, not extending to the base.

14. Perigynia lance-ovate to ovate, usually crimped on the "shoulder," all widely spreading; spikes globose ....47. *C. cristatella*

14. Perigynia lanceolate, not crimped on the "shoulder," at least some of them usually ascending; spikes usually not globose.

15. All but the uppermost spikes remote from each other .... 45. *C. projecta*

15. Spikes crowded, with only 1 or 2 remote spikes, if any .... 43. *C. tribuloides*

6. Perigynia widest above or below the middle, but not at the middle.

16. Perigynia widest a little above the middle; wing of the perigynium not reaching the tip of the perigynium.. 52. *C. albolutescens*

16. Perigynia widest a little below the middle or nearly at the base; wing of the perigynium reaching the tip of the perigynium.

17. Perigynia lanceolate, the wing diminishing above the base of the perigynium .... 45. *C. projecta*

17. Perigynia ovate to orbicular to narrowly ovate, the wing extending all the way to the base of the perigynium.

18. All spikes except sometimes the lowest 1 or 2 crowded..... 49. *C. normalis*

18. Several of the lower spikes usually remote.

19. Sheaths papillose; pistillate scales ovate; beak of the perigynium 0.8–1.2 mm long..50. *C. tenera*

19. Sheaths glabrous; pistillate scales lanceolate to narrowly ovate; beak of the perigynium 1–2 mm long.

    20. Perigynia 2.5–3.5 mm long .... 51. *C. festucacea*

    20. Perigynia 3.5–4.5 mm long .... A10. *C. echinodes*

4. Some or all of the perigynia 2 mm wide or wider.

  21. Pistillate scales aristate to short-awned.

    22. Perigynia ovate, 2.3–3.5 mm long, conspicuously nerved..........46. *C. straminea*

    22. Perigynia rhombic to suborbicular to ovate to elliptic, 4–7 mm long, nerveless or finely nerved.

      23. Perigynia 4.5–7.0 mm long, widest at the middle, elliptic to ovate, the beak 2.0–2.8 mm long; achenes 1.2–1.5 mm wide .... A13. *C. missouriensis*

      23. Perigynia 4–5 mm long, widest above the middle, rhombic to suborbicular, the beak 0.7–1.5 mm long; achenes 0.7–1.0 mm wide.

        24. Wing of perigynium very broad, usually extending to the base of the perigynium; perigynia finely nerved on the ventral face, rounded at the base .... 60. *C. alata*

        24. Wing of perigynium not particularly broad, diminishing before reaching the base of the perigynium; perigynia more or less nerveless, cuneate at the base .... 55. *C. suberecta*

21. Pistillate scales obtuse to acute to acuminate, never aristate nor short-awned.

  25. Perigynia broadest above the middle.

    26. Wing of perigynium diminishing before reaching the base of the perigynium; perigynia more or less nerveless on both faces .... 55. *C. suberecta*

    26. Wing of perigynium extending to the base of the perigynium; perigynia at least finely nerved on both faces.

      27. Perigynia spreading, the wing not reaching the tip of the perigynium .. .. 52. *C. albolutescens*

      27. Perigynia appressed to ascending, the wing reaching the tip of the perigynium.

        28. Perigynia finely nerved, rhombic-orbicular .... 54. *C. cumulata*

        28. Perigynia strongly nerved, obovate .... 53. *C. longii*

  25. Perigynia broadest at or below the middle.

    29. Perigynia less than 4 mm long.

      30. Perigynia broadest below the middle .... 51. *C. festucacea*

      30. Perigynia broadest at the middle.

        31. Perigynia spreading.

          32. Wing of the perigynium not reaching the tip of the perigynium .... 52. *C. albolutescens*

          32. Wing of the perigynium reaching the tip of the perigynium .... 49. *C. normalis*

        31. Perigynia appressed to ascending.................................53. *C. longii*

  29. Some or all of the perigynia 4 mm long or longer.

    33. Perigynia nerveless on the ventral face, orbicular.

      34. Perigynia 2.5–3.5 mm wide..56. *C. brevior*

      34. Perigynia 3.5–5.0 mm wide.................................... 58. *C. reniformis*

    33. Perigynia nerved, at least finely so, on the ventral face, ovate to obovate to suborbicular.

      35. Spikes globose to subglobose.

        36. Sterile shoots present; perigynia 3.0–4.5 mm long, 1.5–2.1 mm wide .... 49. *C. normalis*

        36. Sterile shoots absent; perigynia 4.0–5.5 mm long, 2.0–3.5 mm wide.

37. Perigynia faintly nerved; beak of perigynium 1.0–1.3 mm long; achenes elliptic to narrowly oblong, 1.2–1.3 mm wide .... 57. *C. molesta*

37. Perigynia conspicuously nerved; beak of perigynium 1.3–1.8 mm long; achenes broadly ovate to orbicular, 1.4–1.8 mm wide .... A11. *C. molestiformis*

35. Spikes longer than broad.

38. Perigynia 4.5–7.5 mm long, 2.5–5.0 mm wide; beak of the perigynium 1.0–2.8 mm long.

39. Perigynia translucent, exposing the achene; beak of the perigynium 1.0–1.8 mm long; sheaths papillate........................ ....59. *C. bicknellii*

39. Perigynia opaque, concealing the achene; beak of the perigynium 1.8–2.8 mm long; sheaths glabrous.

40. Beak of perigynium one-half as long as the body of the perigynium; pistillate scales acuminate .... A13. *C. missouriensis*

40. Beak of perigynium one-fourth as long as the body of the perigynium; pistillate scales obtuse .... A12. *C. opaca*

38. Perigynia 2.6–4.5 mm long, 1.5–2.7 mm wide; beak of the perigynium 0.5–1.0 mm long.

41. Perigynia spreading, the wing not reaching the tip of the perigynium... 52. *C. albolutescens*

41. Perigynia appressed to ascending, the wing reaching the tip of the perigynium..53. *C. longii*

Page 29. *Carex stenophylla* Wahlenb. var. *enervis* (C. A. Mey.) Kukenth.

Our plants seem to match up well with the Asian *Carex duriuscula*, although there are a few small differences. Bailey named our plants *Carex eleocharis*, but *C. duriuscula* clearly predates Bailey's binomial. Our plant has also been considered a variety of the European *Carex stenophylla*, but *C. stenophylla* is clearly a different species. Taxonomy of this species follows:

Carex duriuscula C. A. Mey. Acad. Imp. Sci. St.-Petersbourg Divers Savan 1:214. 1831.

Carex eleocharis L. H. Bailey, Mem. Torrey Club 1:6. 1889.
Carex stenophylla Wahlenb. var. *enervis* (C. A. Mey.) Kukenth. in Engelm. Das Pflanzenr. 4 (20):122. 1909.

The illustration on page 30 should be relabeled *Carex duriuscula*.

The following county should be added to the map on page 30: Carroll.

Page 31. *Carex praegracilis* W. Boott. The following counties should be added to the map on page 31: Champaign, Iroquois, Ogle, Vermilion.

Page 33. *Carex sartwellii* Dewey. The following counties should be added to the map on page 36: Ford, Iroquois, Livingston, McDonough.

Page 37. *Carex foenea* Willd. The species that I applied the binomial *C. foenea* to should be called *Carex siccata*. *Carex foenea* is an entirely different species in section Ovales and is not known from Illinois. The taxonomy follows:

Carex siccata Dewey, Am. Journ. Sci. Arts 10:278. 1826.

There are no new records for this species.

The illustration on page 38 should be relabeled *Carex siccata*.

Page 39. *Carex foenea* Willd. var. *enervis* Evans & Mohlenbr. This variety, with perigynia nerveless on the ventral face, should be transferred to *C. siccata*.

Its taxonomy follows:

Carex siccata Dewey var. **enervis** (Evans & Mohlenbr.) Mohlenbr. comb. nov. (basionym: *Carex foenea* var. *enervis* Evans & Mohlenbr.)

Carex foenea Willd. var. *enervis* Evans & Mohlenbr. Trans. Ill. Acad. Sci. 64:270. 1972.

There are no new records for this variety.

Page 39. *Carex retroflexa* Muhl. ex Willd. The following counties should be added to the map on page 41: Franklin, Greene, Montgomery, Shelby.

Page 41. *Carex texensis* (G. S. Torr. ex Bailey) Fern. The following counties should be added to the map on page 43: Montgomery, Ogle. In the Cyperaceae volume of Flora of North America, this species is inexplicably not listed from Illinois, even though there are extant specimens and literature reports for it in Illinois.

Page 43. *Carex rosea* Schkuhr ex Willd. The following counties should be added to the map on page 45: Brown, Greene, Knox, Perry, Richland.

Page 45. *Carex radiata* Wahlenb. The following county should be added to the map on page 47: Gallatin.

Page 47. *Carex socialis* Mohlenbr. & Schwegm. The following county should be added to the map on page 49: Washington.

Page 49. *Carex cephalophora* Muhl. ex Willd. The following counties should be added to the map on page 51: Lee, Mason.

Page 51. *Carex mesochorea* Mack. The following counties should be added to the map on page 53 (top): Macon, Morgan, Winnebago.

Page 53. *Carex leavenworthii* Dewey. The following counties should be added to the map on page 53 (bottom): Alexander, Cass, Jersey, Montgomery, Saline, Sangamon, Union.

Page 54. *Carex muhlenbergii* Schk. ex Willd. var. *muhlenbergii*. The correct spelling for this taxon should be *Carex muehlenbergii* var. *muehlenbergii*. Because the other two varieties of *C. muhlenbergii* recognized in the first edition of *Carex* are now treated as distinct species in this work, a new description for *C. muhlenbergii*, as well as a new illustration, is provided, since the illustration on page 56 is actually of *Carex plana*.

Carex muehlenbergii Schk. ex Willd. Sp. Pl. 4:231. 1805. Fig. A1.

Plants perennial, cespitose, from fibrous roots and short, woody, black rhizomes; culms erect, rough to the touch beneath the inflorescence, up to 1 m tall, light brown at base, longer than the leaves; leaves 5–10, up to 30 cm long, 2.0–4.5 mm wide, confined to the lower fourth of the plant, flat or slightly canaliculate, pale green, rough along the margins and on the upper surface; sheaths open,

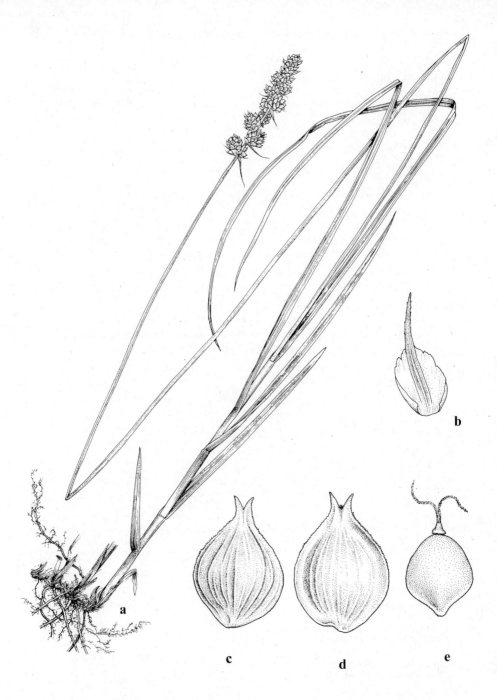

A1. Carex muehlenbergii
(Muhlenberg's sedge).

a. Habit.
b. Pistillate scale.
c. Perigynium, dorsal view.

d. Perigynium, ventral view.
e. Achene.

tight, concave and thickened at the mouth, not as rough as the upper leaf surface, the hyaline ventral band well-developed; ligule as wide as long; inflorescence unbranched, ovoid, up to 4 cm long, up to 1 cm broad, with most of the spikes crowded into a head, but with one or two of the lower spikes separated; bracts setaceous, the lowest ones 1.0–2.0 (–3.5) cm long; spikes 5–10, 5–35 mm long, about as wide, the staminate flowers above the pistillate; pistillate scales ovate, 2.5–3.6 mm long, 1.6–2.2 mm wide, aristate, green or hyaline with a green midvein, a little shorter and narrower than the perigynia; staminate scales narrowly lanceolate, acuminate, pale brown; perigynia up to 20 per spikelet, 3.0–4.5 mm long, 2.0–2.5 mm wide, plano-convex, suborbicular to broadly ovoid, pale green, the ventral surface with 5–9 prominent veins, with a 2-toothed, serrulate beak 0.75–1.00 mm long; achenes 2.0–2.2 mm long, lenticular, ovoid-orbicular, apiculate, jointed to the style, the style swollen at base; stigmas 2, slender, reddish brown.

This species is similar to *C. austrina* by virtue of the conspicuous nerves on the ventral face of the perigynia, but differs by its shorter and narrower perigynia and its smaller achenes.

The following counties should be added to the map on page 55: Clay, Effingham, Fayette, Marion, McLean, Williamson.

Page 57. *Carex muhlenbergii* Schk. var. *austrina* Small. Although often considered a variety of *C. muhlenbergii*, the larger perigynia and achenes are sufficient enough to recognize this as a separate species. Its taxonomy and description follow:

Carex austrina (Small) Mack. Bull. Torrey Club 34:151. 1907. Fig. A2.

Carex muehlenbergii Schk. var. *austrina* Small, Fl. S.E.U.S. 218. 1903.

Plants perennial, cespitose, from fibrous roots and short, woody, black rhizomes; culms erect, rough to the touch beneath the inflorescence, to 1.2 m tall, brown at base, usually a little longer than the leaves; leaves 4–12, up to 35 cm long, 2–5 mm wide, mostly confined to the lower fourth of the plant, more or less flat, green, rough along the margins and usually on the upper surface; sheaths open, tight, concave and thickened at the mouth, usually slightly scabrous, the hyaline ventral band well-developed; ligule as wide as long; inflorescence unbranched, ovoid to oblongoid, up to 4 cm long, up to 1 cm broad, with most of the spikes crowded into a head, but with one or two of the lower spikes separated; bracts setaceous, the lowest ones 1–2 (–5) cm long; spikes 3–10, the staminate flowers above the pistillate; pistillate scales ovate to ovate-lanceolate, acuminate, pale brown; perigynia up to 20 per spikelet, 3.5–4.5 mm long, 2.5–3.0 mm broad, plano-convex, suborbicular to broadly ovoid, pale green, with several conspicuous nerves, with a 2-toothed, serrulate beak to 1 mm long; achenes 2.2–2.5 mm long, lenticular, ovate-orbicular, apiculate, jointed to the style, the style swollen at base; stigmas 2, slender, reddish brown.

Common Name: Southern sedge.
Habitat: Along a railroad (in Illinois).
Range: Indiana to Kansas, south to Texas and Florida.
Illinois Distribution: Known only from along a railroad in Champaign County where it probably is adventive.

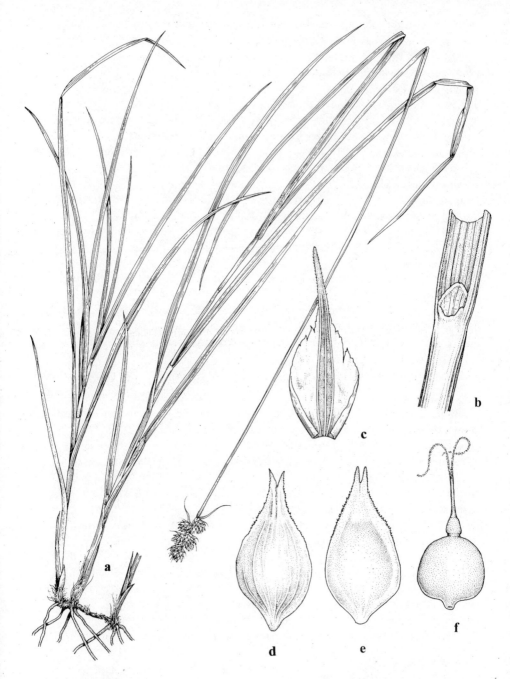

A2. Carex austrina
  (Southern sedge).

a. Habit.
b. Sheath.
c. Pistillate scale.

d. Perigynium, dorsal view.
e. Perigynium, ventral view.
f. Achene.

In the Cyperaceae volume of Flora of North America, this species is inexplicably not listed for Illinois, even though there are specimens and literature reports for it in Illinois.

Specimens that I reported earlier from Jackson and Perry counties are actually *C. gravida.*

Carex austrina flowers in May and June.

Page 57. *Carex muhlenbergii* Schk. var. *enervis* Boott. Illustr. Carex 3:124. 1862. After many years of observing and studying the *C. muehlenbergii* complex in Illinois and other states that I have taught in, I have come to the conclusion that this taxon should be treated as a separate species. Its taxonomy and description follow:

Carex plana Mack. Bull. Torrey Club 50:350. 1923. The illustrations on page 56 of the first edition of *Carex* are of *C. plana* and not *C. muhlenbergii.*

Carex muehlenbergii Schk. var. *enervis* Boott, Illustr. Carex 3:124. 1862.

Plants perennial, cespitose, from fibrous roots and short, woody, black rhizomes; culms erect, rough to the touch beneath the inflorescence, to 80 cm tall, brown at base, longer than the leaves; leaves 5–10, to 25 cm long, 2–4 mm wide, usually confined to the lower fourth of the culm, flat, green, rough along the margins and the upper surface; sheaths open, tight, concave and thickened at the mouth, occasionally septate-nodulose, somewhat scabrous, the hyaline ventral band well-developed; ligule as wide as long; inflorescence unbranched, oblongoid to oblong-ovoid, to 3.5 cm long, to 1.2 cm broad, with most of the spikes crowded into a head, but with one or two of the lower spikes separated; bracts setaceous, 1–2 cm long; spikes 5–15, to 25 mm long and broad, the staminate flowers above the pistillate; pistillate scales 2.0–2.5 mm long, 1.2–1.3 mm wide, lance-ovate, short-awned, green or hyaline with a green midvein, about two-thirds as long as and narrower than the perigynia; staminate scales narrowly lanceolate, acuminate, pale brown; perigynia up to 15 per spikelet, 2.7–3.1 mm long, 1.8–2.2 mm wide, plano-convex, lance-ovoid, green, both surfaces nerveless or with a few obscure nerves, with a 2-toothed, serrulate beak up to 0.75 mm long; achenes 1.8–2.1 mm long, lenticular, ovoid, apiculate, substipitate, jointed to the style, the style swollen at base; stigmas 2, slender, reddish brown.

Common Name: Nerveless sedge.
Habitat: Woods, fields, often in calcareous areas.
Range: Vermont to Kansas, south to Texas and Mississippi.
Illinois Distribution: Scattered in the state; apparently somewhat more common than *C. muehlenbergii.*

This species differs from *C. muehlenbergii* by its nerveless perigynia and its slightly shorter and narrower perigynia. I have observed that most specimens with perigynia less than 3 mm long also have nerveless perigynia.

Mackenzie described two species similar to *C. muehlenbergii* that have nerveless perigynia—*C. plana* and *C. onusta*, the latter mostly from Texas. Most botanists consider these to be synonymous with *C. muehlenbergii*, but I disagree. *Carex*

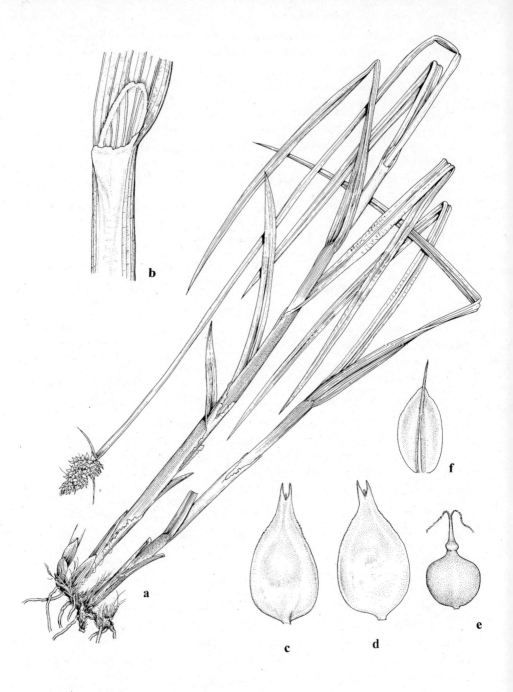

A3. Carex gravida
(Heavy sedge).

a. Habit.
b. Sheath.
c. Perigynium, dorsal view.

d. Perigynium, ventral view.
e. Achene.
f. Pistillate scale.

onusta, which has a shorter stature, is readily recognized by its strongly yellow-brown pistillate scales. It also differs by having perigynia 1.4–1.7 mm wide and a more yellowish perigynium.

Although there is some intergradation between *C. plana* and *C. muehlenbergii*, as one would expect in evolving organisms, there is even more intergradation between *C. austrina* and *C. muehlenbergii*.

Page 57. *Carex spicata* Huds. The following county should be added to the map on page 59: Winnebago.

Page 59. *Carex arkansana* Bailey. The following counties should be added to the map on page 61: Douglas, Jackson, Washington.

Page 61. *Carex gravida* Bailey var. *gravida*. In the first edition of *Carex*, I treated *C. gravida* with two varieties in Illinois. I now believe that var. *lunelliana* should receive species status, particularly because of its very distinctive perigynia. Therefore, a new description of *Carex gravida* follows:

Carex gravida Bailey, Mem. Torrey Club 1:5. 1889. Figs. A3 and 17f.

Plants perennial, cespitose, from fibrous roots and short, dark, woody rhizomes; culms erect, up to 1 m tall, very rough to the touch beneath the inflorescence, light brown at base; leaves 4–6, green, up to 30 cm long, 3–8 mm wide, flat, rough on the upper surface, smooth and septate-nodulose below; sheaths open, loose, pale with green nerves, septate-nodulose, usually convex and slightly thickened at the mouth; inflorescence an elongated, sometimes interrupted head up to 5 cm long, up to 1.5 cm broad; bracts setaceous, the lowest as long as the inflorescence; spikes 8–15, the staminate flowers above the pistillate; pistillate scales short-awned, green or brownish with a conspicuous midvein, about as long as or longer than the perigynia; staminate scales narrow, acuminate, pale brown; perigynia up to 15 per spike, 3.5–5.5 mm long, 2.0–2.5 mm broad, plano-convex, ellipsoid, greenish on the slightly raised margins, shiny, stramineous above the spongy base, faintly nerved on both faces, tapering to a 2-toothed, serrulate beak 1.0–1.5 mm long; achenes 2 mm long, lenticular, yellow-brown, apiculate, jointed to the style, the style swollen at the base; stigmas 2, reddish brown to pale brown.

Common Name: Heavy sedge.
Habitat: Dry prairies, old fields, disturbed meadows.
Range: Ontario to Saskatchewan, south to Wyoming, New Mexico, southern Illinois, and western Ohio.
Illinois Distribution: Scattered throughout the state.

In the first edition of *Carex*, I noted that it is often difficult to distinguish this species from what I then called var. *lunelliana*. However, after a decade of more intense study, particularly in the field, I have become convinced that the variety should be given the rank of species. Although I admit that there is some intergradation, my studies show some good correlating features. More than 85 percent of the specimens I have examined with more ellipsoid perigynia that taper to a beak 1.0–1.5 mm long also have very faint nerves on the perigynia, and 75 percent of these also have short-awned pistillate scales. These are *C. gravida*. Conversely, specimens with ovoid to orbicular perigynia are likely to have conspicuous nerves on the dorsal face of the perigynia, particularly evident in living material. These latter specimens invariably have a more contracted beak 0.5–1.0 mm long and have pistillate scales that are usually cuspidate and not short-awned. These are *C. lunelliana*.

Carex gravida flowers from mid-May through July.

Carex lunelliana Mack. Bull. Torrey Club 42:615. 1915. Figs. 17 a–e.
Carex gravida Bailey var. *lunelliana* (Mack.) F. J. Hermann, Am. Midl. Nat. 17:855. 1936.

Plants perennial, cespitose, from fibrous roots and short, dark, woody rhizomes; culms erect, up to 80 cm tall, rough to the touch beneath the inflorescence, light brown at base; leaves 4–6, light green, up to 20 cm long, 4–8 mm wide, flat, rough on the upper surface, smooth and usually septate-nodulose below; sheaths open, loose, light green, septate-nodulose, truncate and thickened at the mouth; inflorescence an elongated head up to 3.5 cm long, up to 1.5 cm broad, dense; bracts setaceous; spikes 6–12, the staminate flowers above the pistillate; pistillate scales acuminate to cuspidate, rarely with a short awn, hyaline with a 3-nerved green center, shorter than the perigynia; staminate scales narrow, acuminate, tan; perigynia 5–15 per spike, 3.7–5.0 mm long, 3–5 mm broad, plano-convex, ovoid to orbicular, green to stramineous to light yellow-brown, the margins not raised, shiny, the base spongy, nerveless ventrally but with several prominent nerves on the dorsal face, abruptly contracted to a 2-toothed serrulate beak 0.5–1.0 mm long; achenes 2 mm long, lenticular, yellow-brown, apiculate, jointed to the style, the style swollen at the base; stigmas 2, reddish brown.

Common Name: Lunell's sedge.
Habitat: Dry prairies, old fields.
Range: Indiana and southern Michigan to South Dakota, south to New Mexico and Arkansas.
Illinois Distribution: Scattered throughout the state except for the northeastern counties.

While examining specimens that had been collected by Lunell, Mackenzie noted that the species known as *C. gravida* actually included two closely related species. Mackenzie named the other species *C. lunelliana*. For a comparison of this species with *C. gravida*, see the discussion under this latter species.

Carex lunelliana flowers from May to late June.

Page 63. *Carex aggregata*. Mack. The following counties should be added to the map on page 65: Brown, Macoupin, Marshall, Peoria, Tazewell.

Page 65. *Carex cephaloidea* (Dewey) Dewey. The following counties should be added to the map on page 67 (top): Bureau, Clark, Grundy, Jo Daviess, Lake, Ogle.

Page 67. *Carex sparganioides* Muhl. ex Willd. The following counties should be added to the map on page 67 (bottom): Fulton, Greene, Lawrence, Macoupin, Mason, McDonough, Ogle, Tazewell.

Page 69. *Carex vulpinoidea* Michx. In the list of synonyms, remove *Carex setacea* Dewey, *Carex xanthocarpa* Bicknell var. *annectens* Bicknell, and *Carex annectens* (Bicknell) Bicknell. Since I have now removed some of the synonyms I had attributed to *C. vulpinoidea* in the first edition, a newly modified description follows:

Carex vulpinoidea Michx. Fl. Bor. Am. 2:169. 1803.

Carex multiflora Muhl. ex Willd. Sp. Pl. 4:243. 1805.
Carex scabrior Sartw. in Boott, Illustr. Carex 3:125. 1862.
Plants perennial, densely cespitose, from fibrous roots and short, stout, dark rhizomes; culms erect, to 1 m tall, scabrous on the sharp angles at least beneath the inflorescence, shorter than the leaves, dark brown at base; leaves 3–several, up to 1.2 m long, 2–5 mm wide, flat or slightly canaliculate, scabrous along the margins, the upper ones exceeding the culms; sheaths open, tight, with a green midnerve, with a broad, opaque, russet-maculate, often septate-nodulose ventral band, convex at the mouth; ligule broader than long; inflorescence composed of up to 40 spikes in an elongated, interrupted head up to 10 cm long and up to 1.5 cm broad, but not appearing bristly; bracts setaceous, subtending most of the spikes, the lowest one up to 5 cm long; spikes green to stramineous, the staminate flowers above the pistillate; pistillate scales lanceolate, strongly awned, hyaline to greenish brown and with a green midnerve, as long as or longer than the perigynia; perigynia several per spike, ovoid, 2–3 mm long, 1.2–1.8 mm broad, plano-convex, green to stramineous, usually with 2–4 nerves on the dorsal face, with a narrow, corky margin, tapering to a 2-toothed, serrulate or smooth beak 0.8–1.2 mm long; achenes 1.2–1.4 mm long, lenticular, ovoid, red-brown, glossy, apiculate, jointed to the style, the style swollen at base; stigmas 2, reddish brown.

Common Name: Fox sedge.
Habitat: Swamps, wet meadows, low areas, moist open ground.
Range: Newfoundland to British Columbia, south to California, Arizona, Texas, and Georgia.
Illinois Distribution: Common; in every county.

Carex vulpinoidea flowers from May to early August.
Following *Carex vulpinoidea*, the following species should be added to the Illinois flora:

Carex setacea Dewey, Am. Journ. Sci. 9:61. 1825. Fig. A4.

Plants perennial, densely cespitose, from fibrous roots and short, stout, dark rhizomes; culms erect, slender, to 1.2 m tall, scabrous on the sharp angles, at least below the inflorescence, usually longer than the leaves, brown at base; leaves 3–several, up to 80 cm long, 2–6 mm wide, flat, scabrous along the margins, usually shorter than the culms; sheaths open, tight, with a green midnerve, septate, russet-maculate, convex at the mouth; ligule broader than long; inflorescence composed of more than 40 spikes in an elongated, interrupted head up to 20 cm long, up to 1 cm broad; bracts setaceous, subtending all the spikes and many of the spikelets, usually several times longer than the spikelets, appearing densely bristly; spikes green

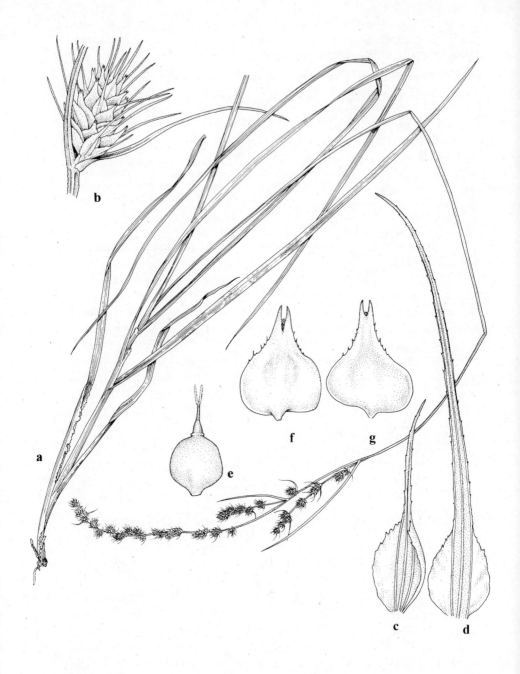

A4. Carex setacea
   (Bristly sedge).

a. Habit.
b. Spikelet.
c, d. Pistillate scales.

e. Achene.
f. Perigynium, dorsal view.
g. Perigynium, ventral view.

to stramineous, ovoid to ovate-oblongoid, 5–10 mm long, the staminate flowers above the pistillate flowers; pistillate scales acuminate to long-attenuate, often awned, greenish brown with a green midnerve; perigynia several per spike, ovate-lanceoloid to lanceoloid, green to stramineous, 2.2–2.8 mm long, 1.2–1.4 mm broad, with 3–5 nerves on the dorsal face, tapering to a 2-toothed, serrulate beak 0.7–1.0 mm long; achenes 1.2–1.6 mm long, lenticular, narrowly ovoid, glossy, red-brown, apiculate, jointed to the style, the style swollen at base; stigmas 2, reddish brown.

Common Name: Bristly sedge.
Habitat: Wet ground.
Range: Vermont to Ontario and Iowa, south to Texas and Kentucky.
Illinois Distribution: Known from Calhoun, Jackson, and Lake counties.

This enigmatic plant was described by Dewey who commented that it had probably been confounded with C. multiflora (= C. vulpinoidea). In its general appearance and in several characters, Dewey noted it is very different from C. multiflora and from any species hitherto described. He remarked that it appears to be intermediate between C. multiflora and C. stipata.

In the twentieth century, most botanists have ignored C. setacea, usually relegating it to an anomalous form of C. vulpinoidea.

The presence of numerous long, setaceous bracts subtending all the spikes and many of the spikelets gives this species a very bristly appearance. In addition, its perigynia are more narrow than those of C. vulpinoidea.

Because its leaves are usually shorter than the culms, it is similar to C. annectens and C. brachyglossa, but these species are not particularly bristly, have broader perigynia, and yellow-brown or golden brown spikes. It also differs from C. fissa because of its densely bristly inflorescences.

Carex setacea flowers in July and August, usually later than the other species mentioned in the last two paragraphs.

Carex fissa Mack. in N. L. Britt. N. Am. Fl. 18:64. 1931. Fig. A5.

Plants perennial, densely cespitose, from fibrous roots and short, stout, dark rhizomes; culms erect, up to 1 m tall, scabrous, at least beneath the inflorescence, longer than the leaves, dark brown at base; leaves 3–several per culm, up to 60 cm long, 2–6 mm wide, flat or slightly caniculate, scabrous along the margins, all of them shorter than the culms; sheaths open, tight, green-nerved, septate, with a broad, opaque, russet-maculate, often septate-nodulose ventral band, rather strongly convex at the mouth; ligule broader than long; inflorescence compound of many spikes in an elongated, interrupted head up to 10 cm long and up to 1.5 cm broad; bracts setaceous, subtending most of the spikes, the lowest up to 5 cm long; spikes many, green to stramineous, the staminate flowers above the pistillate; pistillate scales lanceolate, short-awned, hyaline to greenish brown and with a green midnerve, as long as or longer than the perigynia; perigynia several per spike, ascending, more or less orbicular, 3–4 mm long, 2.0–2.5 mm broad,

A5. Carex fissa
   (Short-leaved fox sedge).

a. Habit.
b. Sheath.
c. Perigynium, dorsal view.

d. Perigynium, ventral view.
e. Pistillate scale.
f. Achene.

plano-convex, green to stramineous, usually with a few nerves on the dorsal face, with a narrow, corky margin, tapering to a 2-toothed, usually smooth beak, the beak strongly notched, 0.8–1.0 mm long; achenes 1.5–2.0 mm long, lenticular, brown, glossy, apiculate, jointed to the style, the style swollen at base; stigmas 2, reddish brown.

Common Name: Short-leaved fox sedge.
Habitat: Swamps, wet meadows, moist open ground.
Range: Illinois to Kansas, south to Texas.
Illinois Distribution: Jackson and Union counties.

This species is similar to *C. vulpinoidea* because of its many-flowered spikes that are pale yellow-green to stramineous, but differs by its leaves that are all shorter than the culms. It is also somewhat similar to *C. setacea* because of its pale yellow-green to stramineous spikelets, but lacks the bristly appearance of the spikes. The short leaves are reminiscent of *C. annectens* and *C. brachyglossa*, but these latter species have yellow-brown and golden brown spikes, respectively.

The flowers of *C. fissa* appear in May and June.

Page 71. *Carex brachyglossa* Mack. When I wrote the first edition of *Carex* in 1999, I treated the *Carex brachyglossa* complex erroneously because I had not thoroughly studied the various entities critically, particularly in the field. During the past decade, I have had a chance to examine thoroughly members of this complex in the field in several parts of their ranges. As a result, I now have a completely different concept of the plants. What I recognized as *C. brachyglossa* in 1999 also includes *C. annectens* which I had erroneously placed in synonymy with *C. vulpinoidea*. Descriptions of *C. brachyglossa* and *C. annectens* follow:

Carex brachyglossa Mack. Bull Torrey Club 20:355. 1923. Fig. 22.

Carex vulpinoidea Michx. var. *xanthocarpa* (E. P. Bicknell) Kukenth. Pflanzenr. IV, 4:148. 1909.
Carex annectens (E. P. Bickn.) E. P. Bicknell var. *xanthocarpa* (E. P. Bicknell) Wieg. Rhodora 24:74. 1922.

Plants perennial, densely cespitose, from fibrous roots and short, stout, dark rhizomes; culms to 75 cm long, longer than the leaves, scabrous below the inflorescence, bluntly triangular, stiff, brown at base; leaves 3–6 per culm, to 60 cm long, 3–6 mm wide, flat or slightly canaliculate, scabrous along the margins; sheaths open, tight, at least the upper with the hyaline ventral band becoming septate-nodulose with age, shallowly convex at summit; ligules as wide as long but not wider; inflorescence of many subcompound spikes in an elongated head up to 8 cm long, dense and usually uninterrupted; lowest bract setaceous, up to 2 cm long, or sometimes absent; spikes 10–15, golden brown, gynecandrous, ovoid, few-flowered; pistillate scales lanceolate, short-awned, not appearing bristly, tawny to yellow-brown, usually shorter than the perigynia; perigynia plano-convex,

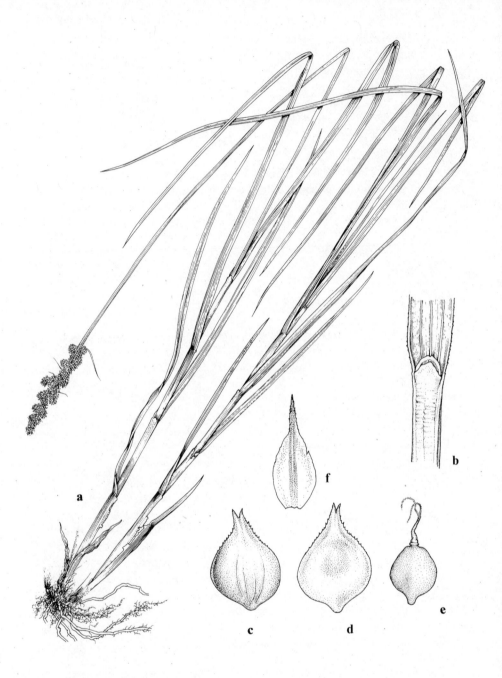

A6. Carex annectens
(Yellow-brown fox
sedge).

a. Habit.
b. Sheath.
c. Perigynium, dorsal view.

d. Perigynium, ventral view.
e. Achene.
f. Pistillate scale.

ascending, 2.2–2.6 mm long, 1.4–1.8 mm wide, deep golden brown, ovate-ellipsoid to subrhomboid, with obscure nerves dorsally, tapering to an obscurely notched, bidentate, serrulate beak 0.3–0.5 mm long, without green margins; achenes lenticular, broadly oblongoid, 1.5–1.8 mm long, red-brown, glossy, apiculate, jointed to the style, the style enlarged at base; stigmas 2, reddish brown.

Common Name: Larger yellow fox sedge.
Habitat: Around ponds and lakes, marshes, fens, old fields, mesic forests, disturbed areas; sometimes as an emergent aquatic.
Range: Maine to Wisconsin, south to Kansas, Missouri, Indiana, and Virginia.
Illinois Distribution: Scattered throughout the state.

For an analysis of the characteristics of this species and *C. annectens*, see the description under *C. annectens*.

This species flowers in May and June.

Carex annectens (E. P. Bicknell) E. P. Bicknell, Bull. Torrey Club 35:492. 1908. Fig. A6.

Carex xanthocarpa E. P. Bicknell var. *annectens* E. P. Bicknell, Bull. Torrey Club 23:22–24. 1903.
Carex vulpinoidea Michx. var. *annectens* (E. P. Bicknell) Farw. Pap. Mich. Acad. Sci. 1:91. 1923.

Plants perennial, densely cespitose, from fibrous roots and short, stout, dark rhizomes; culms to 1 m tall, longer than the leaves, scabrous below the inflorescence, bluntly triangular, stiff, brown at base; leaves 3–6 per culm, to 40 cm long, 2–5 mm wide, flat or slightly canaliculate, scabrous along the margins, none of them exceeding the culms; sheaths open, tight, at least the upper with the hyaline ventral band becoming septate-nodulose with age, strongly convex at summit; ligules broader than long; inflorescence composed of many subcompound spikes in an elongated head up to 7 cm long, crowded or the lowest sometimes separated; lowest bract setaceous, 2–5 cm long; spikes 10–15, yellow-brown, gynecandrous, globose to ovate-oblongoid, many-flowered; pistillate scales lanceolate, long-awned, sometimes appearing slightly bristly, tawny to yellow-brown, longer than the perigynia; perigynia loosely spreading, plano-convex, 2.6–3.2 mm long, 1.7–2.4 mm wide, yellow-brown to greenish, ovoid to suborbicular, with 2–4 conspicuous nerves dorsally, tapering to a conspicuously notched bidentate, serrulate beak, the beak 0.5–0.7 mm long with a greenish margin; achenes lenticular, ovoid-oblongoid to suborbicular, 1.2–1.7 mm long, red-brown, glossy, apiculate, jointed to the style, the style enlarged at base; stigmas 2, reddish brown.

Common Name: Yellow-brown fox sedge.
Habitat: Around ponds and lakes, marshes, fens, old fields, sometimes as an emergent aquatic.
Range: Maine to Wisconsin, south to Texas and Georgia.
Illinois Distribution: DuPage County.

The two species are usually recognized from a short distance away, since *C. annectens* has yellow-brown spikes with the lowest setaceous bract 2–5 cm long and

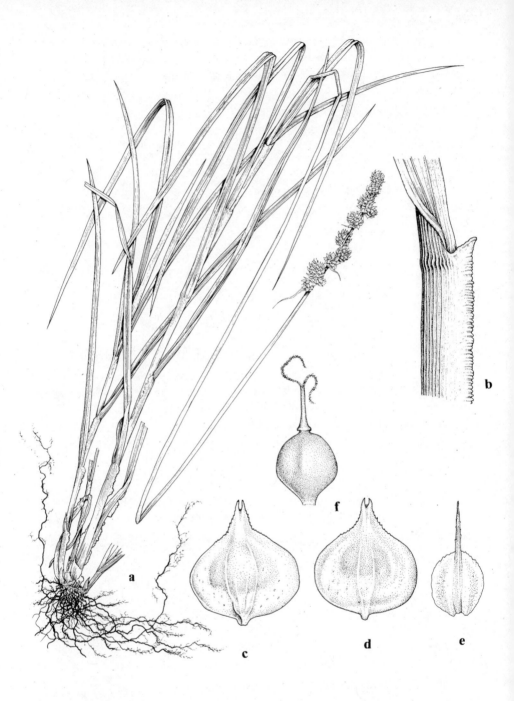

A7. Carex triangularis
(Eastern fox sedge).

a. Habit.
b. Sheath, with ligule.
c. Perigynium, dorsal view.

d. Perigynium, ventral view.
e. Pistillate scale.
f. Achene.

C. brachyglossa has golden brown spikes with the lowest setaceous bract nearly absent or less than 2 cm long. Closer observations of the perigynia, pistillate scales, and achenes show other differences. The differences are summarized in the table below:

Carex annectens	Carex brachyglossa
leaves 2–5 mm wide	leaves 3–6 mm wide
sheath strongly convex at summit	sheath shallowly convex at summit
spikes usually uninterrupted	lowest spike often separated
lowest bract 2–5 cm long	lowest bract up to 2 cm long
spikes yellow-brown	spikes golden brown
spikes many-flowered	spikes few-flowered
pistillate scales long-awned, appearing bristly, longer than the perigynia	pistillate scales short-awned, not appearing bristly, shorter than the perigynia
perigynia loosely spreading	perigynia ascending
perigynia 2.6–3.2 mm long, 1.7–2.4 mm wide	perigynia 2.2–2.6 mm long, 1.4–1.8 mm wide
perigynia ovoid to suborbicular	perigynia ovate-ellipsoid to subrhomboid
dorsal face of perigynium with 2–4 conspicuous nerves	dorsal face of perigynium obscurely nerved
beak of perigynium conspicuously notched, 0.5–0.7 mm long, with greenish margins	beak of perigynium obscurely notched, 0.3–0.5 mm long, without greenish margins
achenes ovate-oblongoid to suborbicular	achenes broadly oblongoid
achenes 1.2–1.7 mm long	achenes 1.5–1.8 mm long

For *C. annectens* and *C. brachyglossa*, across their entire range, where I have been studying and teaching field courses, more than 95 percent of the plants with golden brown spikes also have very short bracts and obscurely nerved perigynia.

Carex annectens flowers in May and June.

Carex triangularis Boeck. Flora 39:226. 1856. Fig. A7.

Carex vulpinoidea Michx. var. *triangularis* (Boeck.) Kukenth. Pflanzenr. IV (20):148. 1909.

Plants perennial, densely cespitose, from fibrous roots and short, stout, dark rhizomes; culms up to 1 m tall, erect, triangular, scabrous on the angles, red-brown to brown at the base; leaves 3–6 per culm, up to 70 cm long, 4–5 mm wide, flat, usually slightly canaliculate, scabrous along the margins, shorter than the culms; sheaths open, tight, convex at the mouth, russet-maculate; ligule broader than long; inflorescence composed of many spikes in an elongated, rather crowded inflorescence up to 6 cm long and up to 1.5 cm broad; lower bracts setaceous, the upper bracts scalelike; spikes 10–15, androgynous; pistillate scales lanceolate, awn-tipped, brown, as long as or longer than the perigynia; perigynia several per spike, broadly ovoid, 2.5–5.0 mm long, 2.5–3.0 mm broad, plano-convex, yellow-brown, russet-maculate, with 3 or 5 nerves, rather abruptly tapering to a bidentate, serrulate beak 0.8–1.2 mm long; achenes about 1.5 mm long, lenticular, ellipsoid, red-brown, dull, apiculate, jointed to the style, the style enlarged at base; stigmas 2, reddish brown.

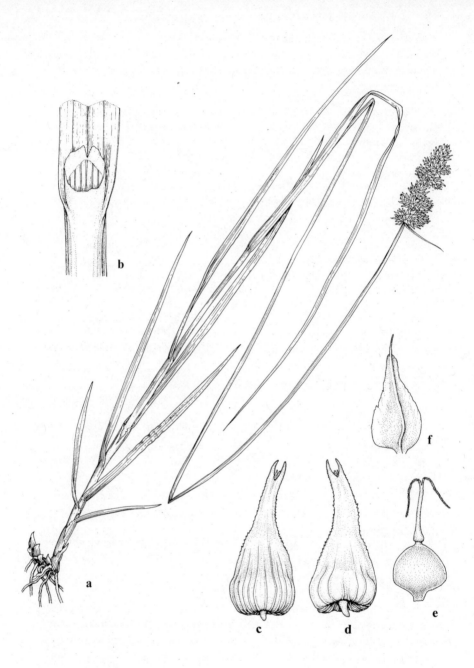

A8. Carex oklahomensis
(False spongy sedge).

a. Habit.
b. Sheath.
c. Perigynium, dorsal view.

d. Perigynium, ventral view.
e. Achene.
f. Pistillate scale.

Common Name: Eastern fox sedge.

Habitat: Wet ditch.

Range: Illinois to Kansas, south to Texas, Mississippi, and Tennessee.

Illinois Distribution: Massac County.

This species is similar in appearance to *C. annectens* because of its yellow-brown perigynia and its leaves that are shorter than the culms. It differs by its larger, broader, red-dotted perigynia and its dull rather than glossy achenes.

Carex triangularis flowers in June.

Page 71. *Carex decomposita* Muhl. The following county should be added to the map on page 74: Pope.

Page 74. *Carex diandra* Schrank. There are no new records for this species.

Page 76. *Carex prairea* Dewey ex Wood. There are no new records for this species.

Page 78. *Carex alopecoidea* Tuckerm. The following county should be added to the map on page 78: Lake.

Page 80. *Carex conjuncta* Boott. The following counties should be added to the map on page 80: Brown, Fulton, Monroe, Pike, Shelby.

Page 80. *Carex stipata* Muhl. in Willd. The following counties should be added to the map on page 83: Adams, Bureau, Carroll, Livingston, Marion, Pike, Pulaski, Shelby, Williamson.

After *Carex stipata*, the following species should be added to the flora of Illinois:

Carex oklahomensis Mack. Torreya 14:126. 1914. Fig. A8.

Carex stipata Muhl. ex. Willd. var. *oklahomensis* (Mack.) Gl. Phytologia 4:22. 1952.

Plants perennial, cespitose, with fibrous roots and short, stout, dark rhizomes; culms up to 80 cm tall, strongly triangular but not easily compressed, glabrous, pale brown at the base, usually overtopped by the leaves; lowest sheaths bladeless; leaves 3–5 per culm, up to 75 cm long, 4–5 mm wide, rather soft but not flaccid, usually pale green, smooth or slightly scabrous along the margins; sheaths open, green but with a hyaline summit, not cross-puckered; ligule scarcely longer than wide; inflorescence elongated, compound, continuous, up to 10 cm long, up to 2.5 cm broad; bracts setaceous; spikes 15–20, the staminate flowers above the pistillate; pistillate scales lanceolate, acute to acuminate, brown with hyaline margins, much shorter than the perigynia; staminate scales narrowly lanceolate, pale brown; perigynia up to 10 per spike, lanceoloid, 4–5 mm long, 1.5–2.0 mm broad, plano-convex, rounded or cordate at the spongy-thickened base, stipitate, pale brown, rather conspicuously nerved, tapering gradually into a 2-toothed, serrulate, greenish beak, 2–3 mm long; achenes 1.5–2.0 mm long, lenticular, ovoid to orbicular, apiculate, substipitate, jointed to the style, the style swollen at base; stigmas 2, reddish brown.

Common Name: False spongy sedge.

Habitat: Bottomland forests, marshes.

Range: Delaware to Illinois to Kansas, south to Texas, Mississippi, and North Carolina.

Illinois Distribution: Williamson County.

Carex oklahomensis is very similar in appearance to *C. stipata*, but its culms are firm and not readily compressed. The perigynia of *C. stipata* have a more long drawn-out beak of the perigynium than does *C. oklahomensis*.

The flowers appear in May and June.

Page 83. *Carex laevivaginata* (Kukenth.) Mack. in Britt. & Brown. The following counties should be added to the map on page 85: Perry, Winnebago.

Carex stipata, *C. laevivaginata*, and *C. oklahomensis* are similar. The following chart summarizes some of the similarities and differences among the three:

character	*C. stipata*	*C. laevivaginata*	*C. oklahomensis*
culms	easily compressed	easily compressed	firm
leaf texture	flaccid	flaccid	not flaccid
leaf width	4–15 mm	3–6 mm	4–5 mm
leaf color	pale green	yellow-green	pale green
sheath at summit	not cartilaginous	cartilaginous	not cartilaginous
sheath	septate-nodulose	septate-nodulose	not septate-nodulose
spikes per culm	15–25	10–15	15–20
perigynium height	3.5–6.0 mm	4.5–8.0 mm	4–5 mm
perigynium color	yellow or pale brown	green or yellowish	pale brown
pistillate scale	acuminate or short awn	acuminate or long awn	acute to acuminate
ratio scale/ perigynium	much shorter	¾ as long as	much shorter

Page 85. *Carex crus-corvi* Shuttlw. in Kunze. The following county should be added to the map on page 87: Madison.

Page 87. *Carex disperma* Dewey. There are no new records for this species. In the Cyperaceae volume of Flora of North America, this species is inexplicably not listed from Illinois, even though there are extant specimens and literature reports for Illinois.

Page 89. *Carex trisperma* Dewey. There are no new records for this species.

Page 91. *Carex canescens* L. There are no new records for this species.

Page 93. *Carex brunnescens* (Pers.) Poir. There are no new records for this species. In the Cyperaceae volume of Flora of North America, this species is inexplicably not listed from Illinois, even though there are extant specimens and literature reports for Illinois.

Page 93. *Carex bromoides* Schkuhr in Willd. There are no new records for this species.

Page 96. *Carex deweyana* Schwein. There are no new records for this species.

Page 98. *Carex interior* Bailey. The following counties should be added to the map on page 98: Iroquois, Jersey, Lee.

Page 98. *Carex sterilis* Willd. The following county should be added to the map on page 101: St. Clair.

Page 101. *Carex echinata* Murray. Add Pope to the map on page 103 and delete Cook and DuPage.

Page 103. *Carex atlantica* Bailey. Add Cook and DuPage to the map on page 105 (top).

Page 105. *Carex muskingumensis* Schwein. The following counties should be added to the map on page 108: Carroll, Franklin, Iroquois, Perry, Vermilion.

Page 108. *Carex scoparia* Schkuhr in Willd. The following counties should be added to the map on page 110: Montgomery, Williamson.

Page 110. *Carex tribuloides* Wahl. The following county should be added to the map on page 112: Bureau.

After *Carex tribuloides*, add the following new species for Illinois:

Carex sangamonensis (Clokey) Mohlenbr. comb. nov. (basionym: *Carex tribuloides* Wahl. var. *sangamonensis* Clokey) Fig. A9.

A9. Carex sangamonensis (Sangamon sedge).

a. Habit.
b. Sheath.
c. Perigynium, dorsal view.

d. Pistillate scale.
e. Achene.
f. Perigynium, ventral view.

Carex tribuloides Wahl. var. *sangamonensis* Clokey, Rhodora 21:84–85. 1919.

Plants perennial, densely cespitose, from short, brown, fibrillose rhizomes; culms to 85 cm tall, triangular, the angles scabrous, soft, erect, usually about as long as the culms, with sterile leafy shoots present; leaves 4–6 (–8) per culm, ascending to spreading, 1.5–4.0 mm wide, the margins scabrous; sheaths tight, with green veins, the summit U-shaped; spikes 4–6 per culm, rarely as many as 8, 6–12 mm long, pale green to stramineous, ascending, obovoid, rounded at the apex, clavate at the base, gynecandrous, overlapping or distinct, forming an inflorescence up to 6 cm long, sometimes flexuous; bracts 0–3, setaceous; pistillate scales lanceolate, acute, two-thirds to three-fourths as long as the perigynia and narrower than the perigynia, hyaline with a green center; perigynia up to 30 per spike, ascending, ovate-lanceolate, plano-convex, 3–4 mm long, 1.2–1.5 mm wide, widest at the middle, less than three times longer than wide, membranous, substipitate, faintly nerved on both faces, the wing abruptly narrowed but not reaching the base of the perigynium, pale green to stramineous, the beak 1.0–1.5 mm long, serrulate, bidentate, the teeth appressed; achenes lenticular, elliptic, 1.0–1.2 mm long, 0.6–0.7 mm wide, dull, stramineous, apiculate, stipitate, weakly continuous with the deciduous style; stigmas 2, slender, elongated, reddish.

Common Name: Sangamon sedge.

Habitat: Rich soil.

Range: Ohio to Kansas, south to Texas, Louisiana, and South Carolina.

Illinois Distribution: Macon and Sangamon counties.

This species has always been considered to be a variety of *C. tribuloides*, but the numerous, often significant differences illustrated in the table below are good evidence of its distinctness. Clokey first discovered this plant in alluvial soil in Macon County, Illinois (*Clokey 2364*, the type specimen).

Carex tribuloides	Carex sangamonensis
culms to 1.2 m tall, rather stiff	culms to 85 cm tall, soft
leaves 3.5–7.0 mm wide	leaves 1.5–4.0 mm wide
spikes 8–15 per culm	spikes 4–6 (–8) per culm
spikes brown, ovoid to globose	spikes pale green to stramineous, obovoid
inflorescence to 8 cm long, not usually flexuotus	inflorescence to 6 cm long, often flexuous
pistillate scales about one-half as long as the perigynia	pistillate scales two-thirds to three-fourths as long as the perigynia
perigynia up to 40 per spike	perigynia up to 30 per spike
perigynia lanceolate	perigynia ovate-lanceolate
perigynia 3.5–5.4 mm long	perigynia 3–4 mm long
perigynia 3–5 times longer than wide	perigynia less than 3 times longer than wide
wing of perigynium extending to below middle of perigynium	wing of perigynium extending nearly to base of perigynium
achenes oblong-ovate	achenes elliptic
achenes 1.2–1.7 mm long, 0.6–0.9 mm wide	achenes 1.0–1.2 mm long, 0.6–0.7 mm wide

To sum up the major differences, *C. sangamonensis* has narrower leaves, fewer spikes per culm, pistillate scales at least two-thirds as long as the perigynia, shorter perigynia that are less than three times longer than wide, the wing of the perigynium extending nearer to the base of the perigynium, and the shorter and narrower achenes. In placing perigynia of *C. tribuloides* and *C. sangamonensis* side by side, the differences are obvious.

This species flowers in May and June.

Page 112. *Carex crawfordii* Fern. There are no new records for this species. In the Cyperaceae volume of Flora of North America, this species is inexplicably not listed from Illinois, even though there are extant specimens and literature reports for Illinois.

Page 114. *Carex projecta* Mack. The following counties should be added to the map on page 114: Cook, Iroquois, Perry, Sangamon.

Page 114. *Carex straminea* Willd. in Schk. The following counties should be added to the map on page 117: Champaign, Iroquois, Macon, St. Clair, Wabash.

Page 117. *Carex cristatella* Britt. in Britt. & Brown. The following counties should be added to the map on page 119: Carroll, Coles, Woodford.

Page 119. *Carex bebbii* Olney. The following counties should be added to the map on page 121: Kankakee, Peoria.

Page 120. *Carex normalis* Mack. The following counties should be added to the map on page 123: Iroquois, Knox, Morgan, Moultrie, Tazewell, Whiteside.

Page 123. *Carex tenera* Dewey. The following counties should be added to the map on page 125 (top): Carroll, Macoupin, Sangamon.

After *Carex tenera*, the following species should be added to the flora of Illinois:

Carex echinodes (Fern.) P. E. Rothrock, Reznicek, & Hipp, Syst. Bot. 34:304. 2009. Fig. A10.

Carex straminea Willd. var. *echinodes* Fern. Proc. Am. Acad. Arts 37:474. 1902.

Plants perennial, densely cespitose, from short, black, scaly, and fibrillose rhizomes; culms to 90 cm tall, sharply triangular, scabrous on the angles just below the inflorescence, stiff, longer than the leaves, brown at base, with the old leaf bases often persisting as stubble; leaves 3–5 per culm, ascending, 1.5–3.5 mm wide, flat or slightly plicate, glabrous except for the scabrous margins; sheaths tight, smooth, with green nerves, septate, the mouth truncate or concave, the lowermost sheaths sometimes purplish brown; spikes 3–8 per culm, gynecandrous, to 14 mm long, green to stramineous, obtuse, somewhat clavate at base, the staminate flowers conspicuous, the spikes arranged in a moniliform inflorescence to 7 cm long, the axis of the inflorescence narrowed and flexed just above the lowest spike; lowest bract usually setaceous, scabrous, the upper bracts scalelike; pistillate scales lanceolate to narrowly ovate, obtuse to acute, reaching or surpassing the base of the beak of the perigynium, the center greenish and 1-(3-)nerved, the margins hyaline; perigynia many per spike, 3.5–4.5 mm long, 1.5–2.0 mm wide, elliptic to narrowly ovate, greenish brown to yellow-brown, widest below the middle, spreading to ascending, less commonly recurved, plano-convex, several-nerved on

A10. Carex echinodes
   (Echinate sedge).

a. Habit.
b. Spikelet.
c. Perigynium, ventral view.

d. Perigynium, dorsal view.
e. Pistillate scale.
f. Achene.

both faces, narrowly and evenly winged to the base, substipitate, the beak 1.2–2.0 mm long, spreading, serrulate, bidentate; achenes lenticular 1.4–1.7 mm long, 0.8–1.2 mm wide, apiculate, stipitate, pale to dark brown, weakly jointed with the deciduous style; stigmas 2, slender.

Common Name: Echinate sedge.
Habitat: Bottomland forests.
Range: Vermont to Manitoba, south to Nebraska, Missouri, Indiana, and Ohio.
Illinois Distribution: Apparently confined to the northern one-sixth of the state: Boone, Carroll, Cook, DeKalb, DuPage, Kankakee, and Winnebago counties.

This species differs from the closely related *C. tenera* by its smooth and not papillose sheaths and the beak of the perigynium spreading and not ascending.

Carex echinodes flowers during May and June.

The following chart summarizes some of the major features of *C. tenera*, and *C. echinodes*:

character	*C. tenera*	*C. echinodes*
leaf width	1.0–2.8 mm	1.5–3.5 mm
sheath	papillose	smooth
spike length	6–10 mm	up to 14 mm
pistillate scale	acute to acuminate	obtuse to acute
perigynium shape	ovate	elliptic to narrowly ovate
beak length	0.8–1.2 mm	1.2–2.0 mm

Page 125. *Carex festucacea* Schk. in Willd. The following counties should be added to the map on page 125 (bottom): Alexander, Edwards, Fulton, Logan, Saline, Schuyler, Scott, Whiteside, Williamson.

Page 127. *Carex albolutescens* Schw. The following county should be added to the map on page 127: Monroe.

Page 127. *Carex longii* Mack. The following county should be added to the map on page 130: Kankakee.

Page 130. *Carex cumulata* (Bailey) Mack. The following county should be added to the map on page 132 (top): Iroquois.

Page 132. *Carex suberecta* (Olney) Britt. There are no new records for this species.

Page 134. *Carex brevior* (Dewey) Mack. The following counties should be added to the map on page 134: Fayette, Jersey.

Page 136. *Carex molesta* Mack. ex Bright. The following counties should be added to the map on page 136: Brown, Carroll, DeWitt, Mercer, Piatt, Saline, Stark, Union, Winnebago, Woodford.

After *Carex molesta*, add the following species that is new for Illinois:

Carex molestiformis Reznicek & P. E. Rothrock, Contr. Univ. Mich. Herb. 21: 300. 1997. Fig. A11.

Plants perennial, cespitose, with usually short, black, fibrillose rhizomes; culms to 1.2 m tall, wiry, triangular, the angles scabrous beneath the inflorescence, with

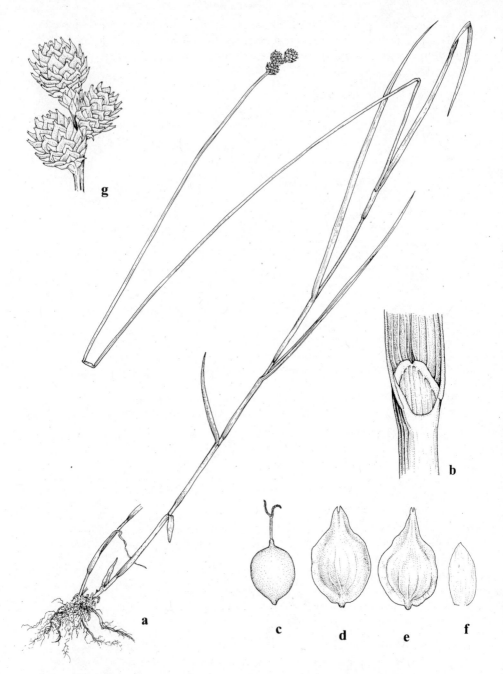

A11. Carex molestiformis
(Molesta-like sedge).
a. Habit.

b. Sheath.
c. Achene.
d. Perigynium, ventral view.

e. Perigynium, dorsal view.
f. Pistillate scale.
g. Spikelets.

old leaf bases sometimes persisting; sterile shoots usually not present; leaves 3–5 per culm, ascending, 1.5–4.0 mm wide, flat, rather thin, light green, scabrous along the margins; sheaths tight, the ventral band more or less truncate near the summit, white-hyaline to pale green, prolonged at the summit, the lowermost brownish; spikes 2–5 per culm, gynecandrous, 6–12 mm long, globose to ovoid, not clavate at base, green or brown, distinct from each other, forming an inflorescence up to 3.2 cm long; bracts scalelike; pistillate scales broadly lanceolate, obtuse to acute, usually reaching only to the base of the beak of the perigynium, the center greenish, usually 3-nerved, the margins hyaline; perigynia 10–45 per spike, 4.0–5.5 mm long, 2.5–3.5 mm wide, broadly ovate to suborbicular, widest near the middle, submembranous, ascending, plano-convex, greenish brown to brown, conspicuously nerved on the outer face, finely nerved on the inner face, winged to the base, substipitate, the beak 1.3–1.8 mm long, serrulate, bidentate, green or brown at tip; achenes lenticular, 1.6–2.0 mm long, 1.4–1.8 mm wide, broadly ovate to orbicular, apiculate, substipitate; stigmas 2, short, reddish brown.

Common Name: Molesta-like sedge.
Habitat: Bottomland forests.
Range: Virginia to southern Illinois and southern Missouri, south to Oklahoma, Arkansas, and Tennessee.
Illinois Distribution: Jackson and Union counties.

This species, in the past considered a part of the variation of *C. molesta*, is very similar to *C. molesta*. *Carex molestiformis* has achenes that are ovate to orbicular, while *C. molesta* has achenes that are elliptic to narrowly oblong. The beak of the perigynium in *C. molestiformis* is slightly longer than the beak of the perigynium of *C. molesta*. *Carex molestiformis* usually has fewer perigynia per spike than *C. molesta*.

Whereas *C. molesta* is found in a variety of habitats in Illinois, from old fields to swamps, *C. molestiformis* appears to be restricted to floodplain forests.

The flowers of *C. molestiformis* appear in May and June.

The following chart summarizes some of the characteristics of *C. molesta*, *C. molestiformis*, and *C. normalis*.

character	C. molesta	C. molestiformis	C. normalis
sterile shoots	not present	not present	present
leaf width	2.0–3.5 mm	1.5–4.0 mm	2.0–6.5 mm
perigynium height	4.0–5.5 mm	4.0–4.5 mm	3.0–4.5 mm
perigynium width	2–3 mm	2.5–3.5 mm	1.5–2.1 mm
perigynium	faintly nerved	conspicuously nerved	distinctly nerved
beak length	1.0–1.3 mm	1.3–1.8 mm	0.8–1.2 mm
achene shape	elliptic to narrowly oblong	broadly ovate to orbicular	oblong
achene width	1.2–1.3 mm	1.4–1.8 mm	1.1–1.3 mm

Page 136. *Carex reniformis* (Bailey) Small. There are no new records for this species.

A12. Carex opaca
(Opaque sedge).
a. Habit.

b. Sheath, back view.
c. Sheath, front view.
d. Perigynium, ventral view.

e. Perigynium, dorsal view.
f. Pistillate scale.
g. Achene.

Page 139. *Carex bicknellii* Britt. in Britt. & Brown. The following counties should be added to the map on page 141 (top): Franklin, Lawrence, Marion.

After *Carex bicknellii*, the following two species should be added that are new for Illinois:

Carex opaca (F. J. Hermann) P. E. Rothrock & Reznicek, Novon 11:223. 2001. Fig. A12.

Carex bicknellii Britt. var. *opaca* F. J. Hermann, Sida 5:49. 1972.

Plants perennial, cespitose, from dark, elongated rhizomes; culms many in a tussock, to 1.5 m tall, triangular, the angles usually scabrous beneath the inflorescence, light brown near the base, with old leaf bases often persisting; sterile shoots few; leaves 3–9 per culm, ascending, 1.5–4.6 mm wide, flat, firm, usually scabrous on the upper surface and the margin, glabrous or nearly so on the lower surface; sheaths tight, smooth, white-hyaline ventrally, not particularly prolonged at summit; spikes 3–10 per culm, 10–20 mm long, ovoid, somewhat tapering or rounded at the base, rounded or sometimes pointed at the apex, golden brown, gynecandrous, crowded or somewhat loosely disposed in the inflorescence, the inflorescence up to 6.5 cm long; lowest bract setaceous, scabrous, the upper bracts scalelike; pistillate scales lanceolate to lance-ovate, acute to obtuse, usually barely reaching the base of the beak of the perigynium, pale brown, the central nerve narrow, yellow-brown, the margins hyaline; perigynia numerous per spike, 6–7 mm long, 3.3–4.8 mm wide, broadly ovate to orbicular, widest at the middle, more or less opaque, appressed to appressed-ascending, flat, yellow-green to yellow-brown, the margins green or brown, not ciliate, conspicuously nerved on both faces, broadly winged to base, sessile or substipitate, the beak 1.8–2.5 mm long, serrulate, bidentate, brown-tipped; achenes lenticular, 1.8–2.5 mm long, 1.3–2.0 mm wide, apiculate, substipitate; stigmas 2; short, reddish.

Common Name: Opaque sedge.
Habitat: Wet prairies, ditches.
Range: Illinois to Kansas, south to Oklahoma and Missouri.
Illinois Distribution: Scattered in the southern counties of Saline, St. Clair, and Washington.

Carex opaca resembles *C. bicknellii*. The most obvious difference is that in *C. bicknellii*, the achene can be seen within the translucent perigynium, while in *C. opaca*, the achene is not visible within the opaque perigynium. The leaf sheaths of *C. opaca* are smooth, while the leaf sheaths of *C. bicknellii* are finely papillose. *Carex opaca* is also similar to *C. missouriensis* by its opaque perigynia, but the pistillate scales in *C. opaca* are acuminate, while those of *C. missouriensis* are awned.

The beak of the perigynium in *C. bicknellii* and *C. opaca* is only about one-fourth the length of the body of the perigynium, while in *C. missouriensis*, the beak is nearly one-half as long as the body of the perigynium.

The following chart displays the distinguishing features of these species:

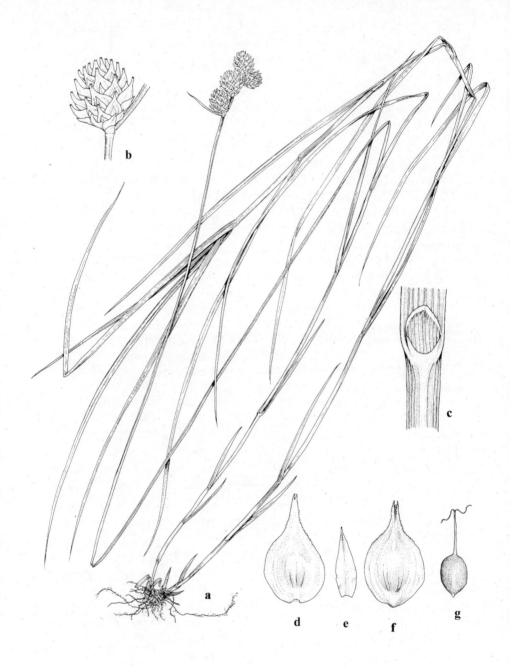

A13. Carex missouriensis
(Missouri sedge).
a. Habit.

b. Spikelet.
c. Sheath.
d. Perigynium, ventral view.

e. Pistillate scale.
f. Perigynium, dorsal view.
g. Achene.

character	C. bicknellii	C. missouriensis	C. opaca
sheath	papillate	glabrous	glabrous
perigynium texture	translucent	opaque	opaque
beak to perigynium body length	¼ as long	½ as long	¼ as long
pistillate scales	acute to obtuse, barely reaching base of perigynium	acuminate to short-awned, reaching base or middle of perigynium	acute to obtuse, barely reaching base of perigynium

Carex opaca flowers from mid-May to mid-June.

Carex missouriensis P. E. Rothrock & Reznicek, Novon 11:280. 2001. Fig. A13.

Plants perennial, cespitose, from dark, elongated rhizomes; culms many in a tussock, to nearly 1 m tall, triangular, the angles usually scabrous beneath the inflorescence, light brown near the base, with old leaf bases often persisting; sterile shoots few; leaves 3–6 per culm, ascending, 2–4 mm wide, flat, firm, usually scabrous on the upper surface and the margin, glabrous or nearly so on the lower surface; sheaths tight, smooth, white-hyaline ventrally, barely prolonged at the summit; spikes 3–8 per culm, 8–15 mm long, globose to ovoid, somewhat tapering or rounded at the base, rounded or somewhat pointed at the apex, brownish, gynecandrous, more or less distinct in the inflorescence, the inflorescence up to 4.5 cm long; lowest bract setaceous or sometimes scalelike; pistillate scales lanceolate to lance-ovate, acuminate or very short-awned, reaching to the base or middle of the beak of the perigynium, usually pale brown with a hyaline margin and yellow-brown center; perigynia numerous per spike, 4.5–7.0 mm long, 2.5–4.0 mm wide, broadly elliptic to ovate, widest at the middle, more or less opaque, ascending, flat, usually yellow-brown, the margin green to brown, not ciliate, nerved on both faces but not usually conspicuously nerved, broadly winged to the base, sessile, the beak 2.0–2.8 mm long, serrulate, bidentate, brown-tipped; achenes lenticular, 1.5–2.2 mm long, 1.2–1.5 mm wide, apiculate, substipitate; stigmas 2, short, reddish.

Common Name: Missouri sedge.
Habitat: Low areas in prairies.
Range: Indiana to Nebraska, south to Kansas and Missouri.
Illinois Distribution: Adams, Bond, Brown, Christian, Clay, Effingham, Fayette, Greene, Macoupin, Madison, Marion, McDonough, Montgomery, and Scott counties.

This species is very closely related to *C. opaca*, differing by its longer beak of the perigynium and its sometimes short-awned pistillate scales. See chart under *C. opaca* for differences between this species and *C. opaca* and *C. bicknellii*.

Carex missouriensis flowers from mid-May to mid-June.

Page 141. *Carex alata* Torr. & Gray. There are no new records for this species.
Page 143. *Carex praticola* Rydb. There are no new records for this species.
Page 143. *Carex leptalea* Wahl. There are no new records for this species.

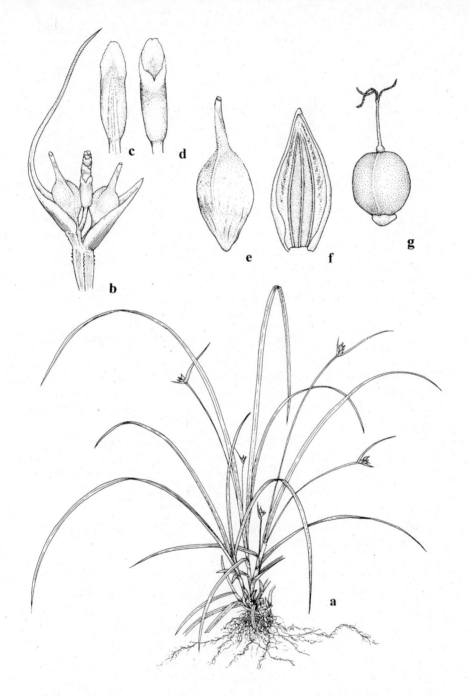

A14. Carex timida (James' look-alike sedge).
a. Habit.

b. Inflorescence.
c. Staminate scale, front view.
d. Staminate scale, back view.

e. Perigynium.
f. Pistillate scale.
g. Achene.

Page 146. *Carex jamesii* Schw. The following counties should be added to the map on page 146 (bottom): Adams, Calhoun, Carroll, Clay, Crawford, DeWitt, La Salle, Montgomery.

After *Carex jamesii*, the following species new to Illinois should be added:

Carex timida R. F. C. Naczi & B. A. Ford, Sida 19:879. 2001. Fig. A14.

Plants perennial, densely cespitose, from short, black rootstocks; culms to 30 cm tall, triangular, usually scabrous on the angles, at least beneath the inflorescence, red-purple near the base, shorter than the leaves; leaves 4–6 per culm, ascending, 1.5–4.0 mm wide, flat, lax, green, scabrous on the margins, occasionally minutely papillate; sheaths tight, nerved, hyaline, the lower brown to red-purple; spikes up to 4 per culm, 3–7 mm long, greenish, androgynous, the uppermost sessile, the lowest on spreading to pendulous peduncles, the staminate part 3.5–6.0 mm long, 0.6–1.0 mm thick and composed of several scales 2.0–3.3 mm long and broadly obtuse to truncate; lowest pistillate scales leaflike, ovate, long-acuminate, up to 5 cm long, the margins hyaline; perigynia 2–3 per spike, 4.5–6.0 mm long, 1.5–2.0 mm thick, turgid, obovoid to ellipsoid, erect, green to brown, nerveless but with two keels decurrent from the base of the beak, more or less spongy at base, with a thick stipe up to 1 mm long, the beak 1.5–2.5 mm long, flat, serrulate; achenes trigonous, 2.0–2.5 mm long, 1.5–2.0 mm broad, broadly stipitate, umbonate at the tip; stigmas 3, rather long, red-brown.

Common Name: James' look-alike sedge.
Habitat: Moist to dry woods.
Range: Ohio to Missouri, south to Oklahoma and Alabama.
Illinois Distribution: Scattered in the southern one-half of Illinois.

This species is very similar to *C. jamesii* and was not segregated from *C. jamesii* until 2001. It differs from *C. jamesii* by its red-purple culm bases, its shorter staminate part of the terminal spike, its somewhat wider staminate scales, and its narrower perigynia that are obovoid to ellipsoid rather than orbicular.

Carex timida flowers in May and June.

Carex jamesii, *C. willdenowii*, and *C. timida* are our only members of Section Phyllostachyae: The following chart summarizes the major characters of these species:

character	*C. jamesii*	*C. willdenowii*	*C. timida*
culm base	green	green	red-purple
spike length	2–10 mm	4–20 mm	3–7 mm
staminate part of spike	3–10 mm	2–8 mm	3.3–6.0 mm
staminate scale length	1–2 mm	1–2 mm	2.0–3.2 mm
perigynium width	2.3–2.5 mm	1.5 mm	1.5–2.0 mm
perigynium shape	orbicular	ellipsoid	obovoid to ellipsoid
achene width	2.5 mm	1.5 mm	1.5–2.0 mm

Page 148. *Carex willdenowii* Schk. in Willd. Add the following county to the map on page 148: Alexander.

Following *Carex willldenowii* are thirteen species forming a complex generally called Section *Montanae*. Since the publication of the first edition of *Carex*, I have made extra effort to study this complex. As a result, I am adding material to each description and am modifying some of the measurements that I had given in the first edition. These additions and changes are reflected below.

When identifying members of Section Montanae, first check to see if some or all of the spikes are hidden near the leaf bases. If so, the plant will be one of the following: *C. umbellata, C. abdita, C. nigromarginata, C. tonsa,* or *C. rugosperma*. In *C. nigromarginata*, there are often spike-bearing culms standing above the leaves, in addition to those hidden near the base of the plant. *Carex nigromarginata* is the only one in this group with the body of the perigynium ellipsoid. *Carex tonsa* is the only one with nearly glabrous perigynia. If all of the spikes are on culms that stand above or equal to the leaves, check to see if the body of the perigynium is ovoid to obovoid or if the body is ellipsoid. If ovoid or obovoid, the plant will be one of the following: *C. pensylvanica, C. lucorum, C. heliophila,* or *C. communis*. If the body of the perigynium is ellipsoid, the plant will be one of the following: *C. communis, C. emmonsii, C. physorhyncha, C. albicans,* or *C. peckii*.

For those species with hidden spikes, the following chart compares and contrasts some of the major characteristics:

character	C. umbellata	C. abdita	C. nigromarginata	C. tonsa	C. rugosperma
culm height	15	20	30	15	15
leaf width (mm)	1.2–2.5	1.0–2.5	1.4–4.0	2.0–4.5	2.0–4.5
leaf color	pale green	pale green	green pale	green	bright green
perigynia/spike	up to 20	up to 20	6–15	up to 20	3–10
perigynium shape	orbicular	orbicular	ellipsoid	orbicular	orbicular
perigynium length (mm)	2.2–3.3	3.2–4.5	3–4	3–5	3.0–4.5
perigynium width (mm)	1.0–1.5	1.2–1.9	1.2–1.5	1.3–1.5	1.0–1.5
beak length (mm)	0.9–1.7	0.8–1.0	0.7–1.0	1–2	2.5–4.0
perigynium color	dull green	dull green	yellow-green	light green	green
ratio beak/body	¾–equal	½–⅔	⅓	⅓–¾	equal

For those species with elevated spikes and ovoid or obovoid perigynia, the following chart compares and contrasts some of the major characteristics:

character	C. pensylvanica	C. lucorum	C. heliophila	C. communis
culm height	45	50	40	60
leaf width	1.5–3.6 mm	1.5–3.5 mm	1.5–4.5 mm	1.8–5.0 mm
perigynia per spike	up to 15	up to 12	up to 15	up to 10
perigynium height	2.2–3.4 mm	3.2–4.6 mm	3.0–4.5 mm	2.5–3.0 mm
perigynium width	1.2–1.7 mm	1.2–1.7 mm	1.5–2.2 mm	1.2–2.6 mm
beak length	0.5–0.9 mm	0.9–1.6 mm	0.4–1.3 mm	0.5–1.8 mm
ratio beak/body	¼	⅔	⅓	⅓
perigynium color	pale green	yellow-green	green	pale green

For those species with elevated spikes and ellipsoid perigynia, the following chart compares and contrasts some of the major characteristics:

character	C. communis	C. emmonsii	C. physorhyncha	C. albicans	C. peckii
culm height	60	40	40	50	45
leaf width (mm)	1.8–5.0	1.0–1.5	1.5–3.0	1.0–2.5	1.0–3.3
leaf color	green	bright green	green	green	pale green
perigynia per spike	up to 10	4–10	4–12	6–12	3–10
perigynium height (mm)	2.5–5.0	2.0–3.3	2.5–3.0	2.5–3.5	3.2–4.2
perigynium width (mm)	1.2–2.6	0.8–1.2	0.8–1.1	0.9–1.1	1.1–1.3
beak length (mm)	0.5–1.8	0.5–1.0	0.7–1.0	0.6–1.0	0.7–1.0
ratio beak/body	⅓	⅓	⅓–½	⅓	⅓
perigynium color	pale green	yellow-green	pale yellow	yellow-green	pale green

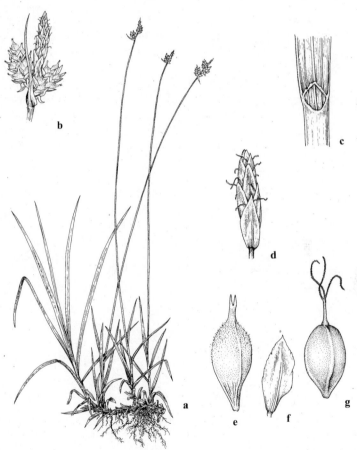

A15. Carex peckii
    (Peck's sedge).
a. Habit.

b. Inflorescence.
c. Sheath.
d. Staminate spikelet.

e. Perigynium.
f. Pistillate scale.
g. Achene

Page 148. *Carex pensylvanica* Lam: Add the following counties to the map on page 151: Greene, Marshall, Montgomery, Tazewell. To the description on page 148, add or change the following data: plants from slender, spreading rhizomes; leaves green, 1.5–3.6 mm wide; perigynia up to 15 per spike, 2.2–3.4 mm long, 1.2–1.7 mm wide, ovoid to obovoid; beak of perigynium 0.5–0.9 mm long, about one-fourth as long as the body.

Page 151. *Carex lucorum* Willd. in Link. There are no new records for this species. To the description on page 151, add or change the following data: plants from slender, spreading rhizomes; leaves green, 1.5–3.5 mm wide; pistillate scales usually longer than the perigynia; perigynia up to 12 per spike, 3.2–4.6 mm long, 1.2–1.7 mm wide, ovoid to obovoid; beak of perigynium 0.9–1.6 mm long, about two-thirds as long as the body. In the Cyperaceae volume of Flora of North America, this species is inexplicably not listed from Illinois, even though there are extant specimens and literature reports for Illinois.

Page 153. *Carex heliophila* Mack. The following county should be added to the map on page 155 (top): McHenry. To the description on page 153, add or change the following data: plants from slender, spreading rhizomes; leaves green, 1.5–4.5 mm wide; perigynia 3.0–4.5 mm long, 1.5–2.2 mm wide, ovoid to obovoid; beak of perigynium 0.4–1.3 mm long.

Page 155. *Carex communis* L. H. Bailey. The following counties should be added to the map on page 155 (bottom): Champaign, Shelby. To the description on page 155, add or change the following data: plants from short, ascending rhizomes; leaves green, 1.8–5.0 mm wide; perigynia 2.5–5.0 mm long, 1.2–1.6 mm wide, obovoid to ellipsoid beak of perigynium 0.5–1.8 mm long, about one-third as long as the body.

Page 157. *Carex emmonsii* Dewey. The following counties should be added to the map on page 157: Fulton, Iroquois. To the description on page 157, add or change the following data: plants from short, ascending rhizomes; culms weak, sprawling; leaves bright green, 1.0–1.5 mm wide; perigynium 2.0–3.3 mm long, 0.8–1.2 mm wide, ellipsoid; beak of perigynium 0.5–1.0 mm long.

Page 157. *Carex albicans* Willd. in Spreng. The following counties should be added to the map on page 160 (top): Brown, Clark, Jo Daviess, Montgomery, Will, Woodford. To the description on page 157, add or change the following data: plants from short, ascending rhizomes; leaves green, 1.0–2.5 mm wide; perigynia 6–12 per spike, 2.5–3.5 mm long, 0.9–1.1 mm wide, ellipsoid; beak of perigynium 0.6–1.0 mm long, about one-third the length of the body.

After *Carex albicans*, the following species new to Illinois should be added:

Carex peckii Howe, Rep. Regents Univ. N. Y. State Museum 47:166. 1894. Fig. A15.

Plants perennial, cespitose, from very long, spreading, reddish purple to brownish rhizomes; culms to 45 cm tall, triangular, slender, rather stiff, erect or ascending, shorter or longer than the leaves, glabrous or somewhat scabrous on the angles beneath the inflorescence, usually reddish purple near the base, with overwintering leaves and leaves of the season present; sterile shoots numerous; leaves

3–8 per culm, up to 1.0–3.3 mm wide, ascending, canaliculate, pale green, papillose on the upper surface, scabrous on the lower surface and along the margins; sheaths tight, reddish brown, concave at the summit; terminal spike staminate, up to 1 cm long, 1.0–1.7 mm wide, sessile or on peduncles up to 2 mm long; staminate scales acute, reddish brown; lateral spikes 2–4, pistillate, approximate, or the lowest not overlapping the one above, up to 7 mm long, sessile; bracts reddish brown, at least at the base, reduced; pistillate scales lanceolate, acute to cuspidate, shorter than the perigynia, reddish brown with hyaline margins; perigynia 3–10 per spike, conspicuous, 3.2–4.2 mm long, 1.1–1.3 mm wide, ellipsoid, trigonous, 2-keeled, membranous, pale green, puberulent, stipitate, tapering into a beak 0.7–1.0 mm long, the beak about one-third the length of the perigynia; achenes trigonous, 2.0–2.5 mm long, 1.0–1.3 mm wide, brown, apiculate; stigmas 3, red-brown.

Common Name: Peck's sedge.
Habitat: Dry woods.
Range: Nova Scotia to Alaska, south to British Columbia, Colorado, Illinois, and New York.
Illinois Distribution: McHenry County.

Carex peckii is similar to both *C. albicans* and *C. emmonsii*, differing by its wider leaves, slightly wider staminate spike, its larger achenes, and its pistillate scales shorter than the perigynia. As a result of this last character, the perigynia are conspicuous. *Carex peckii* also has very long, spreading rhizomes.

The collection of this species was made by George Vasey, presumably from McHenry County, well over a century ago.

Page 160. *Carex physorhyncha* Liebm. ex Steud. The following county should be added to the map on page 160 (bottom): Pope. To the description on page 160, add or change the following data: plants with long spreading rhizomes; leaves green, 1.5–3.0 mm wide; pistillate scales equaling or exceeding the perigynia; perigynia 4–12 per spike, 2.5–3.0 mm long, 0.8–1.1 mm wide, ellipsoid; beak of perigynium 0.7–1.0 mm long, one-third to one-half as long as the body. In the Cyperaceae volume of Flora of North America, this species is inexplicably not listed from Illinois, even though there are extant specimens and literature reports for it in Illinois.

Page 162. *Carex nigromarginata* Schw. The following county should be added to the map on page 162: Alexander. To the description on page 162, add or change the following data: plants from short ascending rhizomes; leaves green, 1.4–4.0 mm wide; pistillate scales as long as or longer than the perigynia; perigynia 6–15 per spike, 3–4 mm long, 1.2–1.5 mm wide, about one-third as long as the body.

Page 162. *Carex umbellata* Schk. in Willd. The following county should be added to the map on page 165 (top): Pope. To the description on page 162, add or change the following data: plants from short ascending rhizomes; leaves pale green, 1.2–2.5 mm wide; pistillate scales as long as or longer than the perigynia; perigynia 2.2–3.3 mm long, 1.0–1.5 mm wide, orbicular; beak of perigynium 0.9–1.7 mm long, three-fourths as or as long as the body.

Page 165. *Carex abdita* Bickn. The following counties should be added to the map on page 165 (bottom): Alexander, Iroquois, Kankakee, Macoupin, Montgomery,

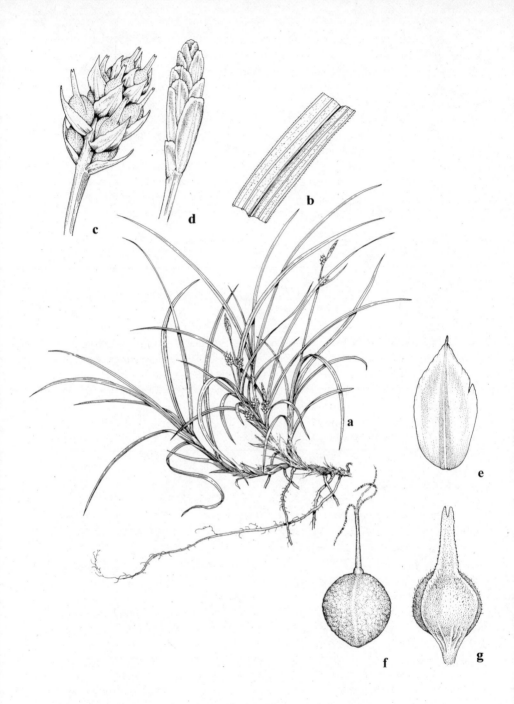

A 16. Carex rugosperma
  (Wrinkle-fruit sedge).
a. Habit.

b. Portion of blade.
c. Pistillate spikelet.
d. Staminate spikelet.

e. Pistillate scale.
f. Achene.
g. Perigynium.

Morgan, St. Clair, Vermilion. To the description on page 165, add or change the following data: plants from short ascending rhizomes; leaves pale green, 1.0–2.5 mm wide; pistillate scales as long as or longer than the perigynia; perigynia 3.2–4.5 mm long, 1.2–1.8 mm wide; beak of perigynium 0.8–1.0 mm long, one-half to two-thirds as long as the body.

Page 166. *Carex tonsa* (Fern.) Bickn. The following county should be added to the map on page 168 (top): La Salle. To the description on page 166, add or change the following data: plants from short ascending rhizomes; leaves pale green, 2.0–4.5 mm wide, smooth along the margins; perigynia 3–5 mm long, 1.3–1.5 mm wide, orbicular; beak of perigynium 1–2 mm long, two-thirds to three-fourths as long as the body.

After *Carex tonsa*, add the following species that is new to Illinois:

Carex rugosperma Mack. Bull. Torrey Club 42:621. 1915. Fig. A16.

Carex tonsa (Fern.) Bickn. var. *rugosperma* (Mack.) Crins, Novon 6:118. 1996.

Plants perennial, densely cespitose, from slender, spreading rhizomes; culms up to 15 cm tall but usually much shorter and hidden among the leaf bases, slender, lax, scabrous on the angles beneath the inflorescence, reddish near the base, fibrillose; leaves 2.0–4.5 mm wide, lax, canaliculate, bright green, scabrous along the margins and often papillate on the surface; sheaths tight, reddish; terminal spike staminate, sessile or nearly so, 8–12 mm long, 1.2–1.8 mm thick, reddish brown; pistillate spikes 2–3, crowded at the base of the plant; bracts setaceous or scale-like; pistillate scales broadly lanceolate to ovate, as long as or a little longer than the perigynia; perigynia 3–10 per spike, 3.0–4.5 mm long, 1.0–1.5 mm broad, the body more or less orbicular, trigonous, 2-keeled, membranous, green, puberulent throughout, long-stipitate, abruptly tapering to a distinct beak 2.5–4.0 mm long, about as long as the body or the perigynium, bidentate; achenes trigonous, 1.7–2.0 mm long, brown, rugulose, apiculate; stigmas 3, red-brown.

Common Name: Wrinkle-fruit sedge.
Habitat: Dry, rocky woods.
Range: Nova Scotia to Minnesota and Saskatchewan, south to Illinois and Georgia.
Illinois Distribution: Pope County.

Sometimes considered a variety of *C. tonsa*, *Carex rugosperma* differs by its densely puberulent perigynia with a longer beak that is about as long as the body of the perigynium. The margins of the leaves of *C. rugosperma* are scabrous, while those of *C. tonsa* are more or less smooth. The bright green leaves of *C. rugosperma* are a good field characteristic.

Carex rugosperma flowers from late April to early June.

Page 168. *Carex pedunculata* Muhl. ex Willd. The following county should be added to the map on page 168 (bottom): Ogle.

Page 170. *Carex richardsonii* R. Br. There are no new records for this species.

Page 173. *Carex hirtifolia* Mack. The following counties should be added to the map on page 173 (bottom): Carroll, Crawford, Macoupin, Montgomery, Sangamon.

A17. Carex gynandra
(Fringed look-alike
sedge).

a. Habit.
b. Perigynium.

c. Achene.
d. Pistillate scale.

Page 175. *Carex garberi* Fern. There are no new records for this species. In the Cyperaceae volume of Flora of North America, this species is inexplicably not listed from Illinois, even though there are extant specimens and literature reports for Illinois.

Page 175. *Carex aurea* Nutt. There are no new records for this species. In the Cyperaceae volume of Flora of North America, this species is inexplicably not listed from Illinois, even though there are extant specimens and literature reports for Illinois.

Page 178. *Carex crinita* Lam. var. *crinita*. The following counties should be added to the map on page 178: Crawford, Johnson, Lawrence, Ogle.

Page 180. *Carex crinita* Lam. var. *brevicrinis* Fern. The following county should be added to the map on page 180: Alexander.

After *Carex crinita* var. *brevicrinis*, add the following species new to Illinois:

Carex gynandra Schwein. Ann. Lyceum Nat. Hist. N.Y. 1:70. 1824. Fig. A17.

Carex crinita Lam. var. *gynandra* (Schwein.) Schwein. & Torrey, Ann. Lyceum Nat. Hist. N.Y. 1:360. 1825.
Carex crinita Lam. var. *simulans* Fern. Portland Soc. Nat. Hist. 2:135. 1897.
Plants perennial, cespitose, from stout rootstocks and slender stolons; culms to 1.5 m tall, triangular, stout, scabrous on the angles, often reddish near the base; leaves 3–6 per culm, up to 1.2 cm wide, somewhat scabrous along the slightly revolute margins, septate between the nerves, the uppermost leaves usually longer than the culms; lowest sheaths red-brown, scabrous, strongly nerved and becoming fibrillose at maturity, the hyaline ventral bands thin; uppermost 1–3 spikes staminate, or with usually some pistillate flowers near the tip, up to 5 cm long, at maturity pendulous on slender peduncles; staminate scales acuminate to awned; lower 2–6 spikes pistillate, narrowly cylindric, up to 10 cm long, pendulous on smooth peduncles; lowest bracts foliaceous; pistillate scales ovate, the apex of the body acuminate, rough-awned, pale brown to reddish brown with a green center, much longer than the perigynia; perigynia numerous, 2–4 mm long, ellipsoid to ovoid, usually pale brown, lustrous, scarcely inflated, nerveless, beakless or with a minute beak up to 0.3 mm long, substipitate; achenes lenticular, 1.5–1.8 mm long, sometimes slightly crimped along the sides; stigmas 2.

Common Name: Fringed look-alike sedge.
Habitat: Swampy woods; edge of marsh.
Range: Nova Scotia to Ontario and Minnesota, south to Illinois, Indiana, and Georgia.
Illinois Distribution: Alexander and Jackson counties.

Although strikingly similar to *C. crinita*, this species differs by its scabrous sheaths and the apex of the body of the pistillate scales acuminate rather than retuse.

Carex gynandra flowers in June and July.

Page 180. *Carex aquatilis* Wahlenb. var. *substricta* Kukenth. in Engler. There are no new records for this variety.

Page 182. *Carex stricta* Lam. The following counties should be added to the map on page 182 (bottom): Brown, Ford, Iroquois, Jo Daviess, Knox, Menard, Montgomery, Schuyler, Shelby.

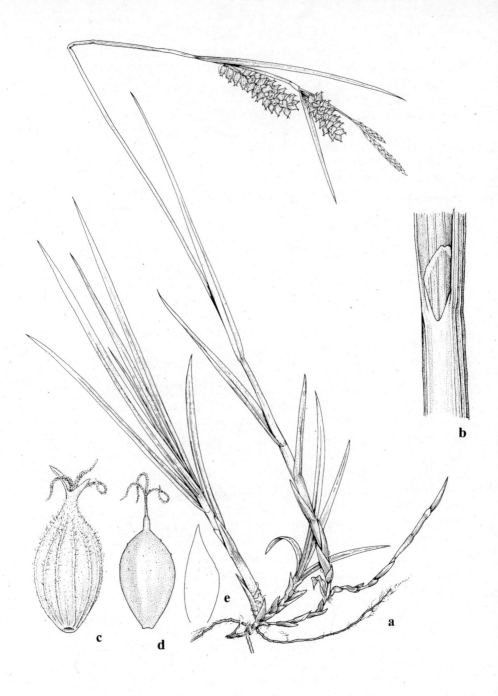

A18. Carex houghtoniana
(Houghton's sedge).

a. Habit.
b. Sheath.
c. Perigynium.

d. Achene.
e. Pistillate scale.

Page 184. *Carex nebrascensis* Dewey. There are no new records for this species.

Page 186. *Carex emoryi* Dewey in Torr. The following counties should be added to the map on page 186: Champaign, Hancock, Kendall, McDonough, Mercer, Woodford.

Page 186. *Carex haydenii* Dewey. There are no new records for this species.

Page 189. *Carex torta* Boott ex Tuckerm. There are no new records for this species.

Page 191. *Carex buxbaumii* Wahlenb. The following county should be added to the map on page 191: Mason.

Page 191. *Carex limosa* L. There are no new records for this species. In the Cyperaceae volume of Flora of North America, this species is inexplicably not listed from Illinois, even though there are extant specimens and literature reports for this species in Illinois.

Page 194. *Carex shortiana* Dewey. The following counties should be added to the map on page 194 (bottom): Alexander, Douglas, McDonough.

Page 194. *Carex* X *deamii* F. J. Herm. There are no new records for this hybrid.

Page 197. *Carex lasiocarpa* Ehrh. var. *americana* Fern. The following counties should be added to the map on page 197 (bottom): Champaign, Iroquois, Will.

Page 199. *Carex pellita* Willd. in Schk. The following counties should be added to the map on page 199: Brown, Carroll, Clay, Knox.

After *Carex pellita*, the following species new to Illinois should be added:

Carex houghtoniana Torr. ex Dewey, Am. Journ. Sci. Arts 30:63. 1836. Fig. A18.

Plants perennial, loosely cespitose, with long, creeping rhizomes; sterile shoots common; culms to 1 m tall, stiff, scabrous on the angles, purple-red at the bases; leaves up to 8.5 mm wide, flat except for the revolute margins, glabrous, septate, with a conspicuous midvein, scabrous toward the tip; lowest sheaths red-purple, fibrillose; upper 1–3 spikes staminate, erect, up to 4 cm long, long-pedunculate; staminate scales light reddish brown; lowest 2–4 spikes pistillate, occasionally with a few staminate flowers at the tip, up to 5 cm long, erect, sessile or subsessile; pistillate scales lanceolate to ovate, acute to short-awned, usually red-brown with a green center; perigynia up to 70 per spike, 4.5–6.5 mm long, broadly ovoid or broadly ellipsoid, sparsely pubescent, conspicuously veiny, with a bidentate beak 1.2–2.0 mm long; achenes trigonous, 1.6–2.0 mm long, yellow-brown, punctate, jointed with the style; stigmas 3.

Common Name: Houghton's sedge.
Habitat: Disturbed sandy area (in Illinois).
Range: Newfoundland to Alberta, south to Minnesota, Illinois, Wisconsin, and New Hampshire.
Illinois Distribution: Whiteside County: Tampico, Geneseo Nursery.

Although obviously related to *C. pellita*, this species is easily distinguished by its sparsely pubescent, veiny perigynia 4.5–6.5 mm long, its longer beak of the perigynium, and its wider leaves.

This plant was collected in 1998 and 1999 in a dry, sandy field planted to native plants in 1997 and mowed for weed control in 1997 and 1998. The plants appeared

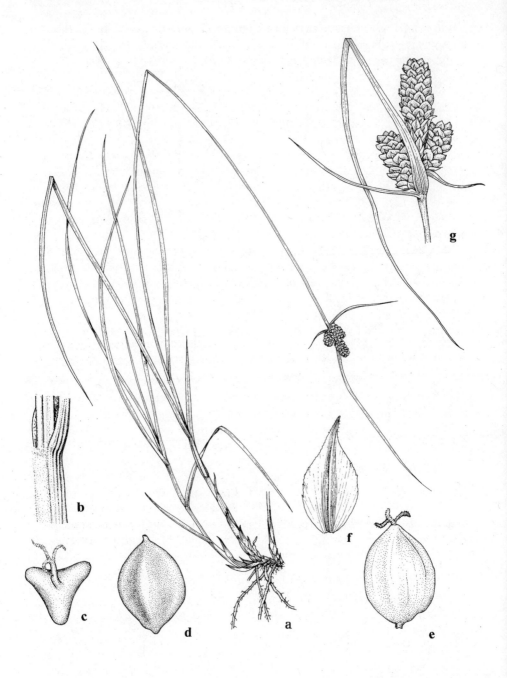

A19. Carex complanata
(Hirsute sedge).
a. Habit.

b. Sheath.
c. Achene, cross-section.
d. Achene, side view.

e. Perigynium.
f. Pistillate scale.
g. Inflorescence.

in a 6- to 8-foot-diameter colony and is aggressively rhizomatous. In personal correspondence with me, Reznicek remarked, "The collector noted that it appeared after the sandy site was planted to native plants, but insisted, when I wrote him and asked the question, that the plant could not have been a contaminant from any seeding practices that went into the site. I'm inclined to agree because it is hard to see how this sporadic early successional species of dry, sandy sites could have gotten into any seed mixtures and, in any event, a 6- to 8-foot-diameter colony could not have appeared one year after planting if it was a seed contaminant. The appearance of the plant from the seed bank after disturbance is consistent with the behavior of *Carex houghtoniana* in the sandy areas of the more northern Great Lakes region, but this southern site is certainly a mystery."

The Illinois specimen was collected on July 2, 1999.

Page 199. *Carex hirta* L. The following counties should be added to the map on page 202 (top): Cook, Iroquois.

Page 202. *Carex pallescens* L. The following county should be added to the map on page 202 (bottom): Montgomery. In the Cyperaceae volume of Flora of North America, this species is inexplicably not listed from Illinois, even though there are extant specimens and literature reports for Illinois.

Page 204. *Carex hirsutella* Mack. The following county should be added to the map on page 204: Montgomery.

Page 206. *Carex caroliniana* Schw. The following counties should be added to the map on page 206: Edwards, Marion. In the Cyperaceae volume of Flora of North America, this species is inexplicably not listed from Illinois, even though there are extant specimens and literature reports for Illinois.

Page 208. *Carex bushii* Mack. The following county should be added to the map on page 208: Schuyler.

After *Carex bushii*, the following species new to Illinois should be added:

Carex complanata Torr. & Hook. Ann. Lyceum Nat. Hist. New York 3:408. 1836. Fig. A19.

Plants perennial, loosely cespitose, with short stolons; culms slender, to 75 cm tall, triangular, sparsely pubescent, more or less scabrous on the angles, usually slightly reddish at the base; leaves up to 6 per culm, up to 4 mm wide, flat except for the revolute margins, pilose on both surfaces, shorter than the inflorescence; sheaths sparsely to densely pilose, reddish brown to green, with a densely pilose, hyaline ventral band, the ligules shorter than wide; terminal spike gynecandrous, erect, up to 2.5 cm long, up to 0.9 cm thick, thick-cylindric with a clavate base, usually on a very short peduncle; staminate scales acuminate to short-awned; lateral spikes 2–4, pistillate, erect, up to 1.5 cm long, up to 4.5 mm thick, sessile or on very short peduncles; pistillate scales ovate, acute to acuminate or short-awned with the awn up to 0.5 mm long, with a green center, shorter than to about equaling the perigynia; perigynia up to 30 per spike, obovoid, trigonous in cross-section, pale or often green above the middle, usually glabrous, 2.0–3.5 mm long, several-nerved, usually short-pointed at the tip; achenes trigonous, 1.6–2.3 mm long, apiculate, rounded or substipitate at the base; stigmas 3.

A20. Carex arctata
   (Drooping woodland
   sedge).

a. Habit.
b. Sheath.
c. Pistillate scale.

d. Achene.
e. Perigynium.

Common Name: Hirsute sedge.

Habitat: Dry woods; roadsides.

Range: Delaware to Iowa, south to Texas and Florida; Mexico.

Illinois Distribution: Lawrence and Williamson counties.

Carex complanata is in a group of closely appearing species that also includes *C. hirsutella*, *C. bushii*, and *C. caroliniana*. Of these, *C. complanata* and *C. hirsutella* have perigynia trigonous in cross-section, while *C. bushii* and *C. caroliniana* have perigynia circular in cross-section. *Carex complanata* differs from *C. hirsutella* by its sparsely pubescent leaves. *Carex bushii* is the only species in the complex that has pistillate scales with an awn longer than 0.5 mm.

This species flowers in May and June.

Carex hirsutella, *C. caroliniana*, *C. bushii*, and *C. complanata* are similar in appearance. The following table summarizes some of their characteristics:

character	C. hirsutella	C. caroliniana	C. complanata	C. bushii
perigynia X-section	trigonous	circular	trigonous	circular
leaf pubescence	dense	+ or – glabrous	sparse	dense
pistillate scales	acute, cuspidate	acute, acuminate	acute or awned	awn 0.5 mm
achene length	1.6–2.0 mm	1.5–2.2 mm	1.6–2.3 mm	2.0–2.5 mm
pistillate scale/ perigynia	shorter	shorter	shorter	longer

Page 208: *Carex virescens* Muhl. ex Willd. The following counties should be added to the map on page 211 (top): Clark, Edgar, Fulton, Gallatin. In the Cyperaceae volume of Flora of North America, this species is inexplicably not listed from Illinois, even though there are extant specimens and literature reports for Illinois.

Page 211. *Carex swanii* (Fern.) Mack. The following counties should be added to the map on page 211 (bottom): Clark, Lee, McDonough, Williamson.

Page 213. *Carex prasina* Wahl. The following counties should be added to the map on page 213: Fayette, Vermilion.

Page 213. *Carex gracillima* Schwein. The following counties should be added to the map on page 216 (top): Crawford, Fulton, La Salle, Vermilion.

After *Carex gracillima*, add the following species new to Illinois:

Carex arctata Boott in W. J. Hooker, Fl. Bor. Am. 2:227. 1830. Fig. A20.

Plants perennial, cespitose, with short rhizomes; culms to 1 m tall, triangular, slender, glabrous but scabrous on the angles, purplish at the base; sheaths glabrous, with the hyaline band red-spotted near the summit; leaves 3–4 per culm, up to 12 mm wide, lax, membranous, septate-nodulose, green, glabrous, sometimes scabrous on the margins; terminal spike staminate, to 4 cm long, 1–2 mm thick, sessile or on peduncles to 2.5 cm long; staminate scales obtuse to very short-awned, hyaline with a green center; lateral spikes 2–5, linear-cylindric, to 8 cm long, to 4 mm thick, on slender, flexuous peduncles to 3 cm long; lower bracts leaflike; pistillate scales oblong-lanceolate, acuminate to short-awned, with the awn to 1.5 mm long, greenish white with a green center, shorter than the perigynia, some of the basal ones empty; perigynia up to 45 per spike, ellipsoid, tapering to both ends, somewhat inflated, 3–5 mm long, 1–2 mm wide, glabrous, membranous, green, often red-dotted, finely nerved, substipitate, with a short, bidentate beak 0.7–1.5 mm long; achenes trigonous, 1.8–2.5 mm long, 1.0–1.8 mm wide, substipitate, jointed with the style; stigmas 3, short.

Common Name: Drooping woodland sedge.
Habitat: Mesic woods.
Range: Newfoundland to Manitoba, south to Minnesota, Illinois, and Pennsylvania.
Illinois Distribution: Wabash County: Beall Woods Nature Preserve.

This species is related to *C. formosa* and *C. gracillima* and occupies similar habitats to those two species. It differs from *C. formosa* by its glabrous sheaths of the leaves, its glabrous culms, its wider leaves, its longer pistillate spikes, its awn of the pistillate scale up to 1.5 mm long, its fewer perigynia per spike, and its longer beak of the perigynium. It differs from *C. gracillima* by its wider leaves, its longer pistillate spikes, its awned pistillate scales, its larger perigynia, and the presence of a beak on the perigynium.

Carex arctata flowers in May and June.

The following table summarizes characteristics of *C. gracillima*, *C. formosa*, and *C. arctata*:

character	*C. gracillima*	*C. formosa*	*C. arctata*
culm	glabrous	slightly pubescent	glabrous
sheaths	glabrous	pubescent	pubescent
leaf width	up to 8 mm	7–8 mm	up to 12 mm
pistillate scale	obtuse to acute	obtuse to acute	acuminate or short awn
pistillate spike	to 6 cm long	to 2.5 cm long	to 8 cm long
perigynia per spike	up to 50	up to 20	up to 45
perigynium height	2.0–3.5 mm	3.5–5.0 mm	3–5 mm
beak	beakless	0.2–0.3 mm long	0.7–1.5 mm long

Page 216. *Carex oxylepis* Torr. & Hook. There are no new records for var. *oxylepis* or for var. *pubescens*.

Page 218. Carex *formosa* Dewey. There are no new records for this species. In the description on page 218, the leaf width should be changed to 7–8 mm.

Page 220. *Carex davisii* Schwein. & Torr. The following counties should be added to the map on page 220: Cass, Union.

Page 220. *Carex debilis* Michx. var. *rudgei* Bailey. After evaluating *C. debilis* var. *debilis* and var. *rudgei*, I have concluded that var. *rudgei* has sufficient diagnostic characters that it should receive species status. At the species level, this plant is known as *Carex flexuosa*. The illustration on page 222 should be changed to *Carex flexuosa*. Following is the taxonomy for this plant at the species level:

A21. Carex debilis
   (White-edge sedge).

a. Habit.
b. Perigynium.

c. Pistillate scale.
d. Achene.

Carex flexuosa Muhl. ex Willd. Sp. Pl. 4:297. 1805. Fig. 107.

Carex tenuis Rudge, Trans. Linn. Soc. London 7:97, 1804, *non* J. F. Gmel. (1791).
Carex debilis Michx. var. *rudgei* L. H. Bailey, Mem. Torrey Club 1:34. 1899.

Plants perennial, cespitose, with short rhizomes; culms up to 1 m tall, triangular, slender, arching, scabrous on the angles, glabrous, purplish at base; overwintering basal leaves up to 7 mm wide; culm leaves up to 3 in number, 2–4 mm wide, flat, soft, glabrous, scabrous along the margins, usually longer than the culms; sheaths glabrous or minutely pubescent (not in Illinois specimens), yellow-brown to red, often red-dotted near the summit ventrally; terminal spike either entirely staminate or with some pistillate flowers at the tip, up to 5 cm long, 1–2 mm thick, very slender, usually on a scabrous peduncle; staminate scales obtuse to acute; lateral spikes 3–4 (–5), entirely pistillate, to 8 cm long, to 4 mm thick, the perigynia barely overlapping, on curving, pendulous, scabrous peduncles; pistillate scales acute to cuspidate, the hyaline margins with reddish streaks, shorter than the perigynia; perigynia up to 25 per spike, lance-ellipsoid, 5–6 mm long, 2–3 mm wide, ascending, green, glabrous or nearly so, strongly 2-ribbed and with several faint nerves, abruptly narrowed to an elongated beak, the beak about one-third as long as the perigynia; achenes trigonous, 2.0–2.2 mm long, apiculate, stipitate; stigmas 3.

Common Name: Rudge's sedge.
Habitat: Mesic or dry woods.
Range: Newfoundland and Nova Scotia to Ontario to Minnesota, south to Illinois, Tennessee, and North Carolina.
Illinois Distribution: Hardin County.

This plant is usually referred to as *C. debilis* var. *rudgei*, but I believe there are enough significant differences to separate it as a different species. As a species, the oldest binomial is *C. tenuis* Rudge, described in 1804, but that binomial had been used in 1791 by J. F. Gmelin for a different species. Therefore, Willdenow's *C. flexuosa*, named in 1805, becomes the next available epithet as a species.

The major distinguishing features of *C. flexuosa* are the shorter and more ellipsoid perigynia that abruptly narrow to a beak about one-third as long as the perigynium.

Carex flexuosa flowers during May and June.

Typical *Carex debilis* is also known from Illinois. Its taxonomy and description follow:

Carex debilis Michx. Fl. Bor. Am. 2:172. 1803. Fig. A21.

Plants perennial, cespitose, with short rhizomes; culms up to 1 m tall, triangular, slender, arching, scabrous on the angles, glabrous, purplish at base; overwintering leaves up to 7 mm wide; culm leaves up to 3 in number, 2–4 mm wide, flat, soft, glabrous, scabrous along the margins, usually longer than the culms; sheaths glabrous or minutely pubescent (not in Illinois specimens), yellow-brown to red, often red-dotted near the summit ventrally; terminal spike either entirely staminate or with some pistillate flowers at the top, up to 5 cm long, 1–2 mm thick, very slender, usually on a scabrous peduncle; staminate scales obtuse to acute; lateral spikes 3–4 (–5), entirely pistillate, to 8 cm long, to 4 mm thick, the perigynia

A22. Carex cherokeensis
(Cherokee sedge).

a. Habit.
b. Sheath.
c. Perigynium.

d. Pistillate scale.
e. Achene.

strongly overlapping, on curving, pendulous, scabrous peduncles; pistillate scales obtuse to acute, the hyaline margins white or pale, shorter than the perigynia; perigynia up to 25 per spike, fusiform, 5.5–8.5 mm long, 1–2 mm wide, ascending, green, glabrous or nearly so, strongly 2-ribbed and with several faint nerves, gradually tapering to an elongated beak nearly half as long as the perigynium; achenes trigonous, 2.0–2.2 mm long, apiculate, stipitate; stigmas 3.

Common Name: White-edge sedge.
Habitat: Mesic woods.
Range: New Jersey to Michigan, south to Missouri, Texas, and Illinois.
Illinois Distribution: Carroll County.

Carex debilis flowers in May and June.
The map on page 107 should be labeled *Carex flexuosa*.
The following table notes the differences between *C. debilis* and *C. flexuosa*:

Carex debilis	Carex flexuosa
perigynia strongly overlapping	perigynia barely overlapping
pistillate scales obtuse to acute, white or pale	pistillate scales acute to cuspidate, with reddish streaks
perigynia fusiform	perigynia lance-ellipsoid
perigynia 5.5–8.5 mm long	perigynia 4–7 mm long
perigynia gradually tapering to an elongated beak nearly half as long as the perigynium	perigynia abruptly narrowed to an elongated beak about ⅓ as long as the perigynium

I have been observing and studying *C. debilis* and *C. tenuis* over much of their ranges for many years and am convinced that the perigynium differences are sufficient enough to merit recognition as a species. It is true that where the ranges overlap, there are a few specimens that appear to be somewhat intermediate. However, all of the perigynia I have examined that are longer than 6 mm are always accompanied by a narrowed beak at least one-half the length of the perigynium.

Page 223. *Carex sprengelii* Dewey in Spreng. The following county should be added to the map on page 223 (bottom): Ogle.

After Carex *sprengelii*, add the following species new to Illinois:

Carex cherokeensis Schwein. Ann. Lyceum Nat. Hist. N.Y. 1:71. 1824. Fig. A22.

Plants perennial, cespitose, from a coarse, fibrous, matted, often tuberlike rhizome; culms to 80 cm tall, triangular, erect, usually scabrous on the angles, glabrous, brownish at the base, the old leaves persistent at the base of the plant but not fibrillose; leaves up to 8 per culm, up to 7 mm wide, flat, scabrous along the margins, shorter than the culms; sheaths glabrous, usually septate, light to dark brown, hyaline near the apex and sometimes red-dotted; terminal spike staminate, up to 6 cm long, up to 4 mm thick, pedunculate and standing above the lateral spikes; staminate scales acute to cuspidate, hyaline with a green center;

lateral spikes up to 12, not crowded, pistillate or the uppermost sometimes androgynous, up to 4 cm long, up to 1 cm thick, on arching or pendulous, scabrous peduncles; pistillate scales elliptic to ovate, acuminate to cuspidate, green or yellow-green with hyaline margins, about as long as the perigynia; perigynia up to 50 per spike, 5.0–6.5 mm long, 1.5–2.5 mm wide, spreading to ascending, the body ovoid, glabrous, 2-ribbed, finely nerved, abruptly tapering to a bidentate beak about half as long as the body of the perigynium; achenes trigonous, up to 2.5 mm long, yellowish, apiculate, substipitate; stigmas 3.

Common Name: Cherokee sedge.
Habitat: Bottomland forests.
Range: Virginia to Missouri, south to Texas and Florida.
Illinois Distribution: Alexander, Jackson, and Union counties.

Carex cherokeensis has the general appearance of *C. sprengelii* by virtue of the inflorescence and thickened fibrous rootstocks. It differs from *C. sprengelii* by its beak about half as long as the finely nerved body of the perigynium and its somewhat wider leaves.

This species flowers during June and July.

Page 225. *Carex granularis* Muhl. ex Willd. After working with *C. granularis* in the field throughout its entire range, I believe it should be divided into two species rather than two varieties. Both species, *C. granularis* and *C. haleana*, occur in Illinois and are included in this second edition. Since the description of *C. granularis* in the first edition of *Carex* also included var. *haleana*, a revised description of *C. granularis* and a description of *C. haleana* are given below:

Carex granularis Muhl. ex Willd. Sp. Pl. 4:179. 1805. Fig. 109 a–g.

Carex granularis Muhl. ex Willd. var. *recta* Dewey in Wood, Class-book 763. 1860.
Plants perennial, cespitose, from short rhizomes; culms slender, spreading to ascending, up to 85 cm tall, triangular, usually slightly scabrous on the angles, brownish at base; overwintering leaves 5–12 mm wide; other leaves up to 5 per culm, 5–12 mm wide, flat, flaccid, more or less glaucous, usually scabrous along the margins, the basal usually longer than the culms; sheaths yellow-brown, russet-maculate on the ventral band; terminal spike 1, entirely staminate, up to 25 mm long, up to 2.5 mm thick, sessile or subsessile; staminate scales acuminate, reddish brown with a green center; lateral spikes 2–5, entirely pistillate, distant or the upper two more or less contiguous, linear-oblong, erect or spreading, 10–23 mm long, 4–6 mm thick, the uppermost sessile or subsessile, the lower ones pedunculate; bracts leaflike, often 10 cm long or longer; pistillate scales usually whitish or reddish, obovate, acuminate to short-awned, a little shorter than the perigynia or occasionally equaling the perigynia; perigynia 10–50 per spike, ascending, obovoid, substipitate, 2.5–4.0 mm long, 1.5–2.5 mm wide, inflated, olive-green to brownish, glabrous, with many strong nerves, with a short, entire, slightly bent or straight beak 0.2–0.4 mm long; achenes trigonous, 1.6–1.8 mm long, yellow-brown, apiculate, stipitate; stigmas 3.

A23. Carex haleana
(Hale's sedge).

a. Habit.
b. Sheath.
c. Pistillate spikelet.

d. Perigynium.
e. Achene.
f. Pistillate scale.

The following counties should be added to the map on page 225: Adams, Brown, Clay, Ogle, Pike, Saline, Shelby, Williamson, Woodford.

Carex haleana Olney, Car. Bot. Am. 6. 1872. Fig. A23.

Carex granularis Muhl. var. *haleana* (Olney) Porter, Proc. Acad. Sci. Phila. 1887:74. 1888.
Carex granularis Muhl. var. *shriveri* Britt. in Britt. & Brown, Ill. Fl. 1:322. 1896.
Carex shriveri (Britt.) Britt. Man. 208. 1901.

Plants perennial, cespitose, from short rhizomes; culms slender, spreading to ascending, up to 75 cm tall, triangular, usually smooth, brownish at base; overwintering leaves 5–10 mm wide; other leaves up to 5 per culm, 5–16 mm wide, flat, not flaccid, slightly glaucous, more or less scabrous along the margins, the basal usually shorter than the culms; sheaths tight, yellow-brown, russet-maculate on the ventral band; terminal spike 1, entirely staminate, up to 25 mm long, up to 2.5 mm thick, sessile or subsessile; staminate scales acuminate, reddish brown with a green center; lateral spikes 2–5, entirely pistillate, distant or the upper two more or less contiguous, linear-oblong, erect or spreading, 7–28 mm long, 3–5 mm thick, the uppermost sessile or subsessile, the lower ones pedunculate; bracts leaflike, often 10 cm long or longer; pistillate scales brown or reddish, ovate, acuminate to cuspidate, much shorter than the perigynia; perigynia 15–50 per spike, ascending, ellipsoid, substipitate, 2.3–3.0 mm long, 1.0–1.5 mm wide, scarcely inflated, green to brownish, glabrous, with several fine nerves, with a short, entire, more or less straight beak about 0.1 mm long, or the beak absent and the perigynia pointed; achenes trigonous, 1.6–1.8 mm long, yellow-brown, apiculate, stipitate; stigmas 3.

Common Name: Hale's sedge.
Habitat: Wet meadows.
Range: Quebec to Saskatchewan, south to Kansas, Missouri, Tennessee, and Virginia.
Illinois Distribution: Scattered throughout the state, but apparently more common in the upper half of Illinois.

Carex haleana differs from *C. granularis* by its usually wider leaves and its smaller ellipsoid perigynia that are scarcely inflated and either beakless or with a minute beak. The perigynia are finely nerved rather than strongly nerved.

Carex haleana flowers in May and June.

Page 227. *Carex crawei* Dewey. There are no new records for this species.

Page 229. *Carex conoidea* Schkuhr ex Willd. There are no new records for this species.

Page 229. *Carex amphibola* Steud. The following counties should be added to the map on page 232 (top): Cook, Macon, Massac, Montgomery, Pulaski, Putnam, Richland, Sangamon. The following citations should be removed from the synonymy of *C. amphibola* since they now belong to a separate species known as *C. planispicata*:

Carex grisea Wahl. var. *rigida* L. H. Bailey, Mem. Torrey Club, 1:56. 1889.
Carex amphibola Steud. var. *rigida* (L. H. Bailey) Fern. Rhodora 44:315. 1942.

After *Carex amphibola*, add the following two species for Illinois:

A24. Carex planispicata
(Flat-spiked sedge).

a. Habit.
b. Spikelets.

c. Perigynium, with scale.
d. Achene.

Carex planispicata Naczi, Journ. Ky. Acad. Sci. 60:37. 1999. Fig. A24.

Carex grisea Wahl. var. *rigida* L. H. Bailey, Mem. Torrey Club 1:56. 1889.
Carex amphibola Steud. var. *rigida* (L. H. Bailey) Fern. Rhodora 44:315. 1942.
Plants perennial, from short rhizomes; culms slender, triangular, to 60 cm tall,
slightly scabrous on the angles, purplish at the base; sterile shoots common; leaves
3–7 mm wide, stiff, flat, scabrous along the margins; sheaths often red-dotted in
the hyaline areas; terminal spike staminate, to 3 cm long, to 2.5 mm thick, on
a glabrous or scabrous peduncle elevated above the pistillate spikes; staminate
scales obtuse to acute, hyaline, with a green center; lateral spikes 2–5, pistillate, to
2.5 cm long, 3–6 mm thick, the lowest often near the base of the plant, with up to
14 perigynia; uppermost bracts leaflike, the sheaths tight; pistillate scales ovate,
awned, hyaline and red-dotted with a green center, shorter than the perigynia;
perigynia distichously arranged, oblongoid to ellipsoid, bluntly trigonous, scarcely
inflated, 3.9–5.0 mm long, dull, glabrous, with numerous impressed nerves, tapering to an entire, straight or slightly curved tip; achenes trigonous, 2.8–3.5 mm
long, yellow-brown, apiculate, stipitate; stigmas 3.

Common Name: Flat-spiked sedge.
Habitat: Moist to wet woods.
Range: New Jersey to Illinois and Missouri, south to Texas and Georgia.
Illinois Distribution: Throughout the southern one-third of Illinois, often with
C. *amphibola*.

This plant, usually considered as synonymous with *C. amphibola*, or as a variety
of it (as var. *rigida*), has been elevated to the rank of species.
The dull perigynia are distichously arranged in *C. planispicata* and shiny and
spirally arranged in *C. amphibola*. The uppermost bracts of *C. planispicata* have
tight sheaths, while the uppermost bracts of *C. amphibola* have loose sheaths.
Carex planispicata flowers in May and early June.

Carex corrugata Fern. Rhodora 44:76. 1942. Fig. A25.

Carex rugata Fern. Rhodora 43:545. 1941, *non* Ohwi (1932).
Plants perennial, cespitose, from short rhizomes; culms slender, to 90 cm tall,
triangular, slightly scabrous on the angles, purple-red at the base; sterile shoots
common; leaves 3.5–8.0 mm wide, rather stiff, flat, scabrous along the margins;
sheaths tight, sometimes red-dotted in the hyaline area; terminal spike staminate,
to 3.5 cm long, 2–3 mm thick, on a glabrous or slightly scabrous peduncle barely
elevated above the pistillate spikes; staminate scales obtuse to acute, hyaline with a
green center; lateral spikes 2–4, pistillate, each with up to 27 perigynia, the spikes
to 3 cm long, to 1 cm thick, the lowest sometimes near the base of the plant; uppermost bracts leaflike, the sheaths more or less loose; pistillate scales ovate, awned,
hyaline along the margins, sometimes red-dotted with a green center, shorter than
the perigynia; perigynia spirally arranged, narrowly ellipsoid to narrowly obovoid,
bluntly trigonous, scarcely inflated, 3.5–4.7 mm long, shiny, with numerous impressed nerves, tapering to an entire, straight tip; achenes trigonous, more or less
round-angled, 2.5–3.3 mm long, yellow-brown, apiculate, stipitate; stigmas 3.

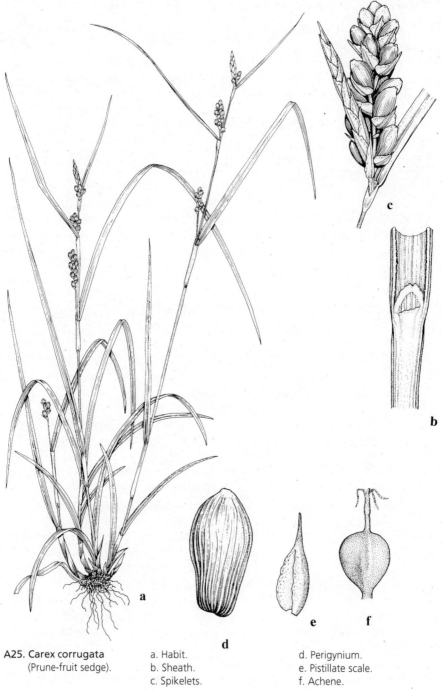

A25. Carex corrugata
(Prune-fruit sedge).

a. Habit.
b. Sheath.
c. Spikelets.

d. Perigynium.
e. Pistillate scale.
f. Achene.

Common Name: Prune-fruit sedge.
Habitat: Floodplain forests.
Range: Maryland to Illinois to Kansas, south to Texas and Florida.
Illinois Distribution: Jackson, Saline, Union, and Wabash counties.

This species often occurs with *C. amphibola* and *C. planispicata*. It differs from *C. amphibola* by its slightly shorter and thicker perigynia with its blunt angles in cross-section. It differs from *C. planispicata* by its shiny perigynia that are spirally arranged.

Carex corrugata flowers in May and June.

The following table compares and contrasts *C. amphibola*, *C. planispicata*, and *C. corrugata*:

character	*C. amphibola*	*C. planispicata*	*C. corrugata*
sheaths	loose	tight	tight
pistillate spike	3–6 mm thick	3–6 mm thick	up to 10 mm thick
perigynia per spike	up to 12	up to 14	up to 27
perigynia arranged	spirally	distichously	spirally
perigynia shape	oblongoid to obovoid	oblongoid to ellipsoid	narrowly ellipsoid
perigynia angles	sharp	blunt	blunt
perigynia	dull	dull	shiny
perigynium height	3.5–4.5 mm	3.9–5.0 mm	3.5–4.7 mm
achene length	2.4–2.5 mm	2.8–3.5 mm	2.5–3.3 mm

Page 232. *Carex grisea* Wahl. The following counties should be added to the map on page 232 (bottom): DeWitt, Johnson, Macon, Monroe, Saline.

Page 234. *Carex flaccosperma* Dewey. There are no new records for this species. In the Cyperaceae volume of Flora of North America, this species is inexplicably not listed from Illinois, even though there are extant specimens and literature reports for Illinois.

Page 234. *Carex glaucodea* Tuckerm. ex Olney. The following county should be added to the map on page 237 (top): Brown.

Page 237. *Carex oligocarpa* Schkuhr ex. Willd. The following county should be added to the map on page 237 (bottom): Gallatin.

Page 237. *Carex hitchcockiana* Dewey. The following counties should be added to the map on page 240 (top): Adams, Brown, Carroll, Champaign, McLean, Pike, Tazewell, Williamson.

Page 240. *Carex meadii* Dewey. The following counties should be added to the map on page 240 (bottom): Jefferson, Johnson, Lee, Marion, Montgomery, Tazewell.

Page 242. *Carex tetanica* Schkuhr. The following county should be added to the map on page 242: Iroquois.

Page 242. *Carex woodii* Dewey. The following county should be added to the map on page 245 (top): St. Clair.

After *Carex woodii*, the following species new to Illinois should be added:

A26. Carex vaginata
(Sheathed sedge).

a. Habit.
b. Staminate spikelet.
c. Perigynium.

d. Pistillate scale.
e. Achene.

Carex vaginata Tausch, Flora 4:557. 1821. Fig. A26.

Plants perennial, colonial, loosely cespitose, with long, slender rhizomes; culms up to 60 cm tall, slender, triangular, smooth except sometimes scabrous just beneath the inflorescence, usually green or brown at base; leaves 3–5 per culm, up to 4 mm wide, green, flat, glabrous; sheaths pale brown, sparsely papillose, the lowest not blade-bearing; terminal spike staminate, to 2 cm long, up to 2.5 mm thick, on a slender, scabrous peduncle; staminate scales obtuse to acute; lateral spikes 2–4, pistillate, on slender ascending peduncles, loosely flowered, to 3 cm long, to 6.5 mm thick; pistillate scales acute or obtuse, purple-tinged; perigynia up to 15 per spike, 3.5–5.0 mm long, 1.5–2.2 mm wide, trigonous, green to dark brown, faintly nerved, glabrous, abruptly contracted to a slender, straight beak up to 2 mm long; achenes trigonous, 2–3 mm long, 1.2–1.5 mm wide, brown, apiculate; stigmas 3.

Common Name: Sheathed sedge.
Habitat: along a road (in Illinois).
Range: Greenland and Labrador to Alaska, south to British Columbia, Montana, Minnesota, Illinois, New York, and New Hampshire.
Illinois Distribution: Lake County: Illinois Beach State Park.

This northern species was found a single time growing along a road in Illinois Beach State Park.

Carex vaginata is distinguished by its loosely flowered pistillate spikes borne on slender, ascending peduncles and by its usually straight beak of the perigynium that arises abruptly from the body of the perigynium.

The Illinois collection was made in late May.

Page 245. *Carex plantaginea* Lam. The following county should be added to the map on page 245 (bottom): La Salle. In the Cyperaceae volume of Flora of North America, this species is inexplicably not listed from Illinois, even though there are extant specimens and literature reports for Illinois.

Page 247. *Carex careyana* Torr. in Dewey. The following counties should be added to the map on page 247: Brown, Effingham, Vermilion.

Page 247. *Carex platyphylla* Carey. The following county should be added to the map on page 250 (top): Mason. In the Cyperaceae volume of Flora of North America, this species is inexplicably not listed from Illinois, even though there are extant specimens and literature reports for Illinois.

Page 250. *Carex digitalis* Willd. The following synonymy citations for *C. digitalis* should be removed since I am recognizing *C. copulata* as a distinct species:

Carex digitalis Willd. var. *copulata* Bailey, Mem. Torrey Club 1:47. 1889.
Carex laxiculmis Schwein. var. *copulata* (Bailey) Fern. Rhodora 8:183. 1906.
Carex copulata (Bailey) Mack. N. Am. Fl. 18:251. 1935.

The following counties should be added to the map on page 250 (bottom): Clark, Cook, Johnson, Winnebago.

After *Carex digitalis*, the following two species new to Illinois should be added:

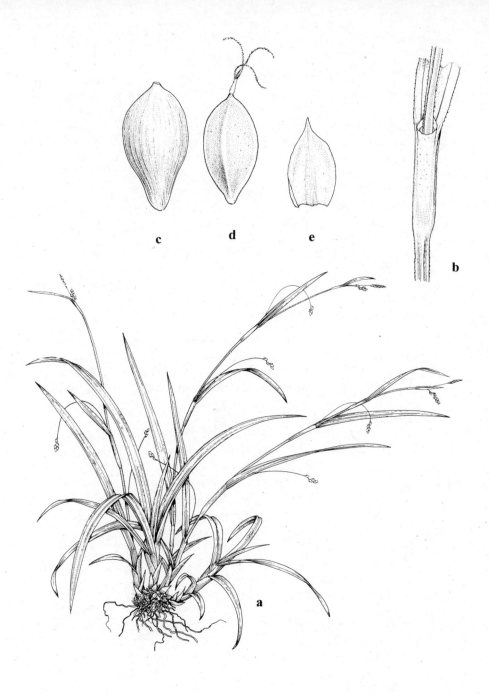

A27. Carex copulata
(Narrow-leaved
spreading sedge).

a. Habit.
b. Sheath.
c. Perigynium.

d. Achene.
e. Pistillate scale.

Carex copulata (L. H. Bailey) Mack. Fl. Prairies & Plains 181. 1932. Fig. A27.
Carex digitalis Willd. var. *copulata* L. H. Bailey, Mem. Torrey Club 1:47. 1889.
Carex laxiculmis Schwein. var. *copulata* (L. H. Bailey) Fern. Rhodora 8:183. 1906.

Plants perennial, cespitose, from short rhizomes; culms slender, to 80 cm tall, spreading to ascending, strongly scabrous, red-brown at the base; leaves up to 8 mm wide, scabrous along the margins, bright green; sheaths pale to dark brown, usually red-dotted, the ventral band hyaline; terminal spike staminate, to 1.8 cm long, to 2.5 mm thick, pedunculate; staminate scales acute to acuminate, brownish with a green center and hyaline margins; lateral spikes 2–4, pistillate, to 2 cm long, 3–4 mm thick, loosely flowered, sometimes with a few staminate flowers at the base, all pedunculate, the lower on spreading to pendulous peduncles up to 8 cm long; pistillate scales ovate, acute to short-awned, brown with a green center and hyaline margins, shorter than the perigynia; perigynia up to 10 per spike, ovoid to oblongoid, 3–4 mm long, trigonous with pointed angles, olive-green, several-nerved, with a short, nearly straight beak up to 1 mm long, substipitate; achenes trigonous, with concave sides, yellow-brown, 2.0–2.4 mm long, apiculate, stipitate; stigmas 3.

Common Name: Narrow-leaved spreading sedge.
Habitat: Mesic woods, wet woods, along streams.
Range: Ontario to Iowa, south to Arkansas, Alabama, and North Carolina.
Illinois Distribution: Occasional throughout the state.

Liberty Hyde Bailey first described this plant as a variety of *C. digitalis*. It subsequently was considered to be a variety of *C. laxiculmis*. Mackenzie proposed it as a possible hybrid between *C. laxiculmis* and *C. digitalis*. Its closest relationship appears to be with *C. laxiculmis*. Following Mackenzie, I am recognizing *C. copulata* as a distinct species, based on its narrower and harshly scabrous bright green leaves and its slightly shorter staminate spikes.

Carex copulata flowers from May until mid-June.

Carex macropoda (Fern.) Mohlenbr. comb. nov. (Basionym: *Carex digitalis* Willd. var. *macropoda* Fern.) Fig. A28.

Carex digitalis Willd. var. *macropoda* Fern. Rhodora 40:400. 1938.

Plants perennial, cespitose, from short rhizomes; culms to 50 cm tall, slender, triangular, somewhat scabrous on the angles, brownish at base; leaves 2.0–2.9 mm wide, rarely wider, scabrous on the margins, dark green; at least the lowest sheaths pale or brownish, the ventral band broad and hyaline; terminal spike staminate, linear to linear-clavate, to 2 cm long, 1.5–2.5 mm thick, on peduncles 8–16 cm long, rarely shorter, surpassing the uppermost bract; staminate scales oblong, acute, usually yellow-brown; lateral spikes 2–5, pistillate, 2–3 cm long,, 3–5 mm thick, loosely flowered, the uppermost stalked, the lower on slender peduncles up to 8 cm long; pistillate scales ovate, acute, keeled, brownish with a green center and hyaline margins, shorter than the perigynia; perigynia up to 12 per spike, obovoid to ovoid, 2.8–3.5 mm long, trigonous, with pointed angles, deep green, with 7–10 nerves, rarely more, very short-beaked to nearly beakless, stipitate; achenes trigonous, with concave sides, 1.0–1.8 mm long, apiculate; stigmas 3.

A28. Carex macropoda
(Southern woodland
sedge).

a. Habit.
b. Perigynium.

c. Achene.
d. Pistillate scale.

Common Name: Southern woodland sedge.
Habitat: Dry or mesic woods.
Range: Pennsylvania to Illinois and Missouri, south to Texas and Florida.
Illinois Distribution: Jackson, Johnson, and Pope counties.

Botanists consider this plant to be a variety of *Carex digitalis*, but there are several significant differences. The table below summarizes these differences:

Carex digitalis	Carex macropoda
peduncle of staminate spike 1–7 cm long, surpassed by uppermost bract	peduncle of staminate spike 8–16 cm long, surpassing uppermost bract
perigynia obovoid to fusiform	perigynia obovoid to ovoid
perigynia 2.5–3.3 mm long	perigynia 2.8–3.5 mm long
perigynia with 11–15 nerves	perigynia with 7–10 nerves
leaves 2.5–5.3 mm wide	leaves 2.0–2.9 mm wide

The long peduncle of the staminate spike that elevates the spike well above the uppermost bract is the most obvious distinguishing feature of *C. macropoda*, but it is not the only significant difference. While the perigynia of *C. digitalis* are sometimes fusiform, the perigynia of *C. macropoda* are never fusiform. The number of nerves on the perigynia of *C. macropoda* are 7–10, rarely more, while the number of nerves on the perigynia of *C. digitalis* are 11–15, rarely fewer. In nearly all of the specimens of *C. macropoda* that I have examined, if the staminate spike is elevated above the uppermost bract, the nerves of at least some of the perigynia are 7–10 in number. In addition, if the staminate spike is elevated above the uppermost bract, at least some of the leaves are not more than 2.5 mm wide.

The overall range of *C. macropoda* is more southern than the range of *C. digitalis*, never occurring north of Pennsylvania and extending south to Texas and Florida. In *C. digitalis*, the range is from Nova Scotia to Wisconsin, south to Missouri and Georgia.

Carex macropoda flowers in May and early June.

Page 252. *Carex laxiculmis* Schwein. The following counties should be added to the map on page 252: McLean, Wayne.

The following table summarizes some of the characteristics of *C. digitalis*, *C. copulata*, and *C. laxiculmis*:

character	C. digitalis	C. copulata	C. laxiculmis
culms	slightly scabrous	harshly scabrous	slightly scabrous
leaf color	dark green	bright green	pale green or glaucous
leaf margin	slightly scabrous	harshly scabrous	slightly scabrous
staminate spike	to 2 cm × 2 mm	to 1.8 cm × 2.5 mm	to 2 cm × 3 mm
staminate spike	short-pedunculate	long-pedunculate	long-pedunculate
pistillate spike	3 cm × 3–4 mm	2 cm × 3–4 mm	2 cm × 4–5 mm
perigynium	deep green	olive-green	olive-green
achene length	2.0–2.2 mm	2.0–2.4 mm	2.2–2.5 mm

After *Carex laxiculmis*, add the following species new to Illinois:

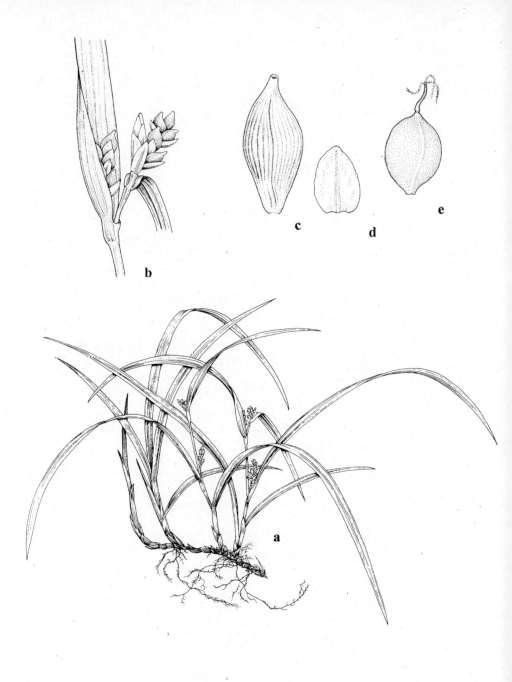

A29. Carex abscondita
    (Thicket sedge).

a. Habit.
b. Spikelets.
c. Perigynium.

d. Pistillate scale.
e. Achene.

Carex abscondita Mack. Bull. Torrey Club 37:244. 1910. Fig. A29.

Plants perennial, cespitose, from short rhizomes; culms slender, to 30 cm tall, erect to ascending, triangular, sometimes scabrous, brownish at the base; leaves up to 9 mm wide, scabrous along the margins, dark green, sometimes recurved, usually longer than the culms; at least the lowest sheaths white to pale brown, the ventral band broad and hyaline; terminal spike staminate, 4–11 mm long, 0.5–1.5 mm thick, sessile or nearly so; staminate scales obtuse, with a green center and hyaline margins; lateral spikes 3–4, pistillate, to 1.2 cm long, 3.5–6.0 mm thick, somewhat congested, the upper sessile or nearly so, the lower sometimes short-pedunculate; pistillate scales ovate, acute, brownish with a green center and hyaline margins, shorter than the perigynia; perigynia up to 13 per spike, obovoid, 2.8–4.0 mm long, trigonous, with pointed angles, green, finely nerved, with short, tapering beak, stipitate; achenes trigonous, ellipsoid, with flat sides, 2.5–4.0 mm long, apiculate; stigmas 3.

Common Name: Thicket sedge.
Habitat: Rich woods.
Range: Connecticut to Missouri, south to Texas and Florida.
Illinois Distribution: Wabash County: Beall Woods Nature Preserve.

Carex abscondita is probably most closely related to *C. digitalis* from which it differs by its shorter and narrower staminate spike, its obtuse staminate scales, its shorter pistillate spikes, and its larger achenes with flat rather than concave sides.

In the first edition of *Carex*, I excluded this species since the only Illinois specimen I had seen was from Beall Woods in Wabash County, but the specimen was too immature to verify. Since then, I have returned to Beall Woods in late May and found mature specimens of *C. abscondita*.

This species flowers in May and early June.

Page 252. *Carex alburisina* Sheldon. The following counties should be added to the map on page 255 (top): Logan, Mercer, Shelby.

After *Carex alburisina*, the following species new to Illinois should be added:

Carex kraliana Naczi & Bryson, Novon 12:520. 2002. Fig. A30.

Plants perennial, cespitose, from short rhizomes; culms to 55 cm tall, triangular, with slightly winged angles, scabrous on the angles; leaves up to 3 cm wide, dark green, usually somewhat scabrous and serrulate on the margins, at least near the tip, with sterile leafy tufts overwintering; sheaths light brown, up to 4 cm long; lower bracts leaflike, to 20 cm long; terminal spike staminate, to 1.5 cm long, to 2 mm thick, often overtopped by the bracts, sessile or subsessile; staminate scales obtuse to acute, with a green center and hyaline margins; lateral spikes 2–4, pistillate, to 2.5 cm long, to 4.5 mm thick, remotely flowered, sessile or on short, ascending peduncles, sometimes concealed by the bract; pistillate scales ovate, acute or short-awned, often whitish throughout, shorter than the perigynia; perigynia up to 14 per spike, ovoid to oblongoid, 2.7–4.5 mm long, trigonous, with rounded

A30. Carex kraliana
(Kral's sedge).

a. Habit.
b. Pistillate spikelet.
c. Perigynium.

d. Pistillate scale.
e. Achene.

angles, glabrous, several-nerved, with a short, curved beak, stipitate; achenes trigonous, with concave sides, 2.0–2.5 mm long, apiculate, substipitate; stigmas 3.

Common Name: Kral's sedge.
Habitat: Mesic woods.
Range: Maryland to Illinois, south to Texas and Florida.
Illinois Distribution: Jackson County: Giant City State Park.

This species, which often grows with *C. albursina* and *C. blanda*, is somewhat intermediate between the two. Its leaves are narrower than those of *C. albursina* but wider than those of *C. blanda*. It is more closely aligned with *C. albursina* by its bracts wider than and often concealing the pistillate spikes. In *C. kraliana*, the pistillate scales are often short-awned and the perigynia are ovoid to oblongoid; in *C. albursina*, the pistillate scales never are short-awned and the perigynia are obovoid.

Carex kraliana flowers in May and June.

The following table summarizes some of the characteristics of *C. albursina*, *C. blanda*, and *C. kraliana*:

character	*C. albursina*	*C. blanda*	*C. kraliana*
culms	slightly wing-angled	not wing-angled	slightly wing-angled
leaf width	to 35 mm	to 12 mm	to 30 mm
leaf color	pale green	pale green to glaucous	dark green
bract	wider than pistillate spike & concealing it	not wider than pistillate spike	wider than pistillate spike & concealing it.
pistillate scale	obtuse to acute	usually awned	acute to short-awned
perigynia	remote	crowded	remote
perigynium shape	obovoid	ellipsoid to obovoid	ovoid to oblongoid
perigynium length	3–4 mm	3.0–4.5 mm	2.7–4.5 mm
perigynia per spike	up to 13	up to 26	up to 14

Page 255. *Carex laxiflora* Lam. The following counties should be added to the map on page 255 (bottom): Cass, Knox.

After *Carex laxiflora*, add the following species new to Illinois:

Carex leptonervia (Fern.) Fern. Rhodora 16:214. 1914. Fig. A31.

Carex laxiflora Lam. var. *leptonervia* Fern. Rhodora 8:184. 1906.

Plants perennial, cespitose, from stout rhizomes; culms to 45 cm tall, decumbent to ascending, triangular, mostly smooth, brown at base; leafy tufts present, without prolonged culms; leaves up to 10 mm wide, often pale green, glabrous or occasionally scabrous along the margins; sheaths brownish, smooth, serrulate; terminal spike staminate, to 1.5 cm long, to 2 mm thick, pedunculate, usually surpassed by the bract; staminate scales ovate, acute, with hyaline margins; lateral spikes 3–4, pistillate, to 2 cm long, to 4 mm thick, more or less crowded, the uppermost sessile or nearly so, the others on ascending peduncles; pistillate

A31. Carex leptonervia
(Few-nerved sedge).

a. Habit.
b. Spikelets.

c. Perigynium, with scale.
d. Achene.

scales ovate, acute to apiculate, with a green center and hyaline margins, usually shorter than the perigynia; perigynia up to 15 per spike, narrowly obovoid to ellipsoid, 2.2–3.5 mm long, trigonous with rounded angles, pale green, with only 2–3 conspicuous nerves, with a short, slightly curved beak up to 1 mm long, stipitate; achenes trigonous, with concave sides, yellow-brown, 1.8–2.5 mm long, apiculate, substipitate; stigmas 3.

Common Name: Few-nerved sedge.
Habitat: Mesic woods.
Range: Labrador to Ontario and Minnesota, south to Illinois, Tennessee, and South Carolina.
Illinois Distribution: Pope County: Belle Smith Springs.

This species, in the complex that contains C. laxiflora, C. blanda, C. granularis, C. styloflexa, and C. striatula, differs from all of these by having perigynia with only 2–3 conspicuous nerves.

The flowers bloom in early April and May.

Page 257. Carex striatula Michx. There are no new records for this species. In the Cyperaceae volume of Flora of North America, this species is inexplicably not listed for Illinois, even though there are extant specimens and literature reports of it for Illinois.

Page 257. Carex blanda Dewey. There are no new records for this species.

Page 260. Carex gracilescens Steud. The following counties should be added to the map on page 260 (bottom): Alexander, Brown, Crawford, Fayette, Jefferson.

Page 260. Carex styloflexa Buckl. The following county should be added to the map on page 263 (top): Perry.

Page 263. Carex cryptolepis Mack. The following county should be added to the map on page 263 (bottom): Kane.

After Carex cryptolepis, add the following species that is new for Illinois:

Carex flava L. Sp. Pl. 2:975. 1753. Fig. A32.

Plants perennial, cespitose, with short rhizomes; culms to 70 cm tall, triangular, smooth, pale brown at base; sterile shoots usually present; leaves up to 6 mm wide, light green, usually scabrous along the margins, usually as long as the culms; sheaths convex at the summit; terminal spike staminate, to 2.2 cm long, to 3 mm thick, sessile or on short peduncles up to 4 mm long; staminate scales obtuse, reddish brown with hyaline margins; lateral spikes 2–5, pistillate, globose or ellipsoid, usually crowded near the tip of the culm, up to 2 cm long, up to 1.2 cm thick, each subtended by a leafy bract; pistillate scales ovate, acuminate, reddish brown, the tip not reaching the base of the beak of the perigynia; perigynia up to 35 per spike, 4.0–6.3 mm long, narrowly ovoid, bright yellow, spreading or the lower ones reflexed, shiny, few-nerved, with a beak over half as long as the body of the perigynium, the beak bidentate, curved, scabrous, 1.8–2.7 mm long; achenes trigonous, with concave sides, nearly black, shiny, 1.3–1.7 mm long, scabrous, short-apiculate; stigmas 3.

Common Name: Yellow sedge.
Habitat: Moist soil.
Range: Labrador to Alaska, south to British Columbia, Idaho, northern Illinois, and New Jersey.
Illinois Distribution: Cook County.

This is a very distinctive species because of the bright yellow perigynia. The very similar *C. cryptolepis* has more yellow-green perigynia. The pistillate scales

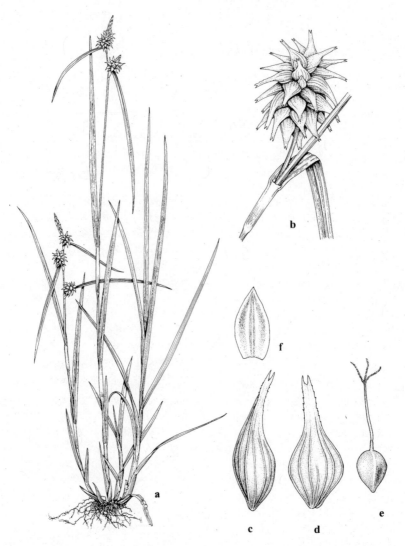

A32. Carex flava
 (Yellow sedge).

a. Habit.
b. Pistillate spikelet.
c. Perigynium, side view.

d. Perigynium, front view.
e. Achene.
f. Pistillate scale.

are reddish brown in *C. flava* and yellow-green in *C. cryptolepis*. The beak of the perigynium is scabrous in *C. flava* and is glabrous in *C. cryptolepis*.

The only Illinois collection is an historic one from Cook County.

Carex flava flowers in early summer.

Page 265. *Carex viridula* Michx. There are no new records for this species.

Carex cryptolepis, C. flava, and *C. viridula* are our only members of Section Ceratocystis. Because they are very similar, I have summarized some of their characteristics in the table below:

character	*C. cryptolepis*	*C. flava*	*C. viridula*
pistillate scales	yellow-green	reddish brown	reddish brown
staminate scales	acute, yellow	obtuse, red-brown	obtuse, red-brown
perigynium color	yellow-green	bright yellow	green or yellow-green
perigynium length	3.2–4.5 mm	4.0–6.3 mm	2–3 mm
beak	glabrous	scabrous	usually scabrous
beak	usually straight	curved	usually straight
beak	= perigynium body	over ½ as long	⅓ as long

Page 267. *Carex frankii* Kunth. The following counties should be added to the map on page 267: DeKalb, Jo Daviess, Lake, La Salle, Madison, Mason, Mercer, Shelby.

After *Carex frankii*, add the following species that is new for Illinois:

Carex aureolensis Steud. Syn. Pl. Glumac. 2:223. 1855. Fig. A33.

Plants perennial, with long, slender rhizomes, usually forming colonies; culms to 85 cm tall, triangular, slightly rough to the touch, occasionally purplish at the base; sterile shoots sometimes present; leaves up to 8 mm wide, deep green, firm, slightly scabrous along the margins and usually on the veins; sheaths tight, septate-nodulose, yellow-brown hyaline at the summit, the lowermost often red-tinged, the ligule about as wide as long; terminal spike staminate, or less commonly with pistillate flowers at the tip, up to 4 cm long, to 3.5 mm thick; staminate scales lanceolate to narrowly ovate, 1.0–1.5 mm wide, hyaline with a green center, awned; pistillate spikes 3–5 sometimes with a few staminate flowers, cylindric, up to 4 cm long, up to 1 cm thick, sessile or short-pedunculate; pistillate scales narrowly lanceolate, 0.4–1.0 mm wide, awned, slightly scabrous, green, much longer than the perigynia; perigynia up to 80 per spike, obconic, glabrous, truncate across the top, 3.0–5.5 mm long, to 2.5 mm broad, olive-green, becoming brownish, many-nerved, abruptly contracted to a bidentate beak 1.2–2.0 mm long; achenes trigonous, obovoid, yellow-brown, 1.5–2.0 mm long, substipitate; stigmas 3.

Common Name: Golden-fruit sedge.

Habitat: Moist woods, wet ditches.

Range: Virginia to Nebraska, south to Texas and Florida.

Illinois Distribution: Alexander, Gallatin, Jackson, Massac, Perry, Pope, Saline, Union, Washington, and Williamson counties.

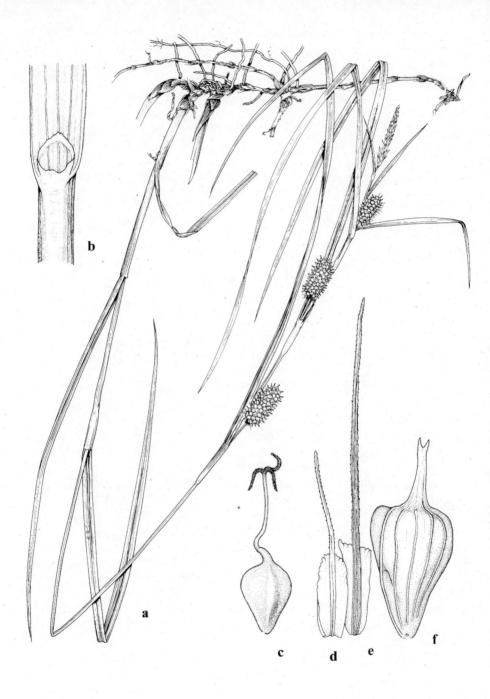

A33. Carex aureolensis
(Golden-fruit sedge).

a. Habit.
b. Sheath.
c. Achene.

d, e. Pistillate scales.
f. Perigynium.

This species looks very much like *C. frankii* but usually has long rhizomes and forms dense colonies. The staminate scales and pistillate scales of *C. aureolensis* are wider than those of *C. frankii*.

Flowers appear from May to September.

The following table summarizes some of the characteristics of *C. frankii* and *C. aureolensis*:

character	*C. frankii*	*C. aureolensis*
rhizomes	short, not colony-forming	long, forming colonies
staminate scale width	0.5–1.0 mm	1.0–1.5 mm
pistillate scale width	0.1–0.5 mm	0.4–1.0 mm
perigynia per spike	up to 100	up to 80
beak length	up to 2.5 mm	1.2–2.0 mm

Page 267. *Carex squarrosa* L. The following county should be added to the map on page 270 (top): McLean.

Page 270. *Carex typhina* Michx. The following counties should be added to the map on page 270 (bottom): Carroll, Cass, Sangamon, Wayne.

Page 272. *Carex lacustris* Willd. The following counties should be added to the map on page 272: Brown, Bureau, Iroquois, Kankakee, Schuyler, Washington.

Page 272. *Carex hyalinolepis* Steud. The following counties should be added to the map on page 275: Brown, Cass, Cook, Henderson, Lake, Macon, Schuyler, Stephenson, Winnebago.

Page 275. *Carex atherodes* Spreng. The following county should be added to the map on page 277 (top): Carroll.

Page 277. *Carex heterostachya* Bunge. There are no new records for this species.

Page 279. *Carex laeviconica* Dewey. The following counties should be added to the map on page 279: Fulton, Sangamon.

Page 279. *Carex trichocarpa* Muhl. ex Schkuhr. The following county should be added to the map on page 283 (top): Woodford.

Page 283. *Carex X subimpressa* Clokey. The following county should be added to the map on page 283 (bottom): DuPage.

Page 285. *Carex comosa* Boott. The following counties should be added to the map on page 285: Lee, St. Clair.

After *Carex comosa*, the following species new to Illinois should be added:

Carex pseudocyperus L. Sp. Pl. 2:978. 1753. Fig. A34.

Plants perennial, cespitose, from short, stout rhizomes; culms to 1 m tall, triangular, scabrous beneath the inflorescence, pale brown at base with several of last year's leaves persistent; leaves up to 13 mm wide, septate-nodulose, rather dark green, flat or W-shaped, usually scabrous on the margins; sheaths pale brown, with hyaline margins, concave at the mouth; ligule longer than wide; lower bracts leaflike; terminal spike staminate, rarely partly pistillate, up to 5 cm long, up to 6 mm thick, sessile or short-pedunculate; lateral spikes 2–5, pistillate, up to 5 cm long, 0.9–1.2 cm thick, at least the lowest ones pendulous or on flexuous peduncles;

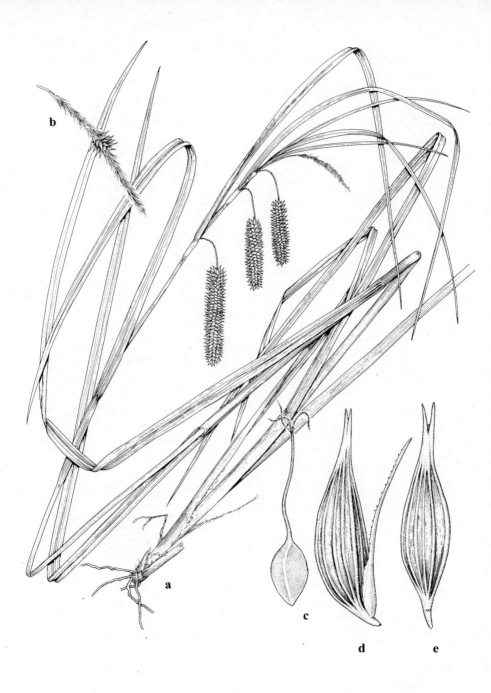

A34. Carex pseudocyperus
(Cyperus-like sedge).
a. Habit.

b. Spikelet, part staminate,
part pistillate.
c. Achene.

d. Perigynium, side view
with scale.
e. Perigynium, front view.

pistillate scales narrowly lanceolate, serrulate-awned, reddish brown with a dark center, one-half to nearly as long as the perigynia; perigynia up to 80 per spike, crowded, divaricately spreading to reflexed, lanceoloid, 3.5–6.0 mm long, up to 1.5 mm wide, not inflated, glabrous, strongly nerved, yellow-green, coriaceous, tapering to a bidentate beak 1–2 mm long, the teeth of the beak stiff, straight or nearly so, 0.7–1.2 mm long; achenes trigonous, 1.5–1.8 mm long, pale brown, continuous with the persistent, flexuous style; stigmas 3, pale brown.

Common Name: Cyperus-like sedge.
Habitat: Along a river (in Illinois).
Range: Nova Scotia to Manitoba, south to Minnesota, Illinois, Ohio, and Massachusetts.
Illinois Distribution: Fulton County: at bridge over Spoon River at Bernadotte.

This species is very similar in appearance to *C. comosa* from which it differs by its shorter perigynia with nearly straight and shorter teeth of the beak of the perigynium, its thinner staminate spike, and its usually narrower leaves.

Carex pseudocyperus flowers during June.

Page 285. *Carex hystericina* Muhl. ex Willd. The following counties should be added to the map on page 288: Adams, Brown, Effingham, Jackson, Livingston, Madison.

The following table summarizes some of the characteristics of *C. comosa*, *C. pseudocyperus*, and *C. hystericina*:

character	*C. comosa*	*C. pseudocyperus*	*C. hystericina*
leaf width	to 17 mm	to 13 mm	to 10 mm
perigynium length	5–7 mm	3.5–6.0 mm	4–6 mm
perigynium width	up to 1.5 mm	up to 1.5 mm	1.5–2.0 mm
perigynium shape	lanceoloid	lanceoloid	lance-ovoid to ovoid
perigynium	not inflated	not inflated	inflated
lateral spikes	3–6	2–5	1–4
lateral spike width	1.7 cm	0.9–1.2 cm	1.0–1.5 cm
achene length	1.7–2.0 mm	1.5–1.8 mm	1.5–1.8 mm

Page 288. *Carex lurida* Wahl. The following counties should be added to the map on page 290 (top): Crawford, Franklin, Iroquois, Monroe, Montgomery, Schuyler, Woodford.

Page 290. *Carex baileyi* Britt. There are no new records for this species. In the Cyperaceae volume of Flora of North America, this species is not listed from Illinois even though there is an extant specimen and a literature report for Illinois.

Page 292. *Carex folliculata* L. There are no new records for this species.

Page 292. *Carex grayi* Carey. The following counties should be added to the map on page 295: Alexander, Iroquois, Lee, Marion, Saline, Schuyler, Shelby, Wayne, Woodford.

Page 295. *Carex intumescens* Rudge. The following counties should be added to the map on page 297: Jersey, Mason, Washington.

Page 297: *Carex louisianica* Bailey. The following counties should be added to the map on page 299: Pulaski, Washington, Wayne.

Page 299. *Carex lupulina* Willd. The following counties should be added to the map on page 302: Brown, Calhoun, Iroquois, Livingston, Mercer, Montgomery, Shelby.

Page 302. *Carex lupuliformis* Sartw. ex Dewey. The following counties should be added to the map on page 304: Crawford, Iroquois, Jersey.

Page 304. *Carex gigantea* Rudge. The following counties should be added to the map on page 306 (top): Lawrence, Massac.

Page 306. *Carex retrorsa* Schw. The following county should be added to the map on page 306 (bottom): Whiteside.

Page 308. *Carex utriculata* Boott in Hook. The following county should be added to the map on page 308: Fulton.

Page 308. *Carex tuckermanii* Boott ex Dewey. The following county should be added to the map on page 311 (top): Henderson. In the Cyperaceae volume of Flora of North America, this species is inexplicably not listed from Illinois, even though there are extant specimens and literature reports for this species for Illinois.

Page 311: *Carex oligosperma* Michx. There are no new records for this species.

Page 313. *Carex vesicaria* L. var. *monile* (Tuckerm.) Fern. The following county should be added to the map on page 313: Iroquois.

Excluded Species

Carex livida (Wahlenb.) Willd. Vasey (1861) reported this species from Illinois, but since Vasey is known to have sometimes placed Illinois on his collection labels even though the plant was collected elsewhere, this fact prevents me from listing this species from Illinois. There is little likelihood that this species would occur in the state.

Carex scabrata Schw. Higley and Raddin (1891) and Pepoon (1927) reported this species from several places in Cook County, and in all probability it may have occurred there. Since there are apparently no specimens from Illinois to substantiate these reports, and the possibility of misidentifications are fairly good, I am excluding this species from the state.

Carex schweinitzii Dewey. This species was reported from Illinois by Kibbe in 1952, but there are no specimens to verify this report, and the range of this plant probably precludes its occurrence in the state.

Glossary

achene. A type of 1-seeded, dry, indehiscent fruit with the seed coat not attached to the mature ovary wall.

acicular. Needlelike.

acuminate. Gradually tapering to a long point.

acute. Sharply tapering to a short point.

androgynous. With staminate flowers above, pistillate flowers below.

antrorse. Projecting forward.

apiculate. Abruptly short-pointed at the tip.

appressed. Lying flat against a surface.

approximate. Close to each other.

aristate. Bearing a short awn.

ascending. Pointing upward.

attenuate. Gradually becoming narrowed.

awn. A bristle terminating a structure.

awned. Bearing an awn.

beak. The narrow terminal tip of some perigynia.

biconvex. Rounded on both the front and back faces.

bidentate. With two teeth.

bract. An accessory structure at the base of a flower, sometimes appearing leaflike, other times setaceous or scalelike.

canaliculate. Grooved.

capillary. Threadlike.

capitate. Forming a head.

cartilaginous. Firm but flexible.

castaneous. Chestnut colored.

cauline. Belonging to a stem or culm.

cespitose. Growing in tufts.

ciliate. Bearing short, marginal hairs.

clavate. Club-shaped.

compressed. Flattened.

concave. Curved on the inner surface; opposite of convex.

conduplicate. Folded together lengthwide.

conic. Cone-shaped.

contiguous. Touching the next one in a series.

convex. Curved on the outer surface; opposite of concave.

convolute. Rolled lengthwise.

coriaceous. Leathery.

cucullate. Hooded.

culm. The stem that terminates in an inflorescence.

cuneate. Wedge-shaped, tapering to the base.

cuspidate. Terminating in a very short point.

cylindric. Shaped like a cylinder.

deciduous. Falling away.

decumbent. Lying flat, but with the tip ascending.

decurrent. Extending beyond the point of attachment.

deltoid. Referring to a solid object that is triangular.

dentate. With sharp teeth, the tips of which project outward.

depressed. Shallowly sunken in.

dilated. Expanded, swollen.

dioecious. Having staminate and pistillate flowers on separate plants.

disarticulating. Breaking off.

distended. Swollen over a structure.

divaricate. Widely spreading.

divergent. Spreading.

dorsal. The outer face of a structure.

ellipsoid. Referring to a solid object that is broadest at the middle, gradually tapering to both ends; narrower than oblongoid.

elliptic. Broadest at the middle, gradually tapering to both ends; narrower than oblong.

emarginate. Having a shallow notch at the tip.

excurrent. Extending beyond the margin of a structure from which it originates.

fibrillose. Bearing numerous fine fibers.

fibrous. Referring to roots borne in tufts.

filament. That part of the stamen supporting the anther.

filiform. Threadlike.

flaccid. Weak, flabby.

flexuous. Flexible; zigzag.

foliaceous. Leaflike.

fusiform. Spindle-shaped.

gibbous. Swollen on one side.

glabrous. Without pubescence or hairs.

granular. Having the surface texture of tiny grains.

glaucous. With a whitish covering that can be wiped off.

globose. Round; globular.

gynecandrous. A spike with pistillate flowers at the tip and staminate flowers below.

hispidulous. With minute stiff hairs.

hyaline. Transparent.

impressed. Sunken into the surface.

inflorescence. A cluster of flowers or spikes.

intervenal. Between the veins.

invaginated. Notched; sunken inward.

involute. Rolled inward.

keeled. Bearing a ridgelike process.

lanceolate. Lance-shaped; broadest near the base, gradually tapering to the narrower apex; narrower than ovate.

lanceoloid. Referring to a solid object that is broadest near the base, gradually tapering to the narrower apex; narrower than ovoid.

lenticular. Lens-shaped.

ligule. The structure at the summit of the ventral side of the sheath where it merges into the blade.

linear. Elongated and uniform in width throughout.

lustrous. Shiny.

maculate. Spotted.

margin. The border of something.

membranous. Membranelike; very thin, often translucent.

moniliform. Beadlike.

mucronate. Possessing a short, abrupt tip.

node. The place on the stem or culm from which leaves and spikes arise.

nodulose. Bearing small knots or knobs.

obconic. Reversely cone-shaped.

oblique. At an angle.

oblong. Broadest at the middle, tapering to both ends; broader than elliptic.

oblongoid. Referring to a solid object that is broadest at the middle, tapering to both ends; broader than ellipsoid.

obovate. Broadly rounded at the apex, becoming narrowed below; broader than oblanceolate.

obovoid. Referring to a solid object that is broadly rounded at the apex, becoming narrowed below.

obtrulloid. Reversely shaped like a brick-layer's trowel.

obtuse. Rounded; blunt.

opaque. Incapable of being seen through.

orbicular. Round.

oval. Broadly elliptic.

ovate. Broadly rounded at the base, becoming narrowed above; broader than lanceolate.

ovoid. Referring to a solid object that is broadly rounded at the base, becoming narrowed above; broader than lanceoloid.

papilla. A small, pimplelike projection.

papillate. Bearing small, pimplelike projections.

peduncle. The stalk of an inflorescence or of a spike.

pedunculate. Bearing a peduncle.

pellucid. Clear; transparent.

pendulous. Drooping; nodding.

perigynium. A saclike covering enclosing the achene in *Carex*.

perennial. Living more than two years.

pilose. Bearing soft hairs.

pistillate. Bearing pistils but not stamens.

prostrate. Lying flat on the substrate.

puberulent. With minute hairs.

pubescent. Bearing some kind of hairs.

punctate. Dotted.

puncticulate. With small dots.

reclining. Lying down.

recurved. Turned backward or downward.

reflexed. Turned downward.

remote. Apart; referring to spikes that do not touch each other.

retrorse. Pointing downward.

retuse. Shallowly notched at a rounded apex.

revolute. Rolled under from the margins.

rhizomatous. Bearing rhizomes.

rhizome. An underground horizontal stem, bearing nodes, buds, and roots.

rhombic. Quadrangular, four-sided.

ribbed. With elevated veins.
rootstock. The underground root system; sometimes synonymous with rhizome.
rudimentary. Poorly developed.
rugulose. Wrinkled-appearing.
russet. Red-brown.

scabrous. Rough to the touch.
scale. A modified leaf that subtends each staminate and each pistillate flower; a structure that is formed on rhizomes.
septate. With dividing walls.
serrate. With teeth that project forward.
serrulate. With very small teeth that project forward.
sessile. Without a stalk.
setaceous. Bristlelike.
sheath. A tubular part of the leaf that surrounds the culm.
spike. The basic unit of the inflorescence in *Carex*.
squarrose. Referring to the tips of perigynia that bend outward or downward.
staminate. Bearing stamens but not pistils.
stigma. The terminal part of the pistil that receives pollen.
stipitate. Possessing a short stalk.
stolon. A slender, horizontal stem on the surface of the ground.
stoloniferous. Bearing stolons.
stramineous. Straw-colored.
striate. Marked with grooves.
style. That part of the pistil between the ovary and the stigmas.

subacute. Nearly tapering to a short point.
subaristate. With an extremely short awn.
subcanaliculate. Nearly forming a canal.
subcoriaceous. Almost leathery.
subcylindric. Nearly cylindric.
subglobose. Almost globe-shaped.
submembranous. Nearly membranelike or thin.
subsessile. With a very short stalk.
substipitate. Possessing a very short stalk.
subterete. Nearly round in cross-section.
subulate. With a very short, narrow point.
subuloid. Referring to a solid object that is long-tapering to the apex.
suffused. Spread throughout; flushed.

terete. Round in cross-section.
translucent. Allowing some light to pass through; intermediate between opaque and transparent.
trigonous. Triangular in cross-section.
trulloid. Shaped like a bricklayer's trowel.
truncate. Abruptly cut across.
tubular. Forming a tube.
turgid. Swollen to the point of bursting.
tussock. A clump.

umbonate. Having a protuberance or rounded elevation.
unisexual. Bearing either staminate or pistillate flowers, but not both.

ventral. The inner face of a structure.

Literature Cited

Brendel, F. 1887. *Flora Peoriana*. Peoria, Illinois.

Crins, W. J., and P. W. Ball. 1989. Taxonomy of the *Carex flava* complex (Cyperaceae) in North America. *Canadian Journal of Botany* 61:1692–1717.

Deam, C. C. 1940. *Flora of Indiana*. Indianapolis: Indiana Department of Conservation.

Engelmann, G. 1843. Catalogue of collections of plants made in Illinois and Missouri by Charles A. Geyer. *American Journal of Science* 46:94–104.

Evans, D. K. 1976. Taxonomy of the *Carex rosea—C. retroflexa* complex in Illinois. Ph.D. diss., Carbondale: Southern Illinois University.

Fernald, M. L. 1950. *Gray's manual of botany*. New York: American Book Company.

Gleason, H. A. 1952. *The new Britton and Brown illustrated flora of the northeastern United States and adjacent Canada*. vol. 1. New York: New York Botanical Garden.

Gleason, H. A., and A. Cronquist. 1963. *Manual of the vascular plants of the northeastern United States and adjacent Canada*. New York: Van Nostrand Reinhold Company.

———. 1991. *Manual of vascular plants of the northeastern United States and adjacent Canada*. 2d ed. New York: New York Botanical Garden.

Higley, W. K., and C. S. Raddin. 1891. Flora of Cook County, Illinois, and a part of Lake County, Indiana. *Bulletin of the Chicago Academy of Science* 2:1–168.

Hoffman, J. B. 1974. A revision of section Lupulinae of the genus *Carex*. Ph.D. diss., Carbondale: Southern Illinois University.

Hutchinson, J. 1959. *Monocotyledons*. Vol. 2, *The families of flowering plants*. Oxford: Clarendon Press.

Kibbe, A. 1952. *A botanical study and survey of a typical mid-western county (Hancock County, Illinois)*. Carthage, Illinois.

Mackenzie, K. K. 1931. Cariceae. *North American Flora* 18:1–168.

———. 1940. *North American Cariceae*. 2 vols. New York: New York Botanical Garden.

Mead, S. B. 1846. Catalogue of plants growing spontaneously in the state of Illinois. *Prairie Farmer* 6:119–22.

Mohlenbrock, R. H. 1970. *The illustrated flora of Illinois. Flowering rush to rushes*. Carbondale: Southern Illinois University Press.

———. 1970a. *The illustrated flora of Illinois. Lilies to orchids*. Carbondale: Southern Illinois University Press.

———. 1972. *The illustrated flora of Illinois. Grasses: Bromus to Paspalum*. Carbondale: Southern Illinois University Press.

———. 1973. *The illustrated flora of Illinois. Grasses: Paspalum to Danthonia.* Carbondale: Southern Illinois University Press.

———. 1975. *Guide to the vascular flora of Illinois.* Carbondale: Southern Illinois University Press.

———. 1976. *The illustrated flora of Illinois. Sedges: Cyperus to Scleria.* Carbondale: Southern Illinois University Press.

———. 1985. *Guide to the vascular flora of Illinois.* revised and enlarged ed. Carbondale: Southern Illinois University Press.

Mohlenbrock, R. H., and D. Ladd. 1978. *Distribution of Illinois vascular plants.* Carbondale: Southern Illinois University Press.

Patterson, H. N. 1876. *Catalogue of the phaenogamous and vascular cryptogamous plants of Illinois.* Oquawka, Illinois.

Pepoon, H. S. 1927. An annotated flora of the Chicago area. *Bulletin of the Chicago Academy of Science* 8:1–554.

Rettig, J. H. 1989. Nomenclatural changes in the *Carex pensylvanica* group (section Acrocystis, Cyperaceae) of North America. *Sida* 13:449–52.

———. 1990. Correct names for the varieties of *Carex albicans*. *Sida* 14:132–33.

Rothrock, P. E. 1991. The identity of *Carex albolutescens, C. festucacea,* and *C. longii* (Cyperaceae). *Rhodora* 93:51–66.

Swink, F., and G. Wilhelm. 1994. *Plants of the Chicago region.* 4th ed. Lisle: Morton Arboretum.

Thorne, R. F. 1968. Synopsis of a putatively phylogenetic classification of the flowering plants. *Aliso* 6:57–66.

Vasey, G. 1861. Additions to the Illinois flora. *Prairie Farmer* 22:119.

Webber, J. M., and P. W. Ball. 1984. The taxonomy of the *Carex rosea* group section Phaestoglochin in Canada. *Canadian Journal of Botany* 62:2058–2073.

Index of Plant Names

Plant names in roman type are accepted names, while those in italics are synonyms and are not considered valid. Numbers in bold refer to pages that contain illustrations.

Robert H. Mohlenbrock is in his sixty-fourth year of plant study. After receiving his doctorate from Washington University in St. Louis, Mohlenbrock taught botany and plant taxonomy at Southern Illinois University (SIU) for thirty-four years, sixteen of which he served as chairman of the department. During his career at SIU Carbondale, he was major professor for ninety graduate students and carried out a vigorous program of research. To date, his research has resulted in the publication of more than sixty books and more than five hundred other publications. Since 1984 he has written "This Land," a monthly column for *Natural History*, published by the American Museum of Natural History in New York. Since retiring from SIU in 1990, Mohlenbrock has been a senior scientist for Biotic Consultants, where he teaches weeklong wetland plant identification classes for government employees and consultants throughout the country, a total of 297 of these in thirty states at the end of 2010.